Artificial Intelligence

Benedikt Paul Göcke, Astrid Rosenthal-von der Pütten (eds.)

Artificial Intelligence

Reflections in Philosophy, Theology, and the Social Sciences

BRILL | mentis

Cover illustration: pixabay.com

Bibliographic information published by the Deutsche Nationalbibliothek

The Deutsche Nationalbibliothek lists this publication in the Deutsche Nationalbibliografie; detailed bibliographic data available online: http://dnb.d-nb.de

Internet: www.mentis.de

Cover design: Anna Braungart, Tübingen
Production: Brill Deutschland GmbH, Paderborn

ISBN 978-3-95743-191-2 (hardback)
ISBN 978-3-95743-748-8 (e-book)

Table of Contents

PART III
Ethics

PART IV
Robots

Introduction

Benedikt Paul Göcke, Astrid Rosenthal-von der Pütten

1. Artificial Intelligence and Machine Learning

Developments in the theoretical and practical sciences over the past decades have brought to light findings which, for the first time in human history, enable us to shape individual and social life in a hitherto unseen way through the use of artificial intelligence. To understand why artificial intelligence has the potential to transform individual and social life, the concept of artificial intelligence must first be rudimentarily clarified.

An artificial intelligence is a virtual machine whose existence is the product of human skill. Like any other machine, artificial intelligence is developed to perform a specific task.[1] The difference to classical mechanical machines is that they solve the tasks for which they are designed in a way that would require consciousness and intelligence in the case of humans. Since intelligence can be understood as the ability to achieve complex goals, artificial intelligence can therefore be understood as a computer program developed by humans that can achieve complex goals.[2]

In research, different and overlapping approaches can be identified how computer programs can be developed that are able to achieve complex goals: In addition to the so-called expert systems, which use their databases to generate a precisely determined output with the help of logical rules, and in addition to evolutionary procedures for identifying the algorithms suitable for a task, there are approaches that construct virtual machines that are based on the way the human brain functions and are therefore referred to as artificial neural networks: Artificial neural networks often operate with probabilities and therefore arrive at their results in a different way than rule-based expert systems, which means that artificial neural networks are generally suitable for

---

1 Cf. Alpaydin (2016: 19): "Artificial intelligence is a part of computer science and our aim is to build useful systems, as in any domain of engineering."

2 Cf. Boden (2005: 1): "Artificial Intelligence (AI) is sometimes defined as the study of how to build and/or program computers to enable them to do the sorts of things that minds can do. Some of these things are commonly regarded as requiring intelligence: offering a medical diagnosis and/or prescriptions, giving legal or scientific advice, proving theorems in logic or mathematics."

 | DOI:10.30965/9783957437488_002

different tasks than expert systems, although, of course, these approaches can be combined.[3]

What the different approaches to AI research have in common is that they are, in principle, capable of implementing an ability necessary for achieving complex goals: the ability to learn from past behaviour. Although the idea of learning machines was already discussed by Alan Turing in the 1950s (cf. Turing 2005) and has been discussed and developed since then, despite some setbacks, significant progress in machine learning could only be achieved in the last decades due to the improved performance of the available processors and the large amounts of available data necessary for some forms of machine learning.

There are at least three ways known from developmental psychology in which an artificial intelligence can learn to perform its task better and better based on given parameters until it reaches its maximum performance (if there is a maximum): supervised, unsupervised and reinforcement learning:

(1) In supervised learning, the artificial intelligence is given positive examples of the desired output and continuous feedback on whether it has fulfilled its task. If, for example, the task is to recognize bicycles, it is first shown pictures of bicycles with the information that bicycles can be seen on the pictures shown, before the artificial intelligence independently checks new pictures to see if bicycles are shown on them. Through continuing human feedback, the recognition process is then optimized further.

(2) The primary purpose of unsupervised machine learning is that artificial intelligence recognizes correlations in the data set provided to it and indicates a probability that a certain characteristic of this data set correlates with a certain other characteristic. For example, based on a woman's shopping behavior, an artificially intelligent system – provided it has access to the necessary data – can indicate the probability that a woman is pregnant if pregnant women in the past showed the same shopping behavior.

(3) Reinforced learning instructs the AI to avoid negative feedback and seek positive feedback, and then evaluates the performance of the system through positive and negative feedback to train the virtual machine to

3 In general, to develop an artificially intelligent system, the following needs to be specified: "1. *Computational theory* corresponds to the goal of computation and an abstract definition of the task. 2. *Representation and algorithm* is about how the input and the output are represented, and about the specification of the algorithm for the transformation from the input to the output. 3. *Hardware implementation* is the actual physical realization of the system." (Alpaydin 2016: 20).

perform its task optimally. Through reinforcement learning, artificial intelligence can, for example, be enabled to practice certain motion sequences that are necessary or sufficient for the fulfilment of its task.

It is the learning ability of artificial intelligence that distinguishes it from previous technologies and constitutes its enormous potential for changing individual and social human life. Depending on the task, artificial intelligence can be implemented as a virtual machine on a computer and specialized in data analysis, or it can take over the physical movements of a machine in cyberphysical form as a control unit that is connected to the environment via sensors. The principle field of application of learning artificial intelligence therefore affects every area of individual and social human life. Since adaptive artificial intelligence is able to learn at a speed that far exceeds the limits and performance of humans due to large amounts of data and processors connected in parallel, the potential effects of this technology on individual and social life can hardly be overrated.

2. Artificial Intelligence – Reflections in Philosophy, Theology, and the Social Sciences

The conference *When Robots Think – Interdisciplinary Views on Intelligent Automation*, held in Münster (Germany) in November 2018, brought together researchers from Germany, the United Kingdom, Switzerland, France, Turkey, Korea, and the USA to discuss artificial intelligence from the perspectives of philosophy, theology, and the social sciences. Although there is a considerable thematic overlap between the contributions from the different disciplines, five core themes could be identified: Consciousness, Superintelligence, Ethics, Robots, and Society.

2.1 *Consciousness*

In his chapter **Conscious, Thinking, and Intelligent Machines? Some Clarifying Remarks in the Field of Artificial Intelligence** *Tobias Müller* addresses the need for conceptual clarification regarding attributions we make to AI. Progress in AI research has led to the relevant literature already assigning many AI systems the ability to think, learn, predict, analyze, decide, know, plan, being autonomous and self-aware. In order to be able to answer the question of whether such attributions are justified, Müller provides considerations from the philosophy of mind and the philosophy of technology, which could be helpful to achieve a clarification of certain terms. For this purpose, he presents various working definitions of "consciousness", "thinking" and "intelligence", and highlights specificities of human consciousness and thinking. He discusses whether these specific qualities can already be found in today's AI systems and whether there are fundamental limits of AI systems.

In **Deus Malignus – The Digital Rehabilitation Of Deception** *Walther Ch. Zimmerli* asks if it makes sense to call machines "intelligent". Zimmerli steps into dialog with Alan Turing's "Computing Machinery and Intelligence" by providing some philosophically orientated footnotes to the main idea of Alan Turing. Building upon a very influential argument in early modern skeptical rationalist philosophy, Descartes' "Deus" or "Genius Malignus" argument, he discusses the neglected implications of the "Turing test" and culminates in a revision and rehabilitation of one of the most abhorred concepts of modern science and philosophy: deception.

Daniel Neumann sees still lots of anthropomorphizing misunderstandings as to what it means for a machine to be intelligent. In his chapter **Artificial Intelligence of the Human Mind. Finding Common Ground Between Intelligent Machines and "Mechanistic" Minds** he suggests reconsidering how human intelligence should be conceived of in mechanistic terms. Neumann takes a closer look at the notion of the spiritual automaton introduced by the 17th century philosopher Baruch de Spinoza who provided a mechanical theory of the mind. Neumann sketches ways in which current methods in AI studies could be seen to reflect aspects of the Spinozistic model of the mind.

In **When Robots (Pretend to) Think** *Davide Ghiglino* and *Agnieszka Wykowska* discuss why artificially intelligent systems are a valuable new tool to learn how humans "work". Based on the assumption that humans necessarily try to understand what other people think and how they feel, and based on the assumption that in order to achieve this goal, humans formulate hypotheses, predictions, expectancies and, more broadly, representations of the others' goals, desires and intentions, they argue that the advent of smart artificial agents is providing researchers with new environments to test mindreading models, thereby helping them to push the cognitive flexibility of the human social brain from the natural domain towards the artificial.

In his **Revisiting the Dancing-Qualia Argument for Computationalism,** *Stefan Reining* argues that according to computationalist theories of the mind, consciousness does not depend on any specific physical substrate – i.e., carbon-based biological material – but instead automatically arises out of the right kind of computational structure. He then challenges David Chalmers' dancing-qualia argument in support of this assumption by discussing experimental results in neurobiology regarding the workings of phenomenal memory that show that the argument is unsound. He also argues, however, that Chalmers' overall case for the possibility of conscious AI can still be vindicated.

2.2 *Superintelligence*

In **Will Superintelligence Take Over? A Critical Reflection on the Apparent and Actual Dangers of AI** *Kilian Karger* provides an overview of the concept of superintelligence and the prerequisites and conditions that are connected with it. He then analyzes whether these are realizable before he critically reflects to what extent such an understanding of artificial intelligence challenges our human self-conception and whether the dangers in the use of AI predicted by advocates of the superintelligence hypothesis are actually the most pressing questions in dealing with AI.

Andreas Theodorou introduces another perspective in his chapter **Why Artificial Intelligence is a Matter of Design.** He first discusses the limitations and bottlenecks of intelligence and argues that an 'all-powerful' artificial general intelligence would not only entail omniscience, but also be impossible. He then reflects upon how culture determines the moral status of all entities, as morality and law, according to Theodorou, are human-made 'fictions' that help us guide our actions. Theodorou concludes that therefore our moral spectrum could be altered to include machines. He argues, however, that this should be avoided.

In **Between Epistemic Progress and Existential Risk Minimization** *Julia Alessandra Harzheim* discusses that future artificial general intelligence (AGI) systems will need to possess a certain degree of autonomy to yield the results we aim for, yet their decisions should preferably not be made arbitrarily – especially when it comes to moral dilemma scenarios. Instead, their actions should be guided by moral maxims we consider valid and do so for epistemically justified reasons. The responsibility to equip AGI systems with an appropriate normative framework and certain premises that prevent them from being a threat to humankind is incredibly high. At the same time, a consensus on moral problems has never been reached; instead, moral pluralism seems to be more prevalent than ever before. Harzheim therefore argues to tackle epistemic and ethical questions first before developing technologically powerful tools in order to guide future AGI systems safely.

2.3 *Ethics*

Same same but different or something completely new? This is the question *Leonie Seng* addresses in **Current Challenges in Ethics of Artificial Intelligence or: Old Wine in New Bottles.** She argues that current discussions on the possibility of moral decision-making by intelligent machines deal with old wine in new bottles: Trying to identify moral values has been part of most ethical considerations since the very beginning of philosophical investigations.

In her chapter **Is Utilitarianism Entirely Useless for Self-Driving Car Ethics? A Critical Reflection on the Rationale for Rule Utilitarianism** *Vanessa Schäffner* discusses the suitability of a rule utilitarian approach in machine ethics. Her comprehensive analysis of the potential that may be unlocked from rule utilitarianism discloses that even though it is unsuitable as the exclusive ethical concept for self-driving car ethics, it still appears to be the most appropriate among utilitarian versions.

In her **The Car's Choice: Illusions of Agency in the Self-Driving Car Trolley Problem** *Rebecca Davnall* identifies another problem concerning the decision-making in autonomous driving. She discusses that the self-driving car trolley problem is unrealistic because it fails to respect well-established truths of vehicle dynamics and misrepresents the information environment that self-driving cars will have to respond to. Davnall also argues that the self-driving car trolley problem induces readers to treat the car as an agent, thereby shielding the agency of car designers and operators from scrutiny.

In **Why Machines That Talk Still Do Not Think, and Why They Might Nevertheless be Able to Solve Moral Problems** *Lukas Brand* argues that the recent success in the development of machines that seem to be in command of a human language through advances in deep learning solve a variety of problems that are crucial for moral decision making. They adapt to the syntax of the spoken language as well as to the unwritten habits of human verbal interaction and perform speech in a way that is meaningful to humans. Even though they still do not think the way humans do, as Searle has shown, they might be able to adapt to human moral decision making alike, learning the written laws as well as the contingent aspects of moral decision making in a way that their action can be evaluated as morally good from an observer's point of view.

2.4 *Robots*

In **Could Artificial General Intelligence be an End-In-Itself?** *Benedikt Paul Göcke* tackles what he regards as the most interesting question in the pervasive discussion of artificial intelligence: Whether artificial intelligence could entail the existence of a conscious entity. He first provides arguments for and against the possibility of constructing artificial general intelligence. He then proposes an argument for the conclusion that artificial general intelligence could indeed be a conscious being and therefore should be considered ethically as an end-in-itself.

Carmen Krämer calls for an exploration on how we want and need to organize our cohabitation with autonomous machines that are entering our everyday worlds. In **Robots and Dignity**, she poses the questions of how robots should treat humans and the world around them, and how humans should treat robots. Considering the Kantian idea that the possession of dignity is based on autonomy, and considering the fact that robots become increasingly autonomous and rational, one of these questions might be if robots have dignity. After a short glimpse on the existing theories, Krämer proposes how to understand dignity and argues that, based on the sketched understanding of dignity, robots cannot be seen as possessors of dignity.

In his contribution *David J. Gunkel* poses **The Right(s) Question: Can and Should Robots Have Rights?** The controversy whether robots might need to be considered "electronic persons" for the purposes of social and legal integration has an underlying important philosophical question: When (if ever) would it be necessary for robots, AI, or other socially interactive, autonomous systems to have some claim to moral and legal standing? Can or should robots ever have anything like rights? Gunkel reviews and critiques current thinking about this subject matter, maps the terrain of the set of available answers that have been provided in the existing literature, and develops an alternative way of responding to it.

Gábor L. Ambrus offers a glimpse into the theological interpretation of artificial intelligence and robotics. In **Image, Servitude, Partnership: What Do the Creation of Human Beings and the Creation of AI Imply?** he provides an overview on theological work already done in the domain of AI and robotics. He concentrates on the "double image" theory, which is the idea that analogously to the way God created humankind in His image, humankind creates artificial intelligence in its own image. He discusses the shortcoming of this theory and suggests a revised "double image" theory to account for the status of artificial intelligence.

Diana Löffler, Swantje Luthe, Jörn Hurtienne and *Ilona Nord* report about human encounters with a robot in a religious context in their contribution **From Experiential to Existential Questions – An Interdisciplinary View on Social Robots in Religious Settings**. In their work, they interweave disciplinary views from interaction studies and religious studies on social robotics in religious communication on the example of the blessing robot BlessU2. They discuss a discursive design study in which the robot expressed Protestant blessings to thousands of visitors on a public exhibition. More than 2,000 comments were analysed regarding experiential aspects such as the acceptability and design features of this social robot as well as how the participants reflected on important existential questions such as the origin of a blessing.

Andreas Bischof and *Arne Maibaum* discuss new problems emerging when robots transition from laboratories into the real world in their chapter **Robots and the Complexity of Everyday Worlds**. Although roboticists have always referred to the idea of robots solving "real world problems", dealing with actual everyday problems has become common in robotics only over the course of the past few years. When transitioning into everyday life, researchers in robotics and human-robot-interaction face a new kind of problem that challenges the established theories, methods and cultures of robotic research and engineering. The authors discuss the nature of these problems as "wicked problems" and their implication for roboticists dealing with everyday worlds.

In **Taught by Technology. Assessing Robots' Social Roles in Early Education** *Scarlet Siebert* raises questions of the adequacy of applying robots in early childhood education because they might deceive children and raise false expectations. Since social robots do not only serve as instructional material for STEM education to teach children how to program software, but social robots themselves teach children subjects they used to learn from humans, e.g. languages and social behavior, they function as social interlocutors rather than tools. Scarlet Siebert therefore discusses different social roles of robots that interact with children for educational purposes and criteria for the assessment of the adequacy of these roles.

2.5 *Society*

In his contribution **Artificial Intelligence and Polygenic Scoring: The Risk of Personal Eugenics** *Johannes Grössl* discusses ethical-social challenges that arise through the application of artificial intelligence. Besides the manipulation of behavior by algorithms (e.g. voting behavior, decision making in online purchases) he identifies a problem in the application of artificial intelligence in the genetic analysis of humans as an area that challenges our self-understanding of being human to a great extent.

In **Could Machines Replace Human Scientists? Digitalization and Scientific Discoveries** *Jan Michel* proposes a new demarcation criterion for science called "the discoverability criterion". By arguing that it is probably an essential feature of science that we can make scientific discoveries, he suggests that artificial intelligence will not be able to replace human scientists as long as artificial intelligence is unable to make scientific discoveries.

In their contribution **Organizational and Societal Implications of the Adoption of Machine Invention Systems** *Dragos-Cristian Vasilescu* and *Michael Filzmoser* propose that machine invention systems have the potential to revolutionize the current standard invention process. They see the need to consider the widespread implications of this possibility on a societal and organizational level. On the organizational side, integration of such machine systems in the largely human-driven innovation process requires careful consideration. Delegation of decisions to human agents and technology assessments seem to be a few of the solutions needed to address the changes incurred by machine invention systems.

3. Acknowledgements

We thank all attendees and keynote speakers who decided to capture the sharp-witted, profound and vivid conversations and discussions, combine this input with their latest research, and contribute to this book. We thank the German Research Foundation (DFG), the Human-Technology Center at RWTH, the Chair for Philosophy of Religion and Philosophy of Science, Catholic Faculty of Theology, Ruhr-Universität Bochum, and the Catholic Academy Franz Hitze Haus in Münster for their support of the conference. And, finally, we thank Max Brunner and Dominik Winter for their assistance in putting this book together.

4. Bibliography

Alpaydin, Ethem. 2016 *Machine Learning*. London: MIT Press. 2016.

Boden, Margaret A. 2005. "Introduction" In: Margaret A. Boden (ed.): *The Philosophy of Artificial Intelligence*. Oxford: Oxford University Press. 1–21.

Turing, Alan. 2005. "Computing Machinery and Intelligence" In: Margaret A. Boden (ed.): *The Philosophy of Artificial Intelligence*. Oxford: Oxford University Press. 40–66.

PART I

Consciousness

Conscious, Thinking, and Intelligent Machines?

Some Clarifying Remarks in the Field of Articial Intelligence

Tobias Müller

Abstract

In recent years modern technology has transformed almost all areas of our lives, some of them fundamentally, through automation and technology. A new dimension of this technical progress can now be seen in "artificial intelligence", that is, in computers whose algorithmic processing is known as "machine learning", the "artificial" generation of knowledge from experience. Progress in AI research has led to the relevant literature already assigning many AI systems the ability to think, learn, predict, analyze, decide, know, and plan. Even autonomy and self-awareness are already attributed to some systems. In order to be able to answer the question of whether such attributions are justified, there is a considerable need for conceptual clarification of this point. In my paper I would like to present some considerations from the philosophy of mind and the philosophy of technology, which could be helpful to achieve a clarification of certain terms. For this purpose, various working definitions of "consciousness", "thinking" and "intelligence" will be presented, whereby specificities of human consciousness and thinking will be highlighted. Subsequently, it will be discussed whether these specific qualities can already be found in today's AI systems and whether there are fundamental limits of AI systems.

1. The Success of Artificial Intelligence Research and the Related Philosophical Questions

Modern technologies influence our daily lives in many ways: Certain facets of communication, social interaction, and production are often already completely digitalized. This was made possible by the use of computer technology, which has proven to be a helpful tool for tasks that previously required a great deal of time and effort. A new dimension of this technical progress is now seen in "artificial intelligence", i.e. in computers whose algorithmic processing is known as "machine learning", which means the "artificial" generation of knowledge from experience. It consists in the fact that an artificial system "learns" from examples and can generalize from them after completing the

 | DOI:10.30965/9783957437488_003

learning phase.[1] A large number of examples are used to train the system to read certain data patterns from a set in order to recognize similar patterns in new case studies. Certain nodes in the artificial neural network are adjusted so that the desired patterns can be recognized quickly and reliably. The adjusted system searches through data sets and tries to identify the prototypes learned from the samples.

Thus, the further development of digitization in the form of "artificial intelligence" achieves an unprecedented dynamic, because now the system not only has a basis of preprogrammed information but is also in a position to recognize new structures independently. The enormous progress in the development of artificial intelligence, which in some areas is already far superior to human cognitive abilities, is the beginning of a far-reaching social change that will have a massive and unpredictable impact on our lifeworld. In the future, AI systems will increasingly be integrated into our everyday and working lives, making them an indispensable part of our day-to-day. Indeed, many experts and scientists therefore see this development as nothing less than a fundamental revolution in our human existence. On the one hand, these technical innovations offer enormous potential for supporting human life (for example in research, communication, and business); on the other hand, these developments also raise fundamental philosophical questions concerning both the "inner life" of AI systems and the way we think about the human mind: What is intelligence? Can machines think? Are intelligent robots conscious? Can the human mind be rightly thought of as a computer?

The question of whether human cognitive abilities can be technically realized was already on the agenda at the beginning of AI research. At a symposium in Dartmouth in 1956, which is regarded as the starting point for efforts towards artificial intelligence, the aim was to describe learning and all other characteristics of intelligence precisely enough that, with these findings, a machine could be built that simulates these processes.[2]

The extent to which the cognitive abilities of AI systems have now been developed can be seen from a wealth of cases in which they surpassed human competitors: A prominent example is the victory of the computer system AlphaGo, which was produced by Google, over a South Korean master player in the board game Go. This millennia-old game represents strategic thinking, human creativity and intelligence, and is considered the most complex

1 Cf. for example Walsh 2017: 61–65.
2 Cf. Ramge 2018: 33.

board game – it has more potential game variants than there are atoms in the universe.[3]

Progress in AI research has led to the fact that the relevant literature already assigns many AI systems the ability to think, learn, predict, analyze, weigh, decide, know, and plan. Even autonomy and self-awareness are already attributed to some systems.[4] In the context of this attribution, one important distinction of great relevance must be pointed out: proponents of so-called weak AI see the goal of AI systems as solving concretely applicable problems by *simulating* human intelligence. It is neither assumed that the computer has mental characteristics nor that the human brain is basically a computer. In contrast, proponents of a so-called strong AI assume that the brain is a computer, such that mental properties and abilities must also be attributed to the AI system in a non-metaphorical sense.[5] In order to answer the question as to whether such attributions are justified, there is a considerable need for conceptual clarification at this point: How do the concepts used differ from concepts of consciousness, thinking and intelligence? Are these related to each other? What criteria are there for attributing these qualities to AI systems?

In my paper I would like to present some considerations from the philosophy of mind and the philosophy of technology which could be helpful in order to achieve some clarification. For this purpose, different working definitions of “consciousness”, “thinking” and “intelligence” shall be presented, whereby specificities of human consciousness and thinking will be highlighted. I will discuss whether these specific qualities can already be found in today’s AI systems, and whether there are principal limits of AI systems.

The starting point for these clarifications is the concept of the Turing machine. According to the generally accepted explanations of the mathematician Alan Turing, the computer is a machine that can manipulate formal symbols according to syntactic rules and thus works algorithmically, i.e. by the step-by-step transformation of a sequence of characters according to a certain scheme. This applies to AI systems as well, even if they can recognize new data patterns through learning processes that were not previously programmed. Attributing consciousness, thinking and intelligence to the computer would mean that these qualities can be traced back to a causal-functional structure, thus allowing them to be represented as algorithmic manipulation of formal characters. This means that in the discussion above, one basic question is of importance:

3 Cf. Walsh 2017: 34–36.
4 Cf. Kurzweil 1990: 19.
5 Cf. Schäfer 1994: 104–105.

Can consciousness, thinking and intelligence be understood as a result or effect of algorithmic processing?

In order to answer this question, it is first necessary to describe the characteristic features of these qualities. In a second step, it will be asked whether these qualities can also be attributed to AI systems.

2. Characteristics of Consciousness, Thinking and Intelligence

Of course, it is not possible to reconstruct here the extensive and disparate discussion of concepts of subjectivity in the philosophy of mind. Instead, I would like to outline the main characteristic of the basic concepts mentioned above, which are inescapable for any meaningful concept of subjectivity. These definitions will also help to distinguish between "consciousness", "thinking", and "intelligence", wherein the first two concepts also will be included under the generic term "mental states".

I would like to begin with the so-called phenomenal consciousness, which has essentially shaped the discussions of recent decades and represents the basic mode of subjectivity, as it is ascribed to all experiencing living beings. The definition of phenomenal experience can be approached with a proposal by David Chalmers:

> We can say that a being is conscious if there is something it is like to be that being [...].[6]

This description goes back to Thomas Nagel, who in his essay "What Is It Like to Be a Bat?" approaches this aspect of consciousness as follows and links it with another aspect:

> But no matter how the form may vary, the fact that an organism has conscious experience at all means, basically, that there is something it is like to be that organism. There may be further implications about the form of the experience; there may even (though I doubt it) be implications about the behavior of the organism. But fundamentally an organism has conscious mental states if and only if there is something that it is like to be that organism – something it is like for the organism.[7]

Here Nagel tries to grasp two basic characteristics of consciousness: One is the phenomenal quality of the conscious experience, namely that it is a certain

6 Chalmers 1996: 4.

7 Nagel 1974: 436.

way to be in this state. This aspect is called the "what-it's-like-ness" dimension of consciousness and always involves another characteristic of phenomenal consciousness which is also mentioned by Nagel: the phenomenal quality is always for a subject who experiences it. Or to put it another way: phenomenal experience is an aspect of an experiencing living being that is capable of having this experience as a subject.[8]

Single qualities of phenomenal consciousness are called "qualia".[9] These include, for example, certain visual and auditory impressions, olfactory and tactile experiences, feelings, etc. Phenomenal consciousness is thus the way in which things present themselves to an experiencing subject.[10] This relation of experiencing living beings to the world through phenomenal consciousness is also a fundamental prerequisite for a thinking subject to generate knowledge about the world. Thus, thinking is a more specific mode of subjectivity that refers to the contents of phenomenal consciousness, which is related to reality.

This relation of human subjectivity to the world is also the basis of the semantic dimension of using signs. This can be illustrated by the so-called semiotic triangle: a sign only has meaning if there is an interpreter (an experiencing and thinking subject) that assigns the sign to an aspect of reality.[11] This means that the use of signs is also linked to the process of producing the meaning of signs by establishing the connection between signs and the associated aspect of reality via an act of thinking. Only in this context can signs represent something qualitatively determined for an interpreter. The semantic dimension thus plays a decisive role for recognizing the world: it links the signs used in thoughts to the world.

A further constitutive feature of human thinking is the ability not only to take given norms into consideration, but also to critically question the associated normative standards themselves and to reflect on their application.[12] Thus, thinking is not restricted to blindly following rules, because the critical examination of whether certain rules should be applied in a concrete case is itself part of rational reflection. Therefore, human thinking has a certain degree of spontaneity. This means that the thinking subject is in principle in a position to freely determine the content and direction of its train of thought.

While phenomenal consciousness and thought form the two basic modes of human subjectivity, the term "intelligence" refers to a practical ability in the

8 Cf. for example Searle 2000: 561.

9 Cf. Searle 2000: 560f. Chalmers uses different tems for this quality: "experience", "what it is like", "subjective experience" and "phenomenal". Cf. Chalmers 1996: 6.

10 Cf. Chalmers 1996: 6–10.

11 Cf. Zwierlein 1990: 352–352.

12 Cf. Stekeler-Weithofer 2012: 184–185.

context of problem solving. Klaus Mainzer has proposed the following definition of intelligence, which I would like to adopt in this context as a working definition:

> A system is called intelligent if it can solve problems independently and efficiently. The degree of intelligence depends on the degree of autonomy, the degree of complexity of the problem, and the degree of efficiency of the problem-solving process.[13]

Already, the formulations "independent" and "problem" assign intelligence to a purposeful rational context, because problem solving relates to a certain task, which is connected with obstacles. Solving a problem thus means that a certain goal is to be achieved with suitable and efficient means and methods.

3. The Attribution of Mental Properties to AI Systems

3.1 *Attempts to Functionalize Phenomenal Consciousness*

The discussion about whether mental properties can be attributed to computers goes back to the 1960s. It started with the question of how mental states can be realized by physical structures, because only then would it be possible to technically build structures that produce artificial mental properties.[14] At the beginning of this discussion, the so-called identity theory dominated the debate, postulating an identity of physical and mental properties. Mental states, according to this early paradigm, are identical to certain classes of physical states (so called type-type theory of identity). However, these positions were confronted with the argument of so-called 'multiple realization', which states that the same mental states can be realized differently in different living beings.[15] The pain of a human being and the pain of an octopus as a phenomenal quality are based on very different physiological structures. Mental states are therefore physically multiply realizable. Criticism paved the way for the so-called functionalist approach, which no longer regards mental states as identical with physical structures, but rather sees mental states as identical

---

13 Mainzer 2016: 3 (translated by the author).

14 The term 'mental' functions as an umbrella term which includes phenomenal consciousness as well as thinking.

15 This line of argumentation goes back to Putnam 1967 and Fodor 1974.

with their causal roles, thus asserting a certain independence of performed function from its physical realization.[16]

A functional state is regarded as the set of all causal relations between mental states and environmental influences as inputs and behavioral reactions as outputs. Ned Block defines functionalism as follows:

> Functionalism is the doctrine that pain (for example) is identical to a certain functional state, a state definable in terms of its causal relations to inputs, outputs, and other mental states. The functional state with which pain would be identified might be partially characterized in terms of its tendency to be caused by tissue damage, by its tendency to cause the desire to be rid of it, and by its tendency to produce action designed to shield the damaged part of the body from what is taken to cause it.[17]

Functionalism is therefore the foundational concept for the position known as strong AI, because if mental states are identical with their causal role, then the respective mental states could also be created through the technical realization of the causal role. However, the assertion of identity between mental states and the causal role in so-called classical functionalism has proved untenable for various reasons. For example, determining what can be regarded as input or output for a specific mental state requires a common point of reference that can no longer be described by a functionalist vocabulary, leading functionalism to not be able to be clear about what respective input or output constellations refer to.[18] In addition, functionalist analysis cannot explain the basic "what-it's-like-ness" aspect of consciousness, because the specification of causal roles does not say anything about phenomenal quality. Thus, the essential characteristic of mental states can no longer be addressed in functionalist approaches.[19] Justifiable doubts have been raised about the possibility that causal roles can provide any unambiguous characterization for mental states. After all, we know from experience that phenomenal states in different people, indeed even in the same person, can be accompanied by completely different behaviors. One can seriously doubt that every phenonemal quality must have a causal role. There are well-known cases in anaesthesia, for example, wherein

16 It should be noted that functionalism is actually ontologically neutral. Nevertheless, it is often combined with a so-called token-token theory of identity, which claims that tokens of mental states are identical to tokens of physical states. Here remains the "hard problem" of consciousness of how a physical state can be identical with a state of phenomenal consciousness.

17 Block 1980: 257.

18 Cf. Hoffmann 2013: 41.

19 Cf. Kim 1998: 126–128.

the patient undergoing a surgery, who has been completely paralyzed by muscle relaxants, nevertheless consciously experiences everything – including the sensation of pain – without this being physically detectable.[20]

These difficulties show that the quality of phenomenal experience is not fully grasped by the description of causal roles. As such, that a computer imitates human behavior does not mean that the computer has a phenomenal consciousness, since there is no necessary connection between the two realms.

3.2 *Attempts to Functionalize Human Thinking*

Similarly, proponents of so-called computer functionalism claim that human thinking can be analyzed in purely syntactical terms. Thinking, then, is understood as a purely syntactic manipulation of symbols, so that on the level of the computer it is conceived as an algorithmic manipulation of a sequence of characters. In order to better deal with the question as to whether the "cognitive abilities" of a computer are identical with human thinking, different levels in the structure of a computer must first be distinguished. This is not usually sufficiently taken into account when equating mind and computer: 1. the physical level of the computer, 2. the syntactic level, and 3. the semantic level. In short, it will be shown that the characteristics of a higher level cannot be brought about by the features of the respective lower level (beginning with the physical level as the lowest level). Additional assumptions have to be made which are not entailed in the lower level. This means that both the syntactic and semantic dimensions of "cognitive" processes of the computer do not lie within the computer itself. This will now be shown by the analysis of the individual levels and their determinations.

On the physical description level, the computer represents a system of physical states which follow each other in time according to causal laws.[21] On this level the computer is nothing else than the sum of physical states. Only the technical and planned realization of the arrangement of certain switching elements and their interaction makes it possible to electronically map functions of Boolean logic to these physical structures by so-called logic gates.[22] This in turn is the condition for the fact that in a second step, different algorithmic structures can be mapped by the programmer on the physical level. It is this intentional mapping that ensures that the physical structures become interpretable syntactic structures which are considered as a set of rules for the

20 Cf. for example Domino, et al. 1999.

21 Cf. Pylyshyn 1980: 113.

22 In principle, logic gates can also be realized physically in a completely different way, but today they are almost all realized electronically.

acceptable combinations of symbols. The physical structures of the computer do not inherently possess this syntactical dimension, because this assignment does not result from the physical properties of the processes alone, but is artificially created by the programmer. The physical structures only have to be arranged in such a way that they also reflect the syntactic manipulations by certain physical manipulations.[23] The syntactic underdeterminedness of the physical structures can be recognized among other things by the fact that there is a kind of multiple realizability here. Syntactic structures can be mapped to physical structures in different ways. From an analysis of the physical structures alone it cannot be concluded which syntax is realized in the concrete case.

In addition to this syntactic level there is now a semantic interpretation by the programmer or user, who connects the syntactic structures with semantic contents. Again, the lower level does not logically entail the interpretive level proceeds from it. No relation to reality follows from a purely syntactic structure. Whether the computer is currently calculating the weather or playing chess cannot be deduced from a purely syntactic analysis, which is due to the fact that pure syntax does not determine the semantics. Only the programmer determines this connection of the different levels, whereby he – as shown – is active on the syntactic as well as the semantic level. Moreover, his movement is freely determined by his decisions. Therefore, it is misleading to understand the computer as a symbol-processing machine, because strictly speaking, the references to semantic interpretations always exist only on the level of a programmer or user which is not accessible to the computer itself and does not have to be accessible for it to complete its task.[24] On this level the program is a physical structure, which in turn generates other physical structures. John Searle has illustrated this distinction in the thought experiment of the so-called Chinese room, through which he clarifies that real understanding does not simply mean following syntactic rules, because an encompassing concept of understanding entails the semantic dimension with which the meaning of the used symbols is determined. This determination finally requires already an experiencing being which is related to reality through its contents of phenomenal consciousness and has the ability to determine autonomously which aspect of reality the symbol stands for.

This can also be illustrated by a simpler example from Klaus Fischer: Imagine a meteorite appears with a surface engraving which looks like this: “<<>><><><<<>>>>><<<<<<><<<”. If one assumes that this engraving is a

---

23 Cf. Fischer 2003: 45.

24 Cf. Fischer 2003: 45.

code and there is no further information about the context or meaning of the engraving, then it is neither possible to decipher this code nor to say whether it is a code at all.[25] The semantics does not result from the syntax (even if we knew the syntactical structure), but are rather the result of a mapping process. If, therefore, computer functionalism regards the human mind as a purely syntactic working machine, the problem of semantics is simply ignored.

This has led to the following problems in the context of computer functionalism: Starting from human epistemological practice, which must be presupposed for every discourse and thus also represents a prerequisite for the construction of computers, thinking is a mode of subjectivity in which real knowledge of reality is gained by generating adequate description in thoughts. Thinking refers primarily to phenomenally experienced things as having certain qualities and thus can be classified into a certain categorical scheme: We recognize something as something which is qualitatively determined. In thinking, we can refer both to phenomenal given, i.e. to those qualities which are given by experience, and to the structure of thinking itself by thematizing its characteristic features and prerequisites. In both cases, the concepts used in certain thoughts must have a qualitative reference to reality that goes beyond pure syntax or formal semantics. In order to determine things qualitatively, it is not sufficient that an abstract pattern is assigned to another abstract pattern, because a semantic underdeterminedness always remains, which must necessarily be overcome in qualitative judgements that refer to the concrete world.[26]

As already outlined above, another difference between human thinking and the "machine thinking" of the computer is that human thinking has in principle the ability to freely determine its thought processes and thus critically question the normative guidelines that guide it.[27] Thinking therefore does not consist in blindly following given rules, but also entails the ability to question the given normative framework including rules, norms, and criteria that determine when certain rules are to be applied.

The argumentation outlined so far also applies to AI systems. For even though AI systems differ from conventionally programmed computers in that their input-output relations are dynamic due to machine learning and the associated training data, they are still Turing machines that algorithmically process character strings in their "learning process".

---

25 Cf. Fischer 2003: 48.

26 A similar problem arises in the logic of the so-called extensionality thesis. Certain intensional correlations cannot be represented solely on the basis of the extensional relations. Cf. Weingartner 1972: 127–178.

27 Cf. Stekeler-Weithofer 1990.

4. A Short Conclusion

The enormous progress in the development of the capabilities of AI systems naturally raises the question as to whether these systems only simulate consciousness, thinking, and intelligence, or whether these features must be attributed to them in a genuine sense. The brief analysis of human phenomenal experience and human thought has revealed some constitutive features of these modes of subjectivity. The technical reconstruction of the computer and a critique of the functionalist paradigm have shown that even AI systems, due to their structure, do not possess these capabilities.

Inasmuch as AI systems are already far superior to human cognitive abilities in certain areas of data analysis, they represent a useful tool that will help us to solve the respective tasks and challenges in many areas where the recognition of data patterns is at stake. In the future, AI systems will have a massive and unpredictable impact on our human lifeworld. However, such systems are not able to be conscious or think in a fuller sense, because the underlying functionalist interpretation of "consciousness" and "thinking" excludes essential aspects of these phenomena and is at the same time the basis of constructing computer systems.

Nevertheless, a form of intelligence in the figurative sense can be assigned to AI systems if one is referring to the definition of intelligence presented above. AI systems are designed through their trained artificial neuronal networks in such a way that they can very effectively and "independently" find new data patterns, which then represent solutions for existing problems. However, it should not be forgotten that the semantic reference of such systems and their "rationality" depend on the engineers, programmers, and users, as well as the associated purposes, and training units.

5. Bibliography

Block, Ned (1980) "Are Absent Qualia Impossible?" *The Philosophical Review* 89, 257–274.

Chalmers, David (1996) *The Conscious Mind. In Search of a Fundamental Theory*. New York / Oxford.

Domino, Karen / Posner K. / Caplan R. / Cheney F. (1999) "Awareness during anesthesia: closed claims analysis." *Anesthesiology* 90, 1053–1061.

Fischer, Klaus: (2003) „Drei Grundirrtümer der Maschinentheorie des Bewusstseins." In: Wolfgang R. Köhler / Hans-Dieter Mutschler (ed.): *Ist der Geist berechenbar?* Darmstadt, 33–57.

Fodor, Jerry (1974) "Special Sciences (Or: The Disunity of Science as a Working Hypothesis)." *Synthese* 28, 97–115.

Hoffmann, Thomas S. (2013) „Hermeneutischer Naturalismus." In: M. Gerhard / C. Zunke (eds.) *Die Natur denken*. Würzburg (2013), 27–55.

Kurzweil, Ray (1990) *The Age of Intelligent Machines*, Cambridge.

Kim, Jaegwon (1998) *Philosophie des Geistes*, Wien / New York.

Lenzen, Manuela (2018) *Künstliche Intelligenz. Was sie kann & was uns erwartet*. München.

Mainzer, Klaus (2016) *Künstliche Intelligenz – Wann übernehmen die Maschinen?* Berlin.

Nagel, Thomas (1974) "What Is It Like to Be a Bat?" *The Philosophical Review* 83(4), 435–450.

Putnam, Hilary (1967) In: W. H. Capitan / D. D. Merrill (eds.), Art, Mind, and Religion. Pittsburgh, pp. 37–48.

Pylyshyn, Zenon W. (1980) "Computation and Cognition: Issues in the foundations of cognitive science" *Behavioral and Brain Sciences*, 3(1), 111–132.

Ramge, Thomas (2018) *Mensch und Maschine. Wie künstliche Intelligenz und Roboter unser Leben verändern*. Ditzingen.

Schäfer Erich (1994) *Grenzen der Künstlichen Intelligenz. John R. Searles Philosophie des Geistes*. Stuttgart / Berlin / Köln.

Searle, John R. (1980) "Minds, Brains, and Programs" *The Behavioral and Brain Sciences* 3, 417–457.

Searle, John (2000) "Consciousness." *Annual Review of Neuroscience* 23, 557–578.

Stekeler-Weithofer, Pirmin (1990) „Verstehen und Begreifen (nach Hegel). Zum Konservativismusproblem technischer Information und Rationalität." In: *Mitteilungen des Deutschen Germanistik-Verbandes* 37/3. Frankfurt/M., 19-25.

Stekeler-Weithofer, Pirmin (2012) *Denken. Wege und Abwege in der Philosophie des Geistes*. Tübingen.

Walsh, Toby (2017) *Android Dreams. The Past, Present, and Future of Artificial Intelligence*. London.

Weingartner, Paul (1972) „Die Fraglichkeit der Extensionalitätsthese und die Probleme der intensionalen Logik." In: R. Haller (ed.) *Jenseits von Sein und Nichtsein*. Graz, 127–178.

Zwierlein, Eduard (1990) „Künstliche Intelligenz und Philosophie. Zur Debatte um J. R. Searle's Einwände gegen harte KI-Versionen." *Journal for General Philosophy of Science* 21, 347–358.

Deus Malignus – The Digital Rehabilitation of Deception

Walther Ch. Zimmerli

Abstract

One of the main questions implied in what we today call "digitalization" is not what happens, *when* computers (or in our case: when robots) think but rather *if* it makes sense to talk of computers, robots, or any kind of machines *as if* they were capable of thinking.[1] Or formulated in a still different way: Does it make sense to call machines "intelligent"? It goes without saying that the *locus classicus* of this question has been Alan Turing's pathbreaking article on "Computing Machinery and Intelligence", published 1950 in *Mind*[2]. What I will be dealing with in what follows will therefore just have the status of some philosophically orientated modest footnotes to the main idea of Alan Turing. These footnotes will be developed in five steps: Before even entering the arena of Robots and Artificial Intelligence[3] I will try to open up a space for thorough reflection which will enable us to discuss these issues by not following the beaten track. In order to do that we will first of all ask whether "bullshit makes sense" by critically dealing with "the Digital", a notion taken for granted by almost everybody (1), and by then asking the seemingly very pedestrian question as to how intelligent the CIA is (2). This then will set the scene for reminding us of a very influential argument in early modern sceptical rationalist philosophy: Descartes' "Deus" or "Genius Malignus" argument (3). From there it will be easy to proceed to an important but often neglected implication of the "Turing test" (4) culminating in a revision and rehabilitation of one of the most abhorred concepts of modern science and philosophy: deception (5), thus launching what I would like to call an "anti-Cartesian experiment".

1 Cf. McCorduck 1979.
2 Turing 1950.
3 Cf. Zimmerli 2014: 11–25.

 | DOI:10.30965/9783957437488_004

1. Does "Bullshit" Make Sense?

Since Harry Frankfurt's famous book the notion of "bullshit"[4] has become nobilitated and transformed into a philosophical term. To put it in the words of the well known Swiss writer Adolf Muschg: "Bullshit doesn't really want to know what it is talking about. It is completely satisfied to behave 'as if'. Bullshit has to sound of something and of so much that nobody even dares asking what is behind it: [...] it is impossible that something that sounds of so much could turn out to be nothing."[5] Or to put it in a less poetic way: Bullshit consists of empty concepts, ideas etc. with no referential meaning. Thus the question is: do they at least make sense?

One of the core tasks of philosophizing consists in distinguishing what makes sense from nonsense, and this task could be called a *bullshit test* accordingly: Especially concepts and ideas which are very popular and frequently used to characterize important features e.g. of the world and/or our present situation in it have to undergo this test. It goes without saying that among these the notion of *the Digital* may be considered one of the (if not *the*) most inflationary used ones and therefore a prime candidate for the bullshit test.

1.1 *Digital vs. Analogue – a "Bullshit"-Alternative?*

Like with respect to quite a few concepts of philosophical interest also the notion of "digital" is to be seen as mutually referring to its counterpart, the notion of "analogue": Digital is what is not analogue *et vice versa*, and at first glance this alternative seems to belong to the group of alternatives "without alternative"; but in a perspective of philosophical criticism it looks promising to apply a bullshit test preferably to those alternatives to which presumably no alternative exists, i.e. in our case to the alternative of "digital vs. analogue".

In doing so we might realize immediately that what first comes to mind in terms of lifeworld examples for the alternative "digital vs. analogue" is not originating from the world of "the Digital", i.e. from IT or computers etc., but from watches, and the explanation would be that digital watches represent time in numbers, whereas analogue watches do so by the relative position of its hands.

But even this simple example at second glance turns out to be much more complicated. On the one hand after a period of time when e.g. "digital" wristwatches were fashionable there has been a renaissance of analogue time measuring instruments, especially in the market segment of luxury wristwatches. On the other hand watches as such are internally working neither digitally nor

4 H. Frankfurt 1986.
5 Muschg 2006: 253 (my transl. WChZ).

analogously, but electronically and/or mechanically. And to further increase the confusion semiotically: The analogous measuring of time is performed by “hands” or “digits” (fingers), reminding us of the original connection between counting and finger calculation. Thus already the etymological findings indicate that “digital or analogue” isn’t a disjunctive alternative (“either-or”), but refers to a rather complex semiotic network. Therefore the seemingly simple idea that “digital” and “analogue” would be mutually excluding each other is rendered obsolete: Not only does the Digital not completely exclude (or for that matter replace) the Analogue, but quite the contrary: The more our world is digitalized the more we need to re-analogize it.[6] Already in this sense the disjunctive alternative “digital vs. analogue” turns out to be misleading. But is it just misleading or in addition even a bullshit alternative?

1.2 *Leibniz in the Supermarket*

As already indicated the literal meaning of “digital” is referring to both fingers and numbers. It is no secret that our decimal system although being far from being unique is connected to the fact that we have ten fingers (lat. “digitus”) with the help of which we are able to perform simple calculations in the elementary numeric space of iterating the number 10.

It isn’t an invention of the 20th or the 21st century that thinking and calculating are somehow related. Since Raimundus Llullus’ “Ars Magna”, i.e. since the end of the 13th century, the idea of an “ars combinatoria” has become notorious, and Thomas Hobbes anticipated three centuries later the pathbreaking insight of G. W. Leibniz that thinking is nothing but calculating (in his words: “computatio”). The obvious question we have to ask here thus is: Why is it that although both Hobbes’ idea that thinking in actual fact is computation, and Llullus’ “ars magna” already existed, the development of digitalization didn’t utilize the whole numeric decimal space, but settled for just two different numeric states: 0 and 1? To better understand not just this but also the whole process called “digitalization” it might turn out helpful to briefly look at *calculization, mechanization*, and *formalization*.[7]

The notion of “calculization” refers to tokens (lat. “calculi” = pebbles or stones used for counting) representing certain quantitative numeric values, e.g. the differently coloured pearls on a calculating frame called “abacus”.

The notion of “mechanization” (gr. “mechane” = instrument) refers to the fact that e.g. on a calculating frame (“abacus”) not only numeric values are represented by differently coloured pearls, but in its very structure also the

6 Cf. W.Ch. Zimmerli 2018b: 38.

7 W. Ch. Zimmerli/S. Wolf, “Einleitung”, in: Zimmerli/Wolf (eds.), *op cit.*, 8–11.

admitted rules according to which these pearls are supposed to be moved representing e.g. the mathematical operations of addition, subtraction, multiplication, and division.

The notion of "formalization", however, is referring to a procedure going at least as far back as to Aristotle who in his logic (gr. "organon" = tool) is operating with variables. i.e. formal symbols representing whole groups or families of tokens and thus allowing the symbols to be moved according to these rules without referring to their content. And although he wasn't the first to realize that all the different admitted moves, i.e. in logic: the different transformations of propositions, had but two values, T(rue) and F(alse), he was the first one to formulate it theoretically.

It goes again without saying that it could be disputed (and would by the way be irrelevant) whom afterwards to attribute priority, but it was without any doubt the philosophical genius of Gottfried Wilhelm Leibniz who made the non-obvious pathbreaking next move: to combine the insight into the two-valued fundamental character of logic with the multiplicity of the states represented by the respective tokens thus opening the avenue to mechanize the discrete states, or to put it differently: to reduce semantics to syntax. The – partly unfinished – project of Leibniz which is relevant here is what he called "characteristica universalis", in our context a universal language capable of calculization, mechanization, and formalization. The main idea is as simple as it is convincing: Given, each and every object as well as each and every state of affairs had been given a proper name which can be expressed by a sequence of 0 and 1, and given a set of rules which also can be reduced to 0 and 1, then the representation of the whole world could be calculated – "characteristica universalis" turning into "mathesis universalis"[8]

Too far-fetched and too abstract? – Just imagine you go to the Supermarket to buy e.g. bananas. You take them off the shelves, put them on one of the automatic scales and punch in the code number for bananas, e.g. 101. The inbuilt printer of the scale then prints a self-adhesive price-tag displaying a bar-code and at least the following information: Item (in this case: bananas), price/kg (in this case: CHF 3.95), netweight (in this case: 0.778 kg), item-price (in this case: CHF 2.68), and date of expiry (in this case 11.02.2019). You bring your bananas to the cashier who draws it over the glass of an automatic scanner, you pay CHF 2.68, get a receipt and leave

According to the digital-analogous-alternative this example could be analyzed as follows:

8 Cf. Mittelstrass 1979: 593–610.

(1) To choose bananas is in itself neither digital nor analogous.
(2) By punching in the respective code-number the first transition to the digital is made.
(3) The printer prints a self-adhesive price-tag in a both machine-readable (digital) and analogous code.
(4) At the cashier's the information on the price-tag is scanned digitally.
(5) The cashier's display shows the respective total in numbers, but not in digital code, and that's why it might be called analogous, although literally it is digital.
(6) The payment is then made either in cash (analogously) or by creditcard (in which case a new cycle of partly digital and partly analogous steps is started).

Although Leibniz even in his wildest dreams wouldn't have thought of the supermarket cashier, his idea of a "characteristica universalis" did indeed represent a major step towards digitalization. This step consisting in a combination of calculization, (electronic) mechanization and formalization and having become possible only because of the complete reduction of semantics to syntax, is to be witnessed in each and every so-called "digital" process – from the first binary calculating machines until today.

1.3 *The Digital – the Discrete Binary*

Leibniz in the Supermarket may teach us one more lesson: Besides demonstrating quite powerfully that "digital" and "analogous" aren't a disjunctive alternative it makes us observe that an argumentative pattern we are familiar with from the discussions on Artificial Intelligence (AI, cf.below 2.2) keeps returning time and again: There can be no doubt that changing the state of affair displayed by machines can be called "behaviour" (and especially "intelligent behaviour") if and only if at some stage something like an attribution of behaviour and intelligence – as a rule by human beings – is happening. And this would hold true even if we, for the argument's sake, would suppose something like an attribution of intelligence to machines by machines, because without a third party's (human observer's) perspective this would remain hidden forever and thus stay an eternal ("metaphysical") secret. To put it differently: Even if machines would attribute behaviour, and especially intelligent behaviour to other machines it would still need a human (intelligent) observer to attribute the quality of (intelligent) behaviour to this process.

So far, however, we haven't even touched the problem of intelligence. What we have done instead is to at least come somewhat closer to an understanding of "digitalization" that is capable of passing the bullshit test: By "digitalization" we understand a process of symbolically representing the whole world in different series of discrete binary states, thus reducing the multiplicity of semantic meanings to the binary syntax of logic. Unless understood this way, the use of "digital" and "analogous" as well as of the alternative "digital vs. analogous" doesn't make sense, i.e. is bullshit.

2. How "intelligent" is the CIA?

As stated above the question as to what happens when machines (or in our case: working machines, i.e. robots) think, is philosophically depending on the other question whether it makes sense to speak of machines/robots *as if* they were thinking, or to put it differently: to attribute intelligence to machines/robots. The very fact that at present we are witnessing a renaissance, or to put it more bluntly: a second hype of "artificial intelligence" isn't in itself proving anything – quite the contrary: it too might be a bullshit word.

2.1 *Of Synonymies und Analogies*

It is a well known fact that the notion of "Artificial Intelligence" when first introduced into academic and public discourse in the middle of the 20th century[9] was quite deliberately chosen by the young mathematician John McCarthy for its neutrality especially over against the paradigmatic fields of automata theory or cybernetics. As will be shown (cf. below 2.2) there are, however, still other reasons connected with the pragmatic meaning of any notion, playing with its ambiguity, in this case especially with respect to the second part "intelligence":

In English the notion "intelligence" refers to collecting and processing of data and information, as in "Central *Intelligence* Agency" (CIA), located in the George Bush Center for *Intelligence* in Langley, VA. In both of these cases it is quite obvious that they don't mean a central agency of extremely intelligent people, i.e. that the notion "intelligence" doesn't refer directly to what we call "intelligence" in ordinary language or in cognitive psychology. And something similar is the case with the first part "artificial" which literally doesn't mean anything but "technical" or "technological". Thus "Artificial Intelligence" could refer to something very unspectacular like technological collecting and processing of data/information.

---

9 After Alan Turing's already mentioned article on "Computing Machinery and Intelligence" (cf. above fn. 2) of 1950 it took more than five years until John McCarthy applied for funding at the Rockefeller Foundation (1955) and then (1956) organized the famous Dartmouth Summer Research Project on Artificial Intelligence (participants: Marvin Minsky, Nathanael Rochester, Caude Shannon, Herbert A. Simon, Allen Newell et al.) which is broadly considered the seminal event of the first wave of the AI-hype.

John McCarthy and most of the members of the worldwide AI-community, however, by using this notion activated one of the most powerful tools connected with human language: the capability of creating and using synonymies, metaphors, and analogies. As Douglas Hofstadter and Emanuel Sander have demonstrated in their encompassing book on analogy this capability is one of the core elements of human thinking and speaking[10] (which in turn sheds another light on the alleged disjunction of "digital vs. analogous"). Already in his proposal for the Dartmouth conference McCarthy was quite explicit about this synonymous use of the notion of "intelligence", drawing on the idea of the so called "Turing machine": "The study is to proceed on the basis of the conjecture that every aspect of *learning* or any other feature of *intelligence* can in principle be so precisely described that a machine can be made to simulate it."[11]

Obviously it was all about *simulating* learning and intelligence by a machine (or a robot); the question as to what the simulated "learning" and "intelligence" is all about was, however, never asked, although already Alan Turing had dealt with it (cf. below 4 and 5).

2.2 *How to Do Things with Words*

So from the very beginning of AI research it has always been this deliberate semiotic ambiguity that was characteristic for the discourse on AI. Therefore, for a philosophical approach to take a stance of criticism of ideology becomes ever more unavoidable: Why is it that the representatives of this then new field of research did not only seem to accept this ambiguity, but to even deliberately look for it?

To answer this question it is helpful to remind ourselves of the "pragmatic turn" in semiotics: "How to Do Things with Words" was the title of the William James lectures John L. Austin gave in 1955 at Harvard University and which, posthumously published as a book in 1962, was together with John Searle's writings to become the most influential founding documents of the so-called "speech act theory"[12], according to which words do not just have a referential, but also a pragmatic meaning: It is not only what they are referring to but it is also what they are capable of doing. The question originating from this approach therefore is: What is it that e.g. McCarthy and all of the other members of the AI-community wanted *to do* by not just accepting but deliberately searching for the ambiguity of the notion "Artificial Intelligence"?

---

10 Hofstadter/Sander 2013.

11 McCarthy/ Minsky/ Rochester 1955 (*Italics* by me, WChZ).

12 Austin 1962; Searle 1969.

Besides the already mentioned motive of searching for neutrality with respect to the dominant paradigms and schools of thought the promise of making the dream of a thinking machine come true seems to be a driving force. And by asking the leading question of any criticism of ideology ("Cui bono?") it becomes obvious that this promise is much more capable of raising funds than the austere idea of technologically collecting and processing data and information. The very notion of "Artificial Intelligence" has a mythical ring to it, it exerts some kind of poetic magic spell capable of opening hearts (and purses!), and this is what can be done with words.

2.3 *The (End of) Singularity is Near*

Since more than half a century people fall for this promise which always follows the same 5 to 10-years pattern: The 5-years-plan of planned economy is eternally recurring in this mythological 5-years promise. It doesn't come as a surprise that one of the most prominent versions of this promise which has attracted a lot of attention over the last decades and which – for whatever reason – is called "singularity" and is linked to another buzzword of these days "posthumanism" has been coined by Vernon Vinge, a mathematician and information scientist who is also science fiction author.[13] Singularity is the point in time when all the advances in technology, particularly in AI, will lead to machines that are in every respect smarter than human beings. Nurtured by the idea of exponential technological growth ("Moore's Law"[14]) Vernon Vinge as early as 1983 wrote:

> We will soon create intelligence greater than our own. When this happens, human history will have reached a kind of singularity, an intellectual transition as impenetrable as the knotted space-time at the centre of a black hole, and the world will pass far beyond our understanding. This singularity I believe, already haunts a number of science-fiction writers

amongst others obviously Vernon Vinge himself. He adds, however – somewhat obscure for a science fiction writer –, that this singularity

> makes realistic extrapolation to an interstellar future impossible.[15]

Much less cautious – as always – is Ray Kurzweil, since 2012 Google's director of Engineering and author of the book "The Singularity is Near"[16], who

---

13 Cf. Vinge 1993.
14 Moore 1965; cf. A. Thackray/D. C. Brock/R. Jones 2015.
15 Vinge 1983; cf. Dooling 2008: 89.
16 Kurzweil 2005.

predicted already in 2005 that in 2029 "an AI will pass a valid Turing Test and therefore achieve human levels of intelligence", and that in 2045 Singularity will be reached "which is when we will multiple our effective intelligence a billionfold by merging with the intelligence we have created"[17].

If a highly visible author like Kurzweil (who by the way has a reputation for not only founding companies by the dozens but also making predictions at an incredibly high frequency as well as for claiming that his accurate prediction rate is close to 86%!) is settling for such a long-term prediction instead of the usual 5 to 10 years it seems to be quite obvious that this again is more of a promise (or a self-fulfilling prophecy) than an accurate prediction. And this is especially true with respect to the fact that these singularity-oriented predictions all have to be seen against the background of the underlying conception of transhumanism believing in various possibilities of human enhancement technologies.

This powerful theoretical model representing a merger of evolutionary thinking and techno-science fiction does have deep roots in the philosophical "overman" tradition of Friedrich Nietzsche. It falls, however, victim to one of the at the same time most common and most detrimental futurologist AI fallacies. To put it in a nutshell: If singularity, i.e. the point in time when machines that are in every respect smarter than human beings, should ever be reached than we humans wouldn't even be capable to notice because we couldn't distinguish it from malfunctioning machines. Epistemologically speaking it would need at least one human being capable of understanding this situation of singularity. But then it by definition wouldn't be singularity and we would still be stuck with humanism (which in my opinion we human beings will be forever). This is what in the tradition of philosophical anthropology was called "conditio humana".[18]

3. "Deus Malignus" – an Anti-Cartesian Experiment

Although the techno-scientific and transhumanist prophecy of the coming Singularity has turned out to be based on a both mistaken and misleading concept, the debate on Artificial Intelligence and digitalization has opened up new avenues of philosophizing as well as put an end to important cornerstone elements taken for granted in traditional philosophy[19]. It goes without saying that this applies especially to the path-breaking philosophical achievements

17 Reedy 2017.
18 Cf. Arendt 1958.
19 For a popular version of what follows cf. Zimmerli 2018: 35.

of René Descartes. I am not talking about his in recent times heavily criticized ontological dualism, but about his two major contributions to the methodology of modern thinking, the "Descartes-Postulate" and the "Descartes-Axiom", both of which have to undergo a thorough scrutiny in the light of digitalization.

3.1 *Doubt – the Descartes-Postulate*

Up to date modern philosophical and scientific methodology seems to follow the postulate of methodical scepticism, *to take nothing for granted, i.e. to doubt everything unless it be perceived clearly and distinctively* ("clare et distincte percipitur"). To put it differently: to avoid deceptions by following the postulate to transform conjectures (scientifically: hypotheses) into insights or rather convictions (scientifically: confirmed knowledge). So the methodologically relevant question is now: How do we go about this process of transformation? And one possible methodological clue is that we might replace "conviction" or "knowledge" by "what has been confirmed beyond any reasonable doubt".

So the Cartesian instrument to find out whether a conjecture or a hypothesis really is capable of being transformed into knowledge is methodical doubt, and Descartes went to great length in order to develop a systematic methodology of doubt in a series of increasingly powerful arguments of doubt, from misperception and trick of the senses via mathematical and logical delusion to the argument of a deceiving genius. As this methodology is crucial for better understanding of what the anti-Cartesian experiment is opposing I will in what follows analyze it in detail:

(1) *The starting point of doubt*: Every kind of methodical and constructive doubt has to begin with destruction of all former beliefs, in Descartes' even more drastic words: "I will devote myself sincerely and without reservations to the general demolition of my opinions" (12:18).[20]

(2) *The generic and pragmatic principle of doubt*: As it is literally impossible to destruct each and every single opinion, the postulated methodical character of doubt is asking for a both generic and pragmatic procedure to reduce the potentially infinite number of possible opinions: "[...] it will not be necessary for me to show that all my opinions are false, which is something I could perhaps never manage. Reason now leads me to think that I should hold back my assent from opinions which are not

20 "[...] serio tandem & libere generali huic mearum opinionum eversioni vacebo", R. Descartes, *Meditations on First Philosophy: With Selections from the Objections and Replies*, ed. and transl. by John Cottingham, Cambridge UP, Cambridge 1996: 12:18 (in what follows the first page number always refers to the English translation by Cottingham, *op cit.*, the second one to the standard edition of the works of R. Descartes by Ch. Adam & P. Tannéry).

completely certain and indubitable just as carefully as I do from those which are patently false. So for the purpose of rejecting all my opinions it will be enough if I find in each of them at least some reason for doubt."[21] This is, what has been called Descartes' "withholding policy"[22].

(3) *Doubting the information given by the senses*: Therefore Descartes decided to "go straight for the basic principles on which all of my former beliefs rested"[23], first of all the information acquired "from senses or through the senses"[24], and here it turns out that sometimes "senses deceive". So according to the pragmatic procedure (2) it seems to be "prudent never to trust completely those who have deceived us even once"[25] (12:18) and thus generally mistrust all information given by the senses.

(4) *Doubting the distinguishability of dream and reality*: Again generalized this also applies to such basic certainties like the one, "that I am sitting here by the fire"[26], because even if it looks "insane" it is possible to doubt this due to the argument of the indistinguishability of dream and reality: "I see plainly that there are never any sure signs by means of which being awake can be distinguished from being asleep"[27] and dreaming.

(5) *Doubting the results of sciences*: "[...] physics, astronomy, medicine, and all other disciplines which depend on the study of composite things, are doubtful"[28].

(6) *Doubting the results of mathematics and logic*: "[...] arithmetic, geometry and other subjects of this kind which deal only with the simplest and most general things regardless of whether they really exist in nature or not, contain something certain and indubitable: For whether I am awake or asleep, two and three added together are five, and a square has no more than four sides. It seems impossible that such transparent truths

21 "Ad hoc autem non erit necesse, ut omnes esse falsas ostendam, quod nunquam fortassis assequi possem; sed quia jam ratio persuadet, non minus accurate ab iis quae non plane certa suntatque indubitata; quam ab aperte falsis assentionem esse cohibendam; satis erit ad omnes rejiciendas; si aliquam rationem dubitandi in unaquaque repero", ibid.

22 "If an opinion appears to be at all uncertain, or if it is possible to doubt it I will withhold assent from it, i.e. I will not believe it.", Wilson 2003: 33.

23 "[...] aggrediar statim ipsa pincipia, quibus illud omne quod olim credidi nitebatur", 12:18.

24 "[...] vel a sensibus, vel per sensus accepi", ibid.

25 "[...] has autem interdum fallere deprehendi, ac prudentiae est nunquam illis plane confidere qui nos vel semel deceperunt",ibid.

26 "[...] ut jam me hic esse, foco assidere[...] ", ibid.

27 "[...] tam plane videonunquam certis iudiciis vigilam a somno posse distingui"; ibid.

28 "[...] Physicam, Astronomiam, Medicinam disciplinasque alias omnes, quae a rerum compositarum consideratione dependent dubias quidam esse", 14:20.

should incur any suspicion of being false."[29] In order to submit even these seemingly self-evident truths to the generalized doubt Descartes had to take recourse to his strongest weapon by asking: "[…] how do I know that God has not brought it about that I too go wrong every time I add two and three or count the sides of a square?"[30] (cf. below 3.3).

3.2 *Ego Cogito – the Descartes-Axiom*

Descartes was quite explicit about what he was up to: his aim was to find the ultimate imperturbable foundation of all knowledge ("fundamentum inconcussum") – and thus not just the truth, but its reflexive mode: certainty.

It is common knowledge that he ended up with a self-referential, recursive mode of his methodological tool: His search for the one indubitable and thus certain truth resulted in the argument, that even if every supposed truth turned out to be dubitable then at least the very fact of doubting was in itself indubitable: It cannot be doubted that I doubt as long as a doubt. And as doubting is – according to his methodology of doubt – the basic mode of thinking and thinking presupposes a thinking subject, it follows that as long as I think I am, "I think, therefore I am" ("Cogito ergo sum"). And this now is the "fundamentum inconcussum", the Descartes-axiom, from which to reconstruct the whole ontology, i.e. the whole world.

Other than the obvious semiotic flaws which have drawn quite some fire of criticism (e.g. mistaking third-person propositions for first-person propositions etc.)[31] the methodology of systematic doubt has survived the centuries and has – in the shape of the methodology of experiments[32] – even reached the status of crucial testing of scientific hypotheses. If anything is characteristic for modern philosophical thinking then this: Only evidence-based results of (scientific) reasoning are accepted in the world of academia.

---

29 ."… atqui Arithmeticam, Geometriam, aliasque ejusmodi, quae nonnisi de simplicissimis & maxime generalibus rebus tractant, atque utrum eae sint in rerum natura necesse, parum curant, aliquid certi atque indubitati continere. Nam sive vigilam, sive dormiam, duo & tria simul juncta sunt quinque, quadratum non plura habet latera quam quatuor; nec fieri posse videtur ut tam perspicuae veritatis in suspicionem falsitatis incurrant", ibid.

30 "[…] ita ego ut fallar quoties duo & tria simul addo, vel numeri quadrati latera", ibid.

31 Cf. already Carnap 1931: 233f.; in a different pragmatic approach Hintikka 1962: 3–32; cf. Kenny 1968.

32 Cf. Galison 1987.

3.3 *Deus Malignus – the Strongest Argument*

Now let us just briefly re-examine this methodology of doubt constitutive for modern science, technology, and philosophy: As pointed out above, its aim was not just truth, but its reflexive mode: certainty. And therefore to come up with something true wouldn't have sufficed. Thus it wasn't about truth, but about guaranteeing truth. Since Aristotle truth was defined as a relation of adequacy between state of affairs and sentences/thoughts, in medieval terminology "adaequatio rei et intellectus". St. Thomas Aquinas knew that in order to answer the question as to the reflexive mode of truth, i.e. certainty, it needed an additional theorem, in his terminology: the theorem of the "analogia entis" which was guaranteed by God himself[33], in the later wording of Spinoza: that "the order and connection of (human) thoughts is the same as the order and connection of things".[34]

Descartes therefore knew that the strongest argument in his methodology of doubt was the argument of a deceiving maligne God whom, however, for different pragmatic reasons he couldn't possibly dare to call God and thus called "*genius* malignus": "I will suppose therefore that not God, who is supremely good and the source of truth, but rather some malicious demon of the utmost power and cunning has employed all his energies in order to deceive me."[35]

The strategy to avoid being misled and deceived by this maligne genius consists in radical scepticism, i.e. in refraining from assenting to any beliefs that hadn't previously passed the test of the Descartes postulate, "so that the deceiver, however powerful and cunning he may be, will be unable to impose on me in the slightest degree"[36].

In order to reach the reflexive status of truth, i.e. certainty, Descartes thought it necessary to fight the strongest counter-argument against the ontological foundation of the adaequatio-relation between the order and connection of things and the order and connection of human thoughts, i.e. against the analogia entis; and this strongest counter argument was the maligne version of God, i.e. a deceiving, maligne genius, in modern terminology: systematic deception.

---

33 Th. Aquin, *De veritate*, q.2, a 11; cf. Przywara 1932; Pannenberg 2007.

34 B. Spinoza, Ethica 2P7, "Ordo & connexio idearum idem est, ac ordo & connexio rerum."

35 "Supponam igitur non optimum Deum, fontem veritatis, sed genium aliquem malignum, eundemque summa potentem & callidum omnem suam industriam in eo posuisse, ut me falleret". (15:22)

36 "[...] nec mihi quidquam iste deceptor, quamtumvis potens, quamtumvis callidus, possit imponere, obfirmata mente cavebo." (15:23)

4. Philosophical Implications of the Turing-Test

One of the results of our philosophical analysis of the process of binary technologisation called "digitalization" has been: Due to the always necessary – analogous – interpretation we human beings are *nolens volens* always fellow players or even partners in the great digital game. And even more: We cannot but be natural DACs (Digital-Analogue-Converters). In spite of all the conceptual pseudo promises like "autonomous driving", "Artificial Intelligence" or "Singularity" there will never be just machines operating in discrete binary steps, but always human being-machine hybrids, and this also applies to robots. But if this is the case then the "paradox of (intelligent) information technology"[37] becomes unavoidable: The more effective and "better" the machine-part of the human being-machine tandem works the less it becomes possible to meet the requirements of the Descartes-postulate: to take nothing for granted unless it be clearly and distinctively perceived ("clare et distincte percipitur").

Although very complex in its implications this paradox can be demonstrated quite easily using simple machine theory: If according to 19th century Hegelian philosopher of technology Ernst Kapp each machine has to be considered some kind of projection of a human organ ("Organprojektionsthese"[38]) then digital ("intelligent") machines are projections of the thinking organ, i.e. the brain. As, however, the very task of every machine is to surpass as far as possible the performance of the human organ it is the projection of, it follows that by definition there is no way to meet the Descartes postulate, i.e. the possibility to control the results. This is even evident with respect to a gadget as simple as a pocket calculator.

In a high tech environment like in the complex IT machinery of an airplane cockpit this problem is dealt with by means of redundancy: a number of additional intelligent machines are in charge with mutually controlling each other "autonomously" as it were. This, however, results in missing the Cartesian goal of direct cognitive control by human beings even further. To put it differently: The Descartes postulate is honoured at the expense of statistically externalising it by binary technology which, however, in fact equals getting rid of literally meeting it.

4.1 *What is the Turing Test Testing?*

The implications of this "anti-Cartesian experiment" (as I would like to call it) becomes obvious especially with respect to our case of Artificial Intelligence

37 Cf. Zimmerli 1986: 294.
38 Kapp 1877: 29 ff.

and Robots. In addition to the problematic ambiguity of the term "intelligence" mentioned above AI and its success story are based on a fundamental thematic shift only now visible: Already Alan Turing had replaced the initial question whether machines are capable of thinking by the description of an experimental arrangement which to his mind was functionally equivalent: the "imitation game":

> I propose to consider the question 'Can machines think?' This should begin with definitions of the meaning of the terms 'machine' and 'think'. [...] Instead of attempting such a definition I shall replace the question by another, which is closely related to it and is expressed in rather unambiguous words. The new form of the problem can be described in terms of a game which we call the 'imitation game'. It is played with three people, a man (A) a woman (B), and an interrogator (C) who may be of either sex. The interrogator stays in a room apart from the other two. The object of the game for the interrogator is to determine which of the other two is the man and which is the woman. [...] We now ask the question, 'What will happen when a machine takes the part of A in this game?' Will the interrogator decide wrongly as often when the game is played like this as he does when the game is played between a man and a woman? These questions replace our original 'Can machines think?'[39]

So from the very beginning, almost 70 years ago, it never was about the question whether machines, resp. the programmes run by them were capable of thinking or were in this sense "intelligent"; the one and only question was whether they could be made to behave in a way that renders an interrogator, i.e. an observer (or even more important: a user) unable to distinguish it from intelligent behaviour or thinking of a human being. With respect to almost everything we today subsume under AI it furthermore doesn't really matter at all, how we would answer the question as to the intelligence of the machine(s) or robots; it is sufficient that they behave (function) the way we want them to behave (function).

Philosophically speaking this shift is of utmost importance: If we recall the argumentative derivation of the Descartes axiom (Ego cogito as the undeniable firm foundation) by meeting the standards of the Descartes postulate (to take nothing for granted that hasn't been perceived clearly and distinctively) we then see that what critical thinking had to fight according to Descartes was the whole methodical sequence of possible deceptions (cf. above 3.1).

So what the Turing Test in actual fact is testing is whether a machine, resp. a program running on a machine is capable to deceive not just one, but any observer/user by performing seemingly cognitive tasks such that he or she will

---

39 Turing 1950: 433.

not be capable of distinguishing it from human cognitive behaviour. In brief: the task of AI is to bring about a situation of complete deception by perfect *indistinguishability*.

4.2 *Doubting Thomas or Deceiving AI*

It is not that easy to thoroughly understand the decisive philosophical change that is taking place with the rise of the Digital. It is not the replacement of analogue by digital devices only; it is not just the development of ever better performing computers or computer networks; and it is not the coming of the age of a whole workforce of robots interacting with human beings (and sometimes replacing them). These all are but visible elements of a huge smokescreen hiding rather than displaying what's behind it.

What is happening is a real fundamental change, the one and encompassing anti-Cartesian experiment undermining and at the same time overturning the very foundations of modern philosophy and science, exemplarily represented by René Descartes. Regardless of whether we go for a rationalist or an empiricist version – the very fundamental characteristic of modern philosophy, science (and technology, as I would like to add!) has since the 17th century always been: to take nothing for granted, unless proven to be evident, following – allegorically speaking – the role model of Doubting Thomas.

The last seven decades, however, initiated by Alan Turing's epochal shift of replacing the question whether machines are capable of thinking by the question of whether a human observer or user would ever be able to distinguish cognitive behaviour displayed by machines from cognitive behaviour displayed by human beings, have turned everything around: Deception doesn't anymore bear the mark of an obstacle or even an enemy ("genius malignus") to be overcome by systematic doubt leading to insights that are not only true, but even certain. On the contrary: We are witnessing something like a *rehabilitation of deception* on the highest possible level; the role model of Doubting Thomas is, mediated by indistinguishability, in the process of being replaced by the role model of simulation (cf. below 5.1) as a new paradigm.

4.3 *Reasonable and Unreasonable Doubt*

As has already been said (cf. above 3.1), Descartes' postulate could also be rephrased as follows: to take for granted only what has been confirmed beyond any reasonable doubt. Therefore we have to ask the self-referential question: Has the rehabilitation of deception been confirmed beyond any reasonable doubt? Or in more general terms: If the rehabilitation of deception as the new paradigm is indeed in the process of replacing (or has already replaced) the

Cartesian paradigm, does then the Cartesian paradigm of testing any opinion according to the reasonable doubt criteria still apply? If the answer is No (as it obviously is supposed to be) the question arises: Is there another set of criteria applicable in order to find out whether claiming the digital rehabilitation of deception makes sense at all?

5. Deception – Revisited and Re-habilitated

Due to the fact that – as already indirectly demonstrated by the "bullshit test" (cf. above 1) – "making sense" is a concept located very much still within the Cartesian paradigm there are other conceptual candidates much more likely to be making sense than the idea of rehabilitating deception. If, however, we remind ourselves of the importance attributed to the pragmatic dimension of semiotics in the same context, we might look for post-Cartesian criteria asking: Could a digital rehabilitation of deception turn out to be *useful*, and if so: in which context?

5.1 *Imitation and Deception*

In order to answer this question we might find it helpful to scrutinize the context of the word "deception" and look for semiotic equivalents. As mentioned above an important first one has already been pointed out by Alan Turing himself and his replacing the question as to whether machines are capable of thinking by a game which he didn't call "deception game", but "imitation game" the goal of which was to design a situation of indistinguishability between the (seemingly cognitive) performance of a machine (computer, robot etc.) and of a human being (or more accurately: two human beings). Keeping in mind that the fourth step of Descartes' methodology of doubt (cf. above 3.1) consisted in the argument as to the indistinguishability between dream and reality we realize the inversely symmetrical structure of the Cartesian paradigm and the anti-Cartesian experiment: Whereas the former used indistiguishability as a reason for increasing doubt ("I am not certain whether I am awake or dreaming") the latter is using it as a *definiens* for the functional equivalent of a positive answer to the question whether machines are capable of thinking. Whereas the former resulted in the attempt to overcome it, indistinguishability is the ultimate goal of the latter: Success in passing the Turing Test consists in permanently failing to distinguish machine performance from human performance. In brief: The *failure* of the former is defining *success* in the latter, as long as we replace "deception" by "simulation".

5.2 *Mundus Vult Decipi*

The old Roman saying that the world wants to be deceived and therefore will be deceived ("mundus vult decipi, ergo decipiatur"[40]) takes on a completely different meaning here: it is one of the undeniable effects of the process called "digitalization" that in almost all realms of reality, from science and technology to the whole lifeworld, deception in the shape of simulation is taking over or has already taken over. Even in our everyday language we quite innocently talk of "Virtual Reality"[41] which under different circumstances would satisfy the conditions of a proverbial wooden iron or a square circle.

Also with respect to the very core of the Cartesian attempt to reach a situation of evidence-based certainty in science, the scientific experiment, simulation is slowly but gradually taking over, and the same applies for different reasons to the whole field of technology (which as a whole is some kind of large-scale real life experiment for scientific theories). As far as the lifeworld is concerned there is almost nothing left which isn't covered by as it were "deceiving" simulation tools like apps, smart-phones, navigation systems, electronic timetables etc.

5.3 *Deception in an "Extra-Moral Sense"*

As demonstrated by these examples we have to get rid of an almost inevitable connotation of the concept of deception, i.e. the automatism of attributing a moral dimension to it. In the semiotics at least of the Cartesian paradigm, deception is usually associated with moral wickedness and cunning up to ideological indoctrination as is clearly visible in the personification of a maligne genius. In order to fully come to terms with the paradigm-shift as represented by the anti-Cartesian experiment we have to develop an understanding of deception in an extra-moral sense following Friedrich Nietzsche's concept of "truth and lie in an extramoral sense"[42].

If interpreted accordingly digitalization is putting an end to the Cartesian paradigm, correcting at least two main features of modern thinking: the conviction that philosophy and science are adversary on the one hand to deception and on the other hand to technology. By means of digitalization a whole new world is being opened to philosophical reflection, based on a new anti-Cartesian insight: that *to know is to make* – but that's another story …

40 "Mundus vult decipi ergo decipiatur" is a very popular Latin phrase, attributed to Scaevola, St. Augustin, S. Franck (and – erranously –) to Petronius.

41 Cf. Rheingold 1991; 1993.

42 Nietzsche 1873; cf. Zimmerli 1999: 253–277.

To summarize: Yes, a digital rehabilitation of deception turns out to be useful once we take a closer look at the changes that have taken place in the whole semiotic environment of the Cartesian paradigm of doubt. And yes, a rehabilitation of deception opens up a whole, non-epistemological new world of phenomena, leaving us philosophers with the task of analyzing and describing it, philosophizing as it were after the end of (Cartesian) philosophy. But no, computers and robots still can't think.[43]

6. Bibliography

Arendt, H. (1958) *The Human Condition*. Chicago/London: University of Chicago Press, Chicago/London.

Austin, J. L. (1962) *How to Do Things with Words*. Oxford: Clarendon. German transl. (1972) *Zur Theorie der Sprechakte*. Stuttgart: Reclam.

Carnap, R. (1931) "Überwindung der Metaphysik durch logische Analyse der Sprache." *Erkenntnis*, 2, 219–241.

Descartes, R. (1996) *Meditations on First Philosophy: With Selections from the Objections and Replies*. Ed. and transl. by John Cottingham, Cambridge: Cambridge UP.

Dooling, R. (2008) *Rapture for the Geeks. When AI outsmarts IQ*. Harmony: Crown.

Frankfurt, H. (2005) *On Bullshit (1986)* Princeton: Princeton UP, German transl. (2006) *Bullshit*, Frankfurt a.M.: Suhrkamp.

Galison, P. (1987) *How Experiments End*. Chicago: University of Chicago Press.

Hintikka, J. (1962) "Cogito, ergo sum: Inference or performance." In: *Philosophical Review* 71, 3–32.

Hofstadter, D. and E. Sander (2013) *Surfaces and Essences: Analogy as the Fuel and Fire of Thinking*, New York: Basic Books, Perseus Book Group. German transl. (2014) *Die Analogie. Das Herz des Denkens*. Stuttgart: Cotta.

Kapp, E. (1877) *Grundlinien einer Philosophie der Technik. Zur Entstehungsgeschichte der Kultur aus neuen Gesichtspunkten*. Braunschweig: George Westermann.

Kenny, A. (1968) *Descartes: A Study of His Philosophy*. New York: Random House.

Kurzweil, R. (2005) *The Singularity is Near. When Humans Transcend Biology*. New York: Penguin. German transl. (2013) *Menschheit 2.0: Die Singularität naht*, Berlin: Lola Books.

43 It is with profound gratitude that I am acknowledging the opportunity to present and discuss this paper on September 21, 2019 in the Fellows Colloquium during my stay as Senior Visiting Fellow at the Institute of Human Sciences/Institut für die Wissenschaften vom Menschen (IWM) in Vienna. Accordingly an abbreviated version of this text will be published in: *IWMPost*, 125, 2020, 20f.

McCarthy, J./M. Minsky/N. Rochester/C. E. Shannon (1955) "A Proposal for the Dartmouth Summer Research Project on Artificial Intelligence". Available at: http://raysolomonoff.com/dartmouth/boxa/dart564props.pdf.

Mittelstrass, J. (1979) "The Philosopher's conception of *Mathesis Universalis* from Descartes to Leibniz." In: *Annals of Science*, 36(6), 593–610.

Moore, G. (1965) "Cramming more components onto integrated circuits." In: *Electronics*. 38(8).

Muschg A. (2006) „Ich ist ein anderer. Respekt in Zeiten der Globalisierung." In: Walther Ch. Zimmerli/S. Wolf (eds.) *Spurwechsel. Wirtschaft weiter denken*. Hamburg: Murmann.

Nietzsche, F. (1873) "Über Wahrheit und Lüge im außermoralischen Sinne" Engl. Transl. (1979) "On Truth and Lie in an Extra-Moral Sense" In: Breazeale, D. Philosophy and Truth: Selections from Nietzsche's Notebooks of the Early 1870's. Atlantic Highlands: Humanities Press, 79–97.

P. McCorduck (1979) *Machines who Think. A Personal Inquiry into the History and Propects of Artificial Intelligence*. San Francisco. German transl. (1987) *Denkmaschinen. Die Geschichte der Künstlichen Intelligenz*, Haar bei München.

Pannenberg, W. (2007) *Analogie und Offenbarung. Eine kritische Untersuchung zur Geschichte des Analogiebegriffs in der Lehre von der Gotteserkenntnis* (1955), Göttingen: Vandenhoeck & Ruprecht.

Przywara, E. (1932) *Analogia entis*. München: Kösel und Pustet.

Reedy, Christianna (2017) "Kurzweil Claims That the Singularity will happen by 2045. Get ready for humanity 2.0." Interview at the SXSW Conference in Austin, Texas, Oct. 5th. https://futurism.com.

Rheingold, Howard (1991) *Virtual Reality*. New York: Summit Books.

Rheingold, Howard (1993) *The Virtual Community. Homesteading on the Electronic Frontier*. Cambridge: MIT Press. German transl. (1992) *Virtuelle Welten. Reisen im Cyberspace*. Reinbek: Rowohlt.

Searle, J. (1969) *Speech Acts: An Essay in the Philosophy of Language*. London: Cambridge University Press. German transl. (1971) *Sprechakte. Ein sprachphilosophischer Essay*. Frankfurt a.M.: Suhrkamp.

Thackray, A. /D. C. Brock/R. Jones (2015) *Moores Law: The Life of Gordon Moore, Silicon Valley's Quiet Revolutionary*, New York: Basic Books.

Turing, A. (1950) "Computing Machinery and Intelligence." In: *Mind* 59, 433–460. German transl. (2002) „Kann eine Maschine denken?" In: W.Ch. Zimmerli/S. Wolf (eds.), *Künstliche Intelligenz. Philosophische Probleme*. 2. unv. Aufl. Stuttgart: Reclam, 39–78.

Vinge, V. (1993) "The Coming Technological Singularity: How to Survive in the Post-Human Era." In: *Proceedings: VISION-21 Symposium*, March 30–31.

Vinge, V. (1983) "op-ed." *Omni magazine*, January 1983.

Wilson, C. (2003) *Descartes' Meditationes. An Introduction*. Cambridge UP.

Zimmerli, W.Ch. (1986) "Who is to Blame for Data Pollution?" In: C. Mitcham/A. Huning (eds.) *Philosophy and Technology II: Information Technology and Computers in Theory and Practice*. Dordrecht: D. Reidel, 291–305.

Zimmerli, W.Ch. (1999) "Nietzsche's Critique of Truth and Science: a Comprehensive Approach." In: B. Babich (ed.) *Nietzsche. Epistemology, and Philosophy of Science*. Dordrecht, 253–277.

Zimmerli, W.Ch. (2014) "From Fiction to Science. A German-Japanese Era-Project." In: G. Funk/B. Irrgang (eds.) *Robotics in Germany and Japan. Philosophical and Technical Perspectives*; Frankfurt a.M.: Lang, 11–25.

Zimmerli, W.Ch. (2018a) "Die Rehabilitierung der Täuschung." In: *Neue Zürcher Zeitung*, 11.09.2018, 35.

Zimmerli, W.Ch. (2018b) „Die Wiederkehr des Analogen im Digitalen." In: *Neue Zürcher Zeitung*, 26.07.2018, 38.

Zimmerli, W.Ch. and Wolf, S. (1994) "Einleitung." In: W.Ch. Zimmerli/S. Wolf (eds.) *Künstliche Intelligenz : philosophische Probleme*. Stuttgart: Reclam, 8–11.

Artificial Intelligence of the Human Mind

Finding Common Ground between Intelligent Machines and "mechanistic" Minds

Daniel Neumann

Abstract

There are still many anthropomorphizing misunderstandings about what it means for a machine to be intelligent. In searching for common ground between human and artificial thinking processes, I suggest reconsidering how human intelligence can be conceived in mechanistic terms. To do this, I take a closer look at the notion of the spiritual automaton of the 17th century philosopher Baruch de Spinoza. In his writings, a mechanical theory of the mind can be found, before the comparison between minds and machines even arises. In a second step I sketch some ways in which current methods in AI studies could be seen to reflect aspects of the Spinozistic model of the mind

1. Spinoza's *spiritual automaton* and the Mechanical Conception of the Mind

The basic idea of this essay is a simple inversion in the concept of artificial intelligence. Instead of asking how the computational power of machines might be compared to the intelligent acts and thoughts of human beings, I want to think about ways in which the human mind itself can be thought of as "machinic", in the sense that conscious thought seems to be determined by outside forces, leaving the thinking person with the experience of an "automatic" thinking process. From today's perspective, this seems to be turning the mind into something less than the free agent it is usually thought of, as though conscious experience were separated from the actual thinking process or computation. That it seems to us this way, is not least the result of the strong sense of subjectivity that philosophers and others have been envisioning, defining and defending since the enlightenment age. To this can be added the concern of the "Philosophy of Mind" to ask what it means for mental events to be based on physical structures. I suspect that this question, standing in the context of neuroscience and related natural sciences, leads us away from a fruitful discussion of the analogy between human and artificial intelligence.

 | DOI:10.30965/9783957437488_005

Instead of trying to assess differences and similarities based on the comparison of hardware and human "wetware", I want to think about how we can conceive of ideas so that they make sense in a human as well as in a machinic environment. In other words, I want to consider not who is thinking, but how one and the same idea can be thought in different instances. This suggestion is in some ways akin to the "multiple realizations" theory which entails that "a single mental kind (property, state, event) can be realized by many distinct physical kinds."[1] My query deviates from this discussion insofar as I am not primarily interested in arguing for or against a reduction of thought to bodily states and I do not discuss thinking in terms of conscious mental states. The artificial intelligence of the human mind I am interested in is of a procedural nature. It is determined by a chain of ideas whose connection creates consciousness as a side effect. The problem is not: Can machines and humans have similar states based on a different computational/neurological architecture? But: Do humans and computers show similar forms of intelligence when processing chains of ideas?

As a source for a mechanical conception of thinking, my choice of the Dutch philosopher Baruch de Spinoza, whose writings predate the authors of enlightenment, is not a coincidence. He is celebrated today not only for his rigorous and idiosyncratic ontology, but is also recognized as having foreshadowed problems concerning artificial intelligence and brain mapping.[2] In this essay, I want to consider the notion, in some ways neglected, of the *spiritual automaton* coined by Spinoza in his "Treatise on the Emendation of the Intellect", written in the 1660s.[3] It has been neglected insofar as it signifies, according to my interpretation, a radical aspect of Spinoza's epistemology (theory of knowledge) which conceives of the human mind as a thinking machine. The *spiritual automaton* means that the everyday stream of thought can be consciously perceived *as* being lawfully determined. According to Spinoza, we can understand and feel that we are determined to think. In coming to terms with this condition, which transcends human thought, lies the chance for a blissful life, which Spinoza tries to sketch in his principal work *The Ethics*.

In a first step, I outline the philosophical concept of the *spiritual automaton*. Secondly, I discuss how Spinoza's philosophy might help in framing questions that are part of current debates about artificial intelligence: What is the nature of human consciousness as opposed to artificial consciousness? How is the ontological status of thought related to the conscious experience of thinking?

1 Bickle 2016.
2 See for instance Damasio 2013.
3 Cf. Spinoza 1985: 37.

How can we find a reasonable and productive way to think of the analogy between the human and the artificial?

The spiritual automaton can be outlined using three main features of Spinoza's epistemology: (1.1) Thought, that is single ideas, does not strictly depend on the human mind for their existence, (1.2) Ideas are of a modular nature, and the same ideas can be found across different minds, (1.3) Consciousness is not defined by a thinking "I", but by the chain of ideas the human mind is composed of.

1.1 *Disembodied Thought*

According to Spinoza's philosophy, ideas are understood not through the properties of the human mind, but through the creative power of the one substance that is the cause of all ideas. This substance, otherwise known as God or Nature, has the power to think infinitely many things in infinitely many ways.[4] The human mind is part of the infinite ways in which this power expresses itself. In other words, my own thoughts are in the first place a result of a power of thinking that does not belong to myself, but to another entity. Minds are modifications of this substance. *If* there is a form of self-reflection in Spinoza's philosophy, then it has to start from this basic fact: My thoughts are not mine, in any strongly subjective or Cartesian sense of the word.

Subsequently, there are two aspects to every idea: ontological and epistemological. Ontologically, this idea exists alongside other ideas and is an entity within a causal network that comprises all the ideas of all minds, including humans, non-humans and the idea of the substance or God or Nature itself. Epistemologically, the idea is an idea *of* something. There is an object that the idea represents. Spinoza says that the human mind is nothing else than the idea of an existing body.[5] That means the human mind consists of ideas that represent this body in terms of feelings, perceptions, as well as hunger and pain. Under these circumstances, the classical requirement for truth can be satisfied, namely that the idea must agree with the object of which it is the idea.

But this is only half the story, because the ontological status of the idea, the fact that the idea *is*, does not depend on the individual mind thinking this idea, but on the divine and natural power which produces all ideas. Spinoza leaves room for interpretation of what the different embeddings of one and the same idea in different minds could mean. What he seems to suggest is that one and the same idea can be part of different chains of ideas, depending on who is

4 Cf. Spinoza 1985: 424.
5 Cf. ibid.: 457.

having it. This opens the questions of how one and the same token idea can mean different things in an artificial and in a human mind. The idea is ontologically the same, but epistemologically different. It means different things to different minds, depending on the connection to other ideas. Now I would like to explore the different shades of objectivity an idea can express according to Spinoza's philosophy.

1.2 *Ideas Are of a Modular Nature*

Ideas are distinguished into adequate and inadequate ideas. An idea is inadequate, if it does not envelop the knowledge of all of its causes. It is one-sided and has a subjective quality. Inadequate ideas present me with a thought that is solely derived from the perspective of my own body. It is a partial representation, insofar as the object of the idea is only perceived by how it affects my own body. Even though this idea only represents my bodily state to me, its objective content is part of a bigger causal network that transcends my body and thus my mind. I can never adequately grasp the nature of my own mind, because there are always other ideas involved that determine me to think this or that, to have inadequate, incomplete ideas.

Adequate ideas on the other hand are those that fully represent their causal place in a network. This network is infinite and goes back to the one divine and natural power of thinking. The condition for the human mind to have an adequate idea is to be its complete cause. Because of the infinite complexity of the network of ideas, only certain aspects of ideas can be understood properly, namely something that is shared across different ideas and different bodies. Much has been made of adequate ideas in terms of a philosophical foundation of the sciences.[6] What I see Spinoza outlining here is much simpler. His description of adequate ideas is in some ways akin to pattern recognition. By extracting certain aspects of ideas, for instance the observation that every mind has an existing body as its object, we can come to an adequate understanding of the world. Spinoza is careful not to conflate this with the process of abstraction, whereby inadequately deduced properties of things are combined to form a random concept.[7]

What makes Spinoza's theory of adequate ideas interesting is that it supposes that there are intelligible aspects of ideas, micro-ideas so to speak, that can be extracted and recombined *without* distorting the macro-ideas themselves. These can be concepts, feelings or even the general properties of ideas and

6 See for instance the concept of "inferential relations" in which adequate ideas can serve as premises to one another. Cf. Perler 2018: 223.

7 Cf. Spinoza 1985: 476.

thinking things themselves. Far from being a random abstraction, adequate ideas point us towards a modular notion of thinking, where ideas can be broken up into self-sufficient bits and spread across different minds. At the same time, the notion of adequate ideas transcends logical formalization. Adequate ideas do not necessarily make up a chain of propositions of inductive or deductive reasoning. They are part of everyday life. These ideas are linked to, enmeshed with, inadequate ideas as part of an individual experience. Let's take a closer look at the chains of ideas that make up the *spiritual automaton.*

1.3 *Chains of Ideas*

The human mind is, according to Spinoza, not defined by a reflexive central instance like the Cartesian cogito, *I think, therefore I am*, but by chains of ideas. Some of these, the adequate ideas, are linked in the same way in different minds, whereas others are related only to the experience of a single body, which are incompatible with the experience of others. But the congruity or incongruity is not a feature of the ideas themselves. As Spinoza says, there is nothing in the ideas themselves which is false.[8] Instead, what makes ideas false or incompatible are the relations between the ideas, which can either follow the order of personal experience or the ontological order. One of the most innovative questions of Spinoza's philosophy derives from this sequential notion of thinking. Namely, how can we bring it about that the flow of our ideas doesn't cause us to be unhappy? How can we find a different relationship to the necessity of thinking this or that? How can we come to terms with the fact that our thinking is determined?

This presupposes some form of reflexivity on the part of the human mind, firstly to acknowledge the fact that the stream of ideas *is* determined and secondly to circumscribe the mental instance that is affected by the chain of ideas, since consciousness follows from the ideas, not the other way around. The doctrine of "ideas of ideas" accounts for reflexivity in Spinoza's philosophy, and it is just as interesting as it is controversial. According to my interpretation, the "ideas of ideas" are indispensable to assess the form of self-awareness that the machine of the human mind is capable of. Since the mind is not the subject of its ideas, but a chain of ideas, we have here, according to Michael Della Rocca something akin to a bundle theory of the human mind.[9] There is no systematic place from where a thinking self can be reflected. Instead, one can conceive of an idea of an idea. That is, one can ask what it means to think something. And here, Spinoza is adamant about distinguishing his account of

8 Cf. ibid.: 472.
9 Cf. Della Rocca 1996: 42.

thinking from others in that he conceives of having ideas as an *activity*, as processing information. In other words, there is a sort of computational power of the human mind to have the ideas (that are at the same time part of God or Nature). To get a sense of this power, the key question is: What is the nature of the idea if its representational content is taken away? What is the idea without its object, as a pure computation so to speak? Spinoza talks about this in a famous note in which he describes what happens during self-reflection:

> For the idea of the mind, i.e., the idea of the idea, is nothing but the form of the idea insofar as this is considered as a mode of thinking without relation to the object. For as soon as someone knows something, he thereby knows that he knows it, and at the same time knows that he knows that he knows, and so on, to infinity.[10]

To know that one knows is the same as to understand the formal quality of an idea. The form of the idea goes back to a divine power of thinking, insofar as God or Nature is the formal cause of the idea. That is to say, every idea depends for its existence on God's power of thinking. An idea is just a mode of thought, a way in which God is understood to be a thinking substance. On the other hand, that same mode of thought is also part of a human mind or constitutes a human mind. The question then is: Who do we attribute the thinking of the idea to? Is there a genuine thinking power proper to the human mind? Or is having ideas a sort of passive participation in the divine power of thinking?

Rather than trying to argue for either option, I would like to think about why Spinoza seemingly chose to leave open the relationship between divine and human thinking. The first question one could ask is: What difference would it make either way? With his "mechanism of the mind", I think Spinoza tries to render this question obsolete. By focusing on ideas as *acts* of thought, the thinking subject is put in brackets. The capacity to think becomes a power defined by the ideas themselves, whereas consciousness is a secondary effect of the idea.[11] What Spinoza establishes is a new means of thinking about the relationship a mind has to having chains of ideas. In this way, he opens up the possibility of conceiving of non-human or artificial thinking processes.

In other words, Spinoza doesn't explicitly tells us who forms our ideas. Between God or Nature as the thinking substance and substratum of the ideas on the one hand and the human mind as something that "has" ideas of an existing body on the other, the creation of ideas remains ambivalent. It is exactly this

10 Ibid.: 467 f.

11 A similar position is taken by Lia Levy, according to whom the concatenation of ideas is necessary to conceive of the mind that has them. Cf. Levy 2000: 291.

ambivalence that I see as a chance to reconsider the analogy between a human and artificial intelligence for two reasons:

(1) Spinoza relieves us from having to look for a central intelligent instance in the case of a thinking machine.

(2) Ideas are not in the first place equated to single conscious mental states but are defined by being part of sequences that minds process.

In this way, we can avoid being confined by unrealistic expectations of the human-machine analogy, a famous example of which would be John Searle's notion of a "strong artificial intelligence" which states that an "appropriately programmed computer really is a mind, in the sense that computers given the right programs can be literally said to understand and have other cognitive states."[12] With Spinoza's epistemology, the need for qualifications such as "actually" and "really" is put into a different perspective. Instead of thinking about the analogy in terms of a simulation of coginition versus real (or natural) cognition, the focus is shifted to performing sequences of operations, something computers as well as human minds can be said to do.

In the next section of my essay, I want to sketch a few ways in which this Spinozistic model of the mind can be related in some general ways to the notion of artificial intelligence.

2. Human and Artificial Algorithms

One way in which artificial intelligence can be contrasted with a human one is in granting to the human mind a sort of master role towards its ideas. The thinking instance, the Cartesian cogito, is not defined by the stream of conscious thought, it is somehow outside of it. This kind of intelligence is usually said to be lacking in machines performing different tasks. As we have seen, Spinoza explicitly rejects this sort of subjectivity because it is an effect, not the foundation, of intelligent thought. For the Dutch philosopher, intelligence is a property inherent in ideas and the relationships between ideas. Intelligence can be found whenever ideas are connected to make up a mind and to represent an object on the one hand and express a divine or natural power of thinking on the other.

Since Spinoza radically separates the capacity to think from any one individual brain or other form of embodiment, ideas become something shared between different minds. Instead of saying that everyone has the same ideas because everyone has the same cerebral capacity to think them, in this case,

12 Searle 1980: 417.

everybody thinks the same because everybody has the same idea tokens, at least in the case of adequate ideas, which, as micro-ideas, contain certain aspects of macro-ideas without distorting them.

Computer-based algorithms could be an example of how adequate ideas are shared between human and artificial minds. Algorithms are not just formalized ways of reasoning that can be executed by both human minds and machines. They determine a mind to think – and 'mind' means any instance where a sequence of operations can be executed. An algorithm satisfies Spinoza's condition for adequate thinking, namely, that the mind is the sole cause of the ideas. Once the steps of the algorithm are mastered, the *spiritual automaton* is able to execute it in a lawful manner. In this framework of a highly formalized thinking process, both humans and machines are able to *use* algorithms. But this usage implies different forms of necessity of the thinking process. When a human mind produces ideas in a lawful fashion without being determined from outside, that is, solely relying on ideas that can be explained through this mind alone, there is a special form of awareness according to Spinoza, the feeling of joy.[13]

Joy is a side product of thinking. It does not reinforce a feeling of self, in the sense that "I am joyous about having come to this or that comprehension", but it points me towards certain truths independent of me, which can be expressed in adequate ideas such as algorithms. Every time the human mind conceives of these kinds of ideas it dehumanizes itself in the sense that these adequate ideas do not depend on a *human* mind to be thought. They can be "thought" by artificial minds as well.

But what does "thinking" mean when talking about computing machines? The way I have presented the analogy up to this point, in terms of algorithms that can be implemented by programming or learning, leaves us with two identical outputs as in the case of an algebraic formula or a logical conclusion that can be accomplished by both a program and a person. In the case of highly formalized operations where a set of rules is given to direct every possible operation, it is fairly straightforward to account for the human-machine analogy.

As stated before, the adequate ideas Spinoza has in mind as being in different minds, far exceed the narrow framework of symbolic manipulation. Current methods of AI research equally transcend simple manipulation of data by rules of induction or deduction. In his book "The Master Algorithm", Pedro Domingos tries to categorize the different ways in which machines can be programmed to perform intelligent tasks. The formalized operations described up to this point would roughly fall under the category of "Symbolism", an AI

13 Cf. Spinoza: 609.

paradigm which is based on rules and decision trees that draw conclusions from large sets of data. The biggest shortcoming of this method is that it is not able to detect more subtle differences within data sets on account of its being based on sets of rules. A typical usage of a symbolic algorithm would be a spam filter, where there is a clear yes/no divide.[14]

Adequate ideas are variations on, and reductions of, inadequate ideas. An adequate idea is a reduced and intelligible form of another idea that is much too complex for the human mind to be understood in every single aspect. This doesn't mean that the only recourse to conceive of an adequate idea is by symbolic manipulation or logical formalization. Domingos talks of another paradigm in AI studies that comes closer to Spinoza's concept of the *spiritual automaton*, namely "analogical reasoning". At the center of the analogy is a simple inference by affinity, recognizing similarities between two things and thereby being able to infer further similarities between them. For the analogy to work, a considerable level of abstraction is needed. In focusing on the similarity between two things, I have to remove my attention from all the other aspects contained in my idea of the things. A new adequate idea results from the new relationship in which the two analogous ideas stand to one another.

Let me give two further examples of how Spinoza imagines the conception of adequate ideas. The first example is the distance from the sun. When we look at the sun, it might appear to us to be much closer than it actually is. This thought is derived from all the impressions contained in the inadequate idea, i.e. the perception, of the sun: the warmth and size for example. The interesting thing is that upon learning that the actual distance is much greater, our sense impression is not changed one bit. When we just look at the sun, our perception of it would still lead us to suppose that it is much closer.[15] The adequate idea of the sun, represented by astronomical knowledge of the distances in our solar system, is at odds with the evidence of the senses. And yet, I can have an adequate and an inadequate idea of the sun at the same time. And while I am not sure if anyone else's perception of the sun is similar to mine, according to Spinoza I can always be sure that the adequate idea is the same, even for different minds.

A second example is the power of association. Today I see for the first time Peter in the morning, Paul at noon and Simon in the evening. If this happens a few times, my associations between the persons I see and the time of day I see them will grow stronger and eventually turn into expectations. This constitutes an inadequate idea of the world based solely on my own experience. Suppose

14 Cf. Domingos 2015: 62.
15 Cf. Spinoza 1985: 473.

then, Spinoza says, that for the next couple of days the persons are switched and I see Simon instead of Peter in the morning and Paul instead of Simon in the evening. This introduces the concept of randomness regarding my notion of past, present and future.[16] It is only now that associations are not just based on my experience, but on the order between the things themselves. When I refrain from understanding things based on the particulars of how they occurred to me, they take on an objective quality. In order to form adequate ideas, I have to bracket everything that is defined only by my individual experience. Like the true distance from the sun, adequate ideas are entangled in the complex inadequate ideas we have.

The method of disentanglement to create order out of complex information, is in some ways akin to how knowledge is extracted within the analogy paradigm of artificial intelligence, as described by Domingos:

> Imagine trying to define what a face is with a decision tree. You could say it has two eyes, a nose, and a mouth, but what is an eye and how do you find it in an image? What if the person's eyes are closed? Reliably defining a face all the way down to individual pixels is extremely difficult, particularly given all the different expressions, poses, contexts, and lighting conditions a face could appear in. Instead, nearest-neighbor takes a shortcut: if the image in its database most similar to the one Jane just uploaded is of a face, then so is Jane's. For this to work, the database needs to contain an image that's similar enough to the new one – for example, a face with similar pose, lighting, and so on – so the bigger the database, the better.[17]

The basic problem we start with in this example is finding the adequate idea of a face, i.e. an idea that enables the mind having this idea to recognize faces. As Domingos makes clear, this cannot be achieved by a symbolist algorithm. There is no logical representation of a face. But there is an analogical method to defining what a face is. It consists simply in comparing a new idea to those already stored in memory. If certain patterns recur in a new picture, it is classified as a face. This method of analogy is in other words based on finding the closest resemblance or nearest neighbor of a current object to those stored in memory.

On the surface, this might seem to be more in line with Spinoza's aforementioned concept of abstraction, where a general idea is formed by abstracting features of random exemplars and putting them together to create an average. But the ideas produced by abstraction are not adequate, because they depend on the singular experiences of a mind and can thus not be found in another

16 Ibid.: 480 f.

17 Domingos 2015: 182.

mind which necessarily has different experiences. Abstraction, in this sense, is induction gone wrong. The case could be made that the nearest neighbor algorithm for face recognition is based on this kind of abstraction. The training sets, containing the examples of different faces, would function as the individual experiences that help to classify new pictures.

This might be true if face recognition was about extrapolating an average face out of all the examples the algorithm has been fed. But the average is of no value, because new data is not compared to the average of all faces, but to those that most resemble it out of all the available exemplars. In other words, in a dynamic process, the algorithm is able to detect specific features in any given example – provided, it is a face. For a face to be a face, a series of variables and their relation to one another are necessary to detect and classify a face in a picture. These variables or nodal points form the basis of a geometry of faces, a dynamic mask that can accommodate different examples.

In retransferring this to the human mind we can see the dynamics of the adequate idea at play. Going back to the examples of the distance from the sun and the associations of daily experience, the adequate idea appears as a form of implicit, automatic knowledge. The adequate idea of the sun informs my concept of it not by changing anything in my perception but by adding to it another perspective, an invisible depth that gives me an idea of the vastness of space. Similarly, the adequate idea doesn't alter my perception of a face but enriches my idea of what a face can be, what the miniscule differences within a change of nodal points, as it were, could mean in terms of emotions and meanings. The adequate idea is not a full-fledged abstraction, an average, but a further layer of experience. The *spiritual automaton* will eventually form several adequate ideas and reproduce them automatically, not unlike an algorithm programmed to recognize faces. The more I "automatically" recognize in a certain face, the better I am equipped to deal with everyday life. For Spinoza, the vanishing line of adequate ideas is a happy and self-sufficient life.

3. Summary

What Spinoza's account offers is, in other words, a broadening of the concept of intelligence and a redefinition of thinking. Taking Spinoza's outlook, we don't have to search for a form of subjectivity of artificial intelligence similar to those of humans. Awareness of oneself, awareness of one's own intelligence is a byproduct of the sequence of ideas. These ideas, provided they are adequate, can be embodied by human and artificial instances. Even though the implementation is very different in both cases, the thinking process is guided by the ideas themselves. This self-sufficient program is what Spinoza calls thinking that is causally dependent solely on a mind.

Lastly, what can Spinoza's philosophy tell us about learning processes beyond the human mind? In short his recipe is: the less thinking depends on the idiosyncrasies of experience and the more a mind is comprised of ideas that can be found in other minds as well, the happier and the more independent someone is. Learning means uncovering connections between ideas that will always be valid, no matter who thinks them under what circumstances. To conceive of these ideas always involves some sort of formalization, be it mathematical, geometrical or ethical. Whoever can process a chain of adequate ideas, becomes an automaton, spiritual or otherwise.

4. Bibliography

Bickle, John (2016) "Multiple Realizability", In: Edward N. Zalta (ed.) *The Stanford Encyclopedia of Philosophy*, available at: https://plato.stanford.edu/entries/multiple-realizability/ accessed November 25, 2018.

Damasio, Antonio (2013) *Looking for Spinoza. Joy, Sorrow and the Feeling Brain*, Harcourt.

Della Rocca, Michael (1996) *Representation and the Mind-Body Problem in Spinoza*, New York.

Domingos, Pedro (2015) *The Master Algorithm. How the Quest for the Ultimate Learning Machine Will Remake Our World*, New York.

Levy, Lia (2000) *L'automate spirituel. La subjectivité moderne d'après L'ethique de Spinoza*, Assen.

Perler, Dominik (2018) "Spinoza on Skepticism." In: *The Oxford Handbook of Spinoza*, Oxford.

Searle, John (1980) "Minds, Brains and Programs." In: *Behavioral and Brain Sciences*, 3(3).

Spinoza, Baruch de (1985) "Treatise on the Emendation of the Intellect." In: *The Collected Works of Spinoza*, Princeton.

When Robots (pretend to) Think

Davide Ghiglino, Agnieszka Wykowska

Abstract

Mental activities are a fascinating mystery that humans have tried to unveil since the very beginning of philosophy. We all try to understand how other people "tick" and formulate hypotheses, predictions, expectations and, more broadly, representations of the others' goals, desires and intentions, and behaviors following from those. We "think" spontaneously about others' and our own mental states. The advent of new technologies – seemingly smart artificial agents – is giving researchers new environments to test mindreading models, pushing the cognitive flexibility of the human social brain from the natural domain to towards the artificial.

1. Introduction

Mental activities are a fascinating mystery that humans have tried to unveil since the very beginning of philosophy. According to the Cambridge dictionary the verb "to think" refers to activities such as "to consider", "to meditate", "to remember", "to wish", "to desire" or "to intend". There is one common feature that links all these verbs: reference to a content, direction toward an object (Brentano 1874), or the feature of "aboutness" (Burgos 2007; Millikan 1984) – something that has been termed by philosophers as "intentionality" and highlighted by some as the key feature of mental states (Brentano 1874), reflected in language structures of "propositional attitudes" (Russell 1905) inherently containing "aboutness" (I believe that ... I worry that ... I hope that ...). Although several authors argue that intentionality might be neither necessary (Husserl, 1900/01) nor sufficient (Davidson, 1980) for a mental act, according to Searle (1980) "aboutness" is what differentiates a natural human thought from a computer program. This idea can be illustrated by the famous "Chinese room" thought experiment: although a computer might be able to perfectly operate on the syntax and symbols of a language, it will not understand the meaning, the "aboutness" of all mental thought of humans.

In terms of etymology, in languages originated from Latin, the verb "to think" is derived from "***pensum***", which used to be the quantity of wool assigned to the weavers in Roman times. It described a simple material, which could be

 | DOI:10.30965/9783957437488_006

transformed into something complex through work and dedication. Whereas for the Latins the activity of thinking was connected with practical and manual activities, the ancient Greeks had already developed and entrenched the concept of "νοῦς" (noûs), that referred to the intellectual perception of reality from the "thinker" perspective. Since Anaxagoras, philosophers, psychologists, neuroscientists and researchers have tried to explain what it means "to think" and what the underpinning mechanisms involved in such an activity are. Reviewing theories of thinking is beyond the aims of this chapter, but it is interesting to point out that after more than two thousand years, the debate on the materialistic or abstract nature of thinking is still heated. According to neuroscientific accounts, the activity of thinking is considered to be the result of the elaboration of information collected by our sensory system, although this does not imply that the act of thinking is physical. Therefore, from a biological point of view, the mind itself can be treated as the result of brain activity, that can be explained in terms of electrochemical signals. On the other hand, certain philosophical perspectives postulate that this explanation seems to be reductive. Various degrees of mind-body dualism or non-reductionism have been proposed since Descartes. Fodor, for example, pointed out that it is impossible to infer high-level properties exhibited by a system just from knowing lower-level properties (Fodor 1975). Similarly, Putnam claims that micro-properties of molecules or atoms are not sufficient to explain macro-properties of behavior (Putnam 1975). The debate about what constitutes thinking and whether the mind is reducible to brain activity continues to be prominent in the philosophy of mind, but it is outside the scope of this chapter.

Despite the lack of agreement about the nature of thinking, attribution of thoughts to others has undeniably practical consequences in everyday life. When we are interacting with others, we probably do not ponder over the etymology, or the meaning of the concept of "thinking", but we still formulate hypotheses, predictions, expectations and, more broadly, representations of the others' goals, desires and intentions, and behaviors following from those. We "think" spontaneously about others' and our own mental states. In order to survive in the complexity of the world in which we are living, our species phylogenetically developed flexible strategies of understanding others' thoughts, strategies that adapt to the situation that is being experienced at a given moment and/or to the interaction partner. In the last fifty years, growing interest in social cognition has led researchers to investigate new questions related to attribution of thoughts to others. Social cognition refers to the mental operations that underlie social interactions, including perceiving, interpreting and generating responses to intentions, disposition and behaviors of others (Green, Horan 2010). The ability to infer and predict intentions, thoughts, desires, intuitions, behavioral reactions, plans and beliefs of other people is a crucial facet of social cognition (Frith, Frith 2012) and is often referred to as "mindreading" or "mentalization". In literature, several definitions of "mentalization" have been put forward, with different connotations across various academic disciplines. However, exploring all facets of this concept goes beyond the scope of this chapter. Thus, we will refer here solely to the (neuro-) psychological definition. In psychological research, mentalization is a concept developed to describe the process of understanding or predicting other people's behaviors in terms of their thoughts, feelings, wishes, beliefs, or desires. It represents all the processes that allow us to understand behavior in terms of underpinning internal states, and making behavior meaningful (Fonagy, 2018). This ability enables individuals to assess or understand other's mental states in a specific situation and, thus, to interpret and anticipate their behaviors (Bèrubè 2013). Mindreading is activated through attribution of mental states to others, mental states that, importantly, may differ from one's own (Korkmaz 2011; Sabbagh 2004). Some authors postulated that such ability is driven by innate neural mechanisms dedicated to mental state reasoning (Fodor 1983; Sholl, Leslie 1999; Karmiloff-Smith et al. 1995). Other authors have claimed that reasoning about others' mental states requires the capability of emphasizing with someone else, and that this depends on personal experience (Lillard, Kavanaugh 2014; Taylor, Carlson 1997). The study of these mechanisms constantly requires the revision of models at multiple levels, which are valuable in so far as each captures different phenomena, and no single level can be eliminated (Schaafsma et al. 2015).

The advent of new technologies – seemingly "smart" artificial agents, such as Apple's Siri, Amazon's Alexa or Google Home assistant is giving researchers new environments in which to test mindreading models, pushing the cognitive flexibility of the human social brain from the natural domain to towards the artificial. Such technologies seem to display a certain degree of "artificial thinking" that resembles (but only to a very limited extent) human's thought. But do artificial agents lead people to treat them like "thoughtful" agents? The artificial intelligence that these agents seem to display might enhance humans' tendency to anthropomorphize agents for which they lack specific knowledge and, as a cascade effect, this might lead towards mentalizing. Whether it is possible, or whether it will be possible in the future to attribute proper mental states toward artificial agents is still an open question. However, addressing this point is important not only for theoretical purposes, but also for practical and ethical consideration about design and development of social artificial agents.

Already in the seventies, Dennett proposed an account addressing how humans understand and predict the behavior of various systems (Dennett 1971).

The author postulated the existence of different strategies, called *stances*. According to Dennett, when we are interacting with inert objects, such as a leaf falling from a tree, we explain the behavior we might see in terms of physical laws. A leaf falling to the ground is just an effect of the gravity pull. Thus, we adopt the *physical stance* in order to understand the behavior of a physical system and to predict its consequences. The very same model can be applied to a ball rolling down the street, whose behavior is explained in terms of acceleration, gravity pull, friction, etc. However, when the object we are perceiving is a complex artifact, such as an airplane landing at the airfield or a car moving through a street, relying on physical information might not be an efficient (or accessible) strategy to explain and predict the behavior we are observing. In order to efficiently predict the behavior displayed by such entities we might need to rely on their functionality and design, using what Dennett calls the *design stance*. The fact that a car is moving is not explicable just in terms of inertial motion, but it becomes clearer as we start thinking about the design of the motor or the brake. When we are interacting with very complex agents, such as humans, neither the *physical* nor the *design* stance can provide an efficient model of explanation. For example, if a person is talking to us and repetitively moves her head in the direction of a clock, neither physical proprieties of the agent nor the anatomy underpinning the neck movement might be reliable sources of information to make efficient predictions regarding the person's behavior. In the depicted scenario, the behavior displayed by the agent might be due to her boredom or her intention to leave us as soon as possible. Therefore, in order to understand and predict her behavior, we need to include in our model a representation of the agent internal states, adopting what Dennett calls *intentional* stance. Adopting the *intentional* stance means to treat the acting agent as an *intentional* being, who is aware of their own behavior and who acts in order to maximize the likelihood of achieving a pre-set goal. In other words, adopting this attitude allows us to treat the behavior as a consequence of a mental act, which is directed toward an object. Here, the term "intentionality" reminds one of Brentano's conceptualization, since adopting the intentional stance implies understanding an immanent interconnection between mental phenomena and objective contents (see Mayer-Hillebrand 1951; Jacquette 1991). The intentional stance allows us to deal with unknown entities as well, or with artifacts whose behavior is ambiguous or impenetrable. The adoption of these three stances depends on the knowledge, the experience and the representation that humans build around the acting agent. This means that when such knowledge and representation changes, also the strategy we are adopting toward the agent might change. This flexibility is essential, since our cognitive system needs to constantly deal with an overwhelming amount of information coming from the environment, relying on a limited pool of resources. Therefore, when a certain explanation model is inefficient, resources need to be allocated toward another model that can grant rapid adaptation of the entire system. On the other hand, when a simpler and less cognitively demanding model is sufficient to explain and predict an agent's behavior, resources are shifted toward this latter model. It is important to point out that the adoption of such strategies does not necessarily imply the ascription of a mind to the perceived agent, unless one claims that adoption of intentionality is sufficient for mind attribution (Dennett, 2009). For example, when we are interacting with agents that display artificial intelligence, the most efficient way to interact with them might be the adoption of the intentional stance. When we need to ask our smart assistant the weather condition or the latest news, the easiest way to communicate our request is treating it *as if* it was an intentional being. This does not imply that we represent it as a "*true believer*" (Dennett 1981), equipped with mental states.

2. Empirical Approaches to Studying the Intentional Stance Towards Artificial Agents

As mentioned above, new technologies provided researchers with new tools to explore in more complex interactive scenarios the question of the attribution of intentionality to artificial agents. In particular, robots are of interest, as they are embodied, thus introducing physical (and potentially social) *presence* in the environment, they can manipulate the environment, thus introducing artificial *agency*. Yet, they still offer excellent experimental control (as they can be programmed to behave in an exactly specified and controlled way) in contrast to experimental protocols where another human is introduced as an interaction partner. In this context, it is therefore promising to examine whether and when humans adopt the intentional stance towards robots.

Some authors claim that implicit social signals lead humans to perceive automatically an agent's behavior as the reflection of mental operations (for a review see Frith, Frith 2008). We speculate that designing artificial agents that can closely imitate humans' behavior, even at the level of implicit social signals would induce humans to perceive them as intentional agents. Some recent studies suggest that the perception of the intentionality of an action[1] might be an automatic and immediate process. In 2007, Terada et al. investigated

1 Note that the assumption of direct perception of intentionality is arguable, as intentionality might be inferred from the perception of action.

whether such automatic process happens when participants are exposed to reactive movements of non-humanoid artificial agents (a chair or a cube with wheels). In their experimental setup, participants were observed while they were freely interacting with an artificial agent in an empty room. The movement of the artificial agent was remotely controlled and could be periodic or reactive. While in the first case, the movement of the artificial agent was random, in the reactive condition the movement was happening immediately after the movement of a participant. At the end of the interaction, participants were asked to fill questionnaires in order to evaluate whether the movement of the agent was perceived as intentional or not. Results showed that the physical appearance of the agent (chair vs cube) did not affect the attribution of intentionality, while the movement profile (periodic vs reactive) played a significant role in such attributions. In their experiment, the authors demonstrated that when the behavior of an artificial agent is contingent on human behavior, participants tend to evaluate it as goal-directed. In their study, the authors claim that enhancing the sense of intentionality perceived toward a robot is the key for smoothing communication between these agents and humans. The same authors demonstrated in a follow-up experiment (Terada et al. 2008) that other factors might affect intentionality attribution toward artificial agents. For example, context and participant's expectations might have an impact on the attribution of intentionality. In line with this, Wiese, Wykowska et al. (2012) showed that what mattered in evoking engagement in joint attention with a robot avatar was the belief that participants held regarding its agency, rather than the actual physical characteristics of the agent. The authors manipulated the instruction that they had given to the participants of their study. In one condition, participants were instructed that the robot face they were observing was controlled by a human, while in another condition, they received an instruction that it was just a pre-programmed automaton. Joint attention (measured in the form of gaze cueing effects, Friesen & Kingstone 1998; Driver et al. 1999) was evoked to a larger extent when participants believed the robot's behavior was controlled by a human, relative to the belief that it was only pre-programmed. Interestingly, the physical features of the observed avatar were identical across conditions, which may suggest that there is no direct perception of intentionality. In a follow up EEG study (Wykowska, Wiese et al. 2014), the authors showed that this effect is reflected at early stages of attentional processing while Özdem et al. (2017) identified – in an fMRI study – neural correlates of such belief manipulation.

In a recent study, Thellman et al. (2017) asked a large number of people (N=93) with various backgrounds to interpret and explain behaviors displayed by two different agents (i.e. a human and a humanoid robot) involved in everyday activities (such as cleaning the floor or cooking). Specifically, the authors presented to their participants pictures depicting the two agents doing exactly the same actions and proposing, for each scenario, a possible explanation of the behavior. All explanations were referring to mental states both when the depicted agent was a robot, and when the depicted agent was a human. After observing each scenario, participants were asked to rate the explanation in terms of plausibility, and to rate the general behavior displayed by the agent in terms of intentionality, desirability and controllability. Results of their experiment showed that there was no difference between the two agents in terms of plausibility, intentionality and desirability ratings between the conditions. However, when the behavior was displayed by the human agent, participants evaluated that the behavior was controlled by the agents themselves. On the contrary, when the behavior was displayed by the artificial agents, the behavior seemed controlled by external causes or was evaluated as programmed. Results of this experiment suggest that humans might adopt the intentional stance to some extent towards artificial agents as well. Yet, the existence of some differences in ratings supports the idea that different strategies are used to interpret artificial agents and human behaviors. Recently, Kamide et al. (2015) used the "Anshin" questionnaire to explore the reactions of more than nine hundred Japanese participants to videos depicting several robots. They discovered that aspects like perceived humanness, proficiency and comfort with a robot changed from robot to robot and from behavior to behavior. Specifically, they noticed that usually the more the humanoid resembled a human in physical appearance, the higher the participants were evaluating the comfort of the interaction and the proficiency of behaviors. In some situations, they found that exposure to more machine-like robots made humans feel stressed, but their results confirm that attributions towards artificial agents can fluctuate according to several factors.

It is important to point out, however, that both Thellmann and colleagues as well as Kamide and colleagues used pictures or videos as stimuli in their experiments. Such "offline" experiments might not be sufficient to capture the social mechanisms involved during "live" interactions. In this context, it is crucial to address the question of whether behavioral parameters of an observed agent in more "live" naturalistic protocols affect the attribution of intentionality. In line with this hypothesis, Weiss and Bartneck (2015) pointed out that evaluations of likeability, anthropomorphism and human-likeness of a robot using their Godspeed questionnaire are dramatically influenced by the interaction with a robot rather than by its physical appearance.

In a recent study, Wykowska et al. (2015) designed two experiments that might help clarify this. In the first experiment, the authors showed to their participants a non-humanoid robot (mechanical arms attached to a picture depicting a face of a human). Participants were asked to observe the behaviors of the robot, trying to identify which one of those behaviors was pre-programmed and which one was remotely controlled by a confederate located in another room. In a second experiment, the same design was adopted, but the artificial agent was a humanoid robot NAO (SoftBank Robotics). In addition to the explicit task of detecting human control of the robot behavior, authors implemented in both experiments a visual letter discrimination paradigm, in order to collect an implicit measure of engagement. While participants were observing the behavior of the robot and rating its agency, they were also asked to discriminate between two letters ("T" or "F") appearing on a screen located behind the robot. Results of both experiments suggest that humans are quite sensitive in detecting subtle differences between movements displayed by an artificial agent. Indeed, participants were able to discriminate accurately when a movement was remotely controlled by a confederate and when it was pre-programmed. Furthermore, this study provides information about implicit aspects of social interaction with artificial agents. The results suggested that when artificial agents show a certain degree of anthropomorphism, the physical appearance of the robot might play a role in implicit social processes of engagement (in joint attention, in the case of this study).

In our own lab, we designed an experiment in a "live" interactive scenario, in which human participants were exposed to two different behaviors of a humanoid robot, one completely programmed and one pre-recorded from a human. After seeing both behaviors, participants were asked to explicitly state whether they detected any differences in the behavior displayed by the robot in the two conditions. For the design of the experiment, and the human behavior specifically, we needed to obtain data from human participants first. We focused only on eye and head movements in an attentional capture setup. We selected attentional capture (being distracted by salient events in an environment), because being distracted is common to several animal species and can be considered a powerful defense mechanism that allows adaptation to environmental changes. Attentional capture is very common in everyday life as well. It might happen that we are engaged in a conversation and our attention gets caught by something we see or hear, causing an abrupt interruption of the conversation. However, when we are interacting with artificial agents, we might not expect that they would be distracted. According to their desired functionality and the way they should be designed, they should be focused on a given task, with no distractibility. Hence, we reasoned, implementing a "human-like" behavior of being distracted on the robot should make it appear human-like.

To obtain the human-like behavior, we recorded reactions (head and eye movements) of twenty-two participants during an attentional capture paradigm. Participants were seated in front of a notebook and engaged in a solitary game, while continuous videos were being played on a different screen positioned in the periphery. Every once in a while a louder sound (i.e. a phone ring, a woman's laughter, a door slam, etc.) would be displayed on that peripheral screen. For the entire duration of the session, behaviors of the participants were recorded using an inertial sensor and an eye tracker. A first qualitative analysis of the results highlighted that our participants' reactions were extremely heterogeneous. Some participants were extremely reactive to distractions while some others were extremely focused on the game, thus ignoring the peripheral stimuli. Analyzing in more depth the collected reactions, we realized that there were no identical reactions, both in terms of trajectory of the movement and in terms of velocity profile of the movements. Kinematic patterns of humans are extraordinarily variable, and even the most sophisticated techniques of emulation might fail to be representative of a human behavior if we do not keep the variability in mind when developing artificial systems.

In order to investigate whether some components of the behavior were more pertinent for the evaluation of human-likeness, reactions to the most salient stimuli were implemented (Ghiglino et al. 2018) in an iCub simulator (Metta et al. 2008). This allowed us to search for some common patterns related to head movements within this large human variability. Results showed that the most important feature for attribution of human-likeness was time spent on the distraction, but not necessarily the amplitude of the movement. We used the results found with this study to select one participant whose reactions were most representative of the entire sample in terms of distractibility, and whose behavior could maximize the attribution of human-likeness toward the artificial agent. Our recordings were not limited to the head movements, but extended to the eyes. Therefore, saccades and fixations that we recorded from the human participant were implemented in the human-based behavior of the robot as well. This behavior was then implemented on our iCub robot (Metta et al. 2008; Natale et al. 2017).

As a "mechanistic" condition, we decided to generate a programmed stereotypical behavior. We hypothesized that in contrast with the huge variability recorded from the human participants, a programmed reaction should be identical across all trials and distractions. Therefore, we computed the average reaction profile we collected during our participants' reactions (in terms of amplitude of the movement and time to perform the sequences of the movement), and we implemented it in the humanoid robot. Every time there was a distraction (i.e. a salient sound), the head of the robot was programmed to turn toward a fixed point in space, every time exactly in the same way. In terms of saccades, the robot in the mechanistic condition always displayed exactly the same horizontal eye movements (from point A to point B and then back to point A).

In the experiment proper, we compared participants' evaluations of a robot behaving like one of our participants (human-like condition) with the robot behaving in a stereotypical, programmed way. We reproduced the attentional capture setup, where an iCub robot was seated in front of a computer screen. The aims of this experiment were to explore whether our participants would be able to detect differences in the behaviors of the humanoid and, most importantly, whether changing the robot behavior from mechanistic to human-like would affect the attribution of intentionality and/or other human-related characteristics. In order to test whether participants adopt the intentional stance towards the iCub or not, and how this is modulated by its behavior, we used a new tool, the InStance questionnaire (Marchesi et al., 2019) developed also by our team. In that tool, we created thirty-five fictional scenarios, depicting the iCub robot engaged in different daily activities. Each scenario consisted in three sequential frames, representing a storyline. After seeing each scenario, two alternative explanations of the story were presented: a mechanistic explanation and a mentalistic explanation. While the first was always referring to the design, functionality and mechanisms underpinning the robot behavior, the latter was always referring to mental states, such as desires, intentions and motivations. These two alternatives were located on the opposite sides of a slider. Participants were asked to move the slider toward the explanation that they believed more plausible in explaining the story depicted in the photographs. This would allow for identifying whether participants adopt the design stance or the intentional stance when explaining a robot behavior. Furthermore, we administered the Anshin questionnaire (Kamide et al. 2015) and the Godspeed questionnaire (Bartneck et al. 2009) to gain more insights into attributions participants made towards our iCub robot. Regarding the Godspeed questionnaire, we implemented in our experiment three subscales of the questionnaire: anthropomorphism, animacy and likeability. According to the authors of the questionnaire, anthropomorphism refers to the attribution of a human characteristics or behavior, while animacy refers to perceived lifelikeness and likeability refers to positive impressions felt toward a non-human being.

Regarding the Anshin questionnaire, the authors identify five components that might affect the attitude towards artificial agents:

- Comfort, which refers to the safety perceived during the interaction with the robot;
- Performance, which is defined as the perceived proficiency in the robot's behavior;
- Peace of mind, which refers to the feeling of being free from uneasiness and stress during the interaction with the robot;
- Controllability, which identifies the fear that robots might physically take over control from humans;
- Robot-likeness, which refers to how much the robot seems conventionally robot-like in terms of appearance, or motion.

Forty participants were recruited in our experiment. They were located in front of the iCub robot, which was behaving either in a human-based or in a pre-programmed way, as described above. Participants were asked to observe the humanoid while it was engaged in a generic task on a screen, in two different randomized sessions. In one of them, the robot was emulating the behavior recorded from the human we selected with the previous experiment. In the other one, the robot was acting stereotypically, as programmed. All participants were exposed to both behaviors of the robot, and after each session they were asked to fill out the Godspeed questionnaire, the Anshin questionnaire and half of the InStance questionnaire. In order to keep participants' attention on the face of the robot, a "cover task" was given: Participants were asked to count the number of times that robot's eyebrows changed position during each session. This task was fairly easy, since changes were very rare, in order to avoid excessive cognitive load. Before debriefing the participants, we asked them to determine explicitly if they detected any difference between the two sessions and to elaborate on these differences. Eventually, participants that detected a difference in the behaviors were asked which one of the two behaviors displayed by the robot resembled more a human being to them.

Surprisingly, almost half of our sample (N=17) was not able to detect any difference, and between the ones that spotted a difference (N=23), only fourteen attributed human-likeness to the robot emulating the behavior recorded from the human participant. This result was quite striking, given the seemingly very noticeable difference between conditions. In order to investigate the possible effect of the experimental manipulation, several analyses on the InStance, Godspeed and Anshin ratings were performed. Regarding the InStance questionnaire, the average score of participants' ratings collected after the exposure to the human-like behavior of the robot was compared with the average score of participants' ratings after the exposure to the machine-like condition and no significant difference between the two experimental conditions was found. Interestingly, also no difference was found when comparing the scores with a "baseline" score obtained in a previous experiment (Marchesi et al., 2019). We investigated also whether the other two questionnaires that we implemented as implicit measures of attitude towards the robot were more sensitive to our experimental manipulation. We conducted within-comparisons for all the subscales of the Godspeed and the Anshin questionnaires, and no significant results were found. In order to have a clearer idea of our participants' answers, we also decided to analyze separately participants that detected no difference between the conditions, participants that attributed human-likeness to the machine-like behavior and participants that attributed human-likeness to the human-based behavior. Despite no significant results being found on comparing these groups, a constant pattern repetitively appeared across our analyses. Participants' explicit evaluation of human-likeness seemed to have been predicted by the attribution of intentionality in the InStance questionnaire and from all Godspeed subscales. Specifically, participants that spotted no differences between the behavior of the robot tended to have identical scores after both sessions of the experiment. In contrast, participants who spotted a difference and attributed human-likeness to the machine-like behavior tended to attribute slightly more intentionality, anthropomorphism, animacy and likeability to the robot after the machine-like session. On the contrary, participants that perceived more human-likeness in the human-based behavior tended to attribute slightly more intentionality, anthropomorphism, animacy and likeability to the robot after the human-based session. This shows that what matters for attribution of intentionality or human-likeness is the belief or subjective impression rather than objective reality, a finding in line with literature (Wiese, Wykowska et al. 2012; Wykowska, Wiese et al. 2014; Wykowska et al. 2015). Another interesting insight arises from the qualitative data collected after our experiment. It seems that participants who perceived no difference between the two behaviors seemed to be too involved in the cover task and paid no attention to the behaviors of the robot. Participants evaluating the human-based condition as more human-like tended to make this decision based on the variability of the behavior displayed by the robot. On the other hand, the same variability was interpreted as a result of a "failure" in the robot program by participants that evaluated the machine-like behavior as more human-like. Some of them reported the saccadic movement of the robot as a completely unnatural behavior that they could not explain. This might suggest that expectations and preconceptions humans have towards an agent heavily affect the way we interpret the behaviors displayed by that agent, presumably over-shadowing empirical experience. This highlights how challenging it is to model human behaviors on a robot in order to create a human-like behavioral appearance, considering not only the large heterogeneity in such even simple behaviors, but also how humans interpret this variability.

The key observation from our study is that our participants' perception of the robot behavior seems to be extremely heterogeneous. This might suggest that besides the physical appearance of the robot, its behavior and the specific characteristics of the interaction, some other factors related to individual differences might play a role in the attitude participants adopt toward an artificial agent. In order to explore this possibility, we investigated possible correlations between sociodemographic information of our participants and their questionnaire ratings. We found a negative correlation between level of education and InStance questionnaire scores collected after both the human-based ($r=-0.43$, $p<0.01$) and the machine-like ($r=-0.33$, $p<0.05$) sessions of the experiment. This is an extremely interesting result and might suggest that the higher the participants' education (and presumably knowledge about technology), the less likely participants are to adopt intentional stance towards artificial agents. This result seems to be perfectly aligned with the idea of adopting the intentional stance both ontogenetically and phylogenetically. From a developmental perspective, some authors claim that already one-year-old children can represent, explain and predict goal-directed action using rudimental strategies that might be precursors of advanced mentalizing abilities. According to these theories, children adopt a sort of "teleological stance" which establishes teleological relation among actions, goals and situation (Gergely, Gergely 1998; Gergely, Gergely 2003; Kelemen 1999). Therefore, although the development of a proper intentional stance seems to occur only around the fourth year of age (Perner 1991), the adoption of goal-directed model to explain other agent's behaviors seems to develop already at an early age. Interestingly, the development and consolidation of design stance seems to occur only between five and seven years of age, as demonstrated by several authors (German, Johnson 2002; Matan, Carey 2001). This suggests that comprehension of the functioning and the design underpinning an action might require additional knowledge to be developed. This can be paralleled by non-scientific common observations of children's behavior: they seem to adopt the intentional stance towards teddy bears, dolls and various toys. Only with time, and as their knowledge grows, the intentional stance becomes increasingly substituted with the design stance, which requires information and knowledge about how things are designed to behave. Coherently with this notion, the more we are exposed to technology and the more we learn about physics, the more we can rely on an explanation that refers to the design and physical stances when observing behavior of a system.

Similarly, from an anthropological perspective, the limited knowledge our ancestors developed around natural phenomena, like the rain or the earthquakes, might be the reason why they attributed such phenomena to the "God's mind". When knowledge has become more developed and accessible, such interpretations have been substituted with physical explanations of the phenomena.

Thus, it seems that when we are interacting or observing behavior of an unknown (or not understandable) entity or phenomenon, we refer to the intentional model, as this might be the most available and default model that we possess. In fact, we are probably most experienced with intentional explanations, rather than scientific or technical explanations of various systems we observe in the environment. However, when our knowledge about an entity changes, the strategy we adopt to understand its behaviors changes as well.

A similar reasoning can be applied to robots: if a person with no formal education, who had never seen a robot before is exposed to a robot, s/he might be more likely to adopt the default intentional stance towards the robot, as compared to a person who is experienced with robotics, or has acquired sufficient level of general education (physics and engineering included). This suggests that it is important to take into consideration not only levels of education and experience with robots but also various age groups when testing attitudes towards robots.

3. Speculations

In addition to education and experience, other factors might play a crucial role in modulating humans' attitude towards robots. In human-human interactions, countless factors affect the way we perceive and understand our conspecifics' behavior, from personality traits to mood (Lopes et al. 2005; Cozolino 2014). For example, it has been widely demonstrated that introverted people show difficulties in communicating (Sallinen-Kuparinen et al. 1991; Schoemaker, Kalverboer 1994) and that people in a depressed mood tend to avoid social interaction (Young, Leyton 2002; Erber et al. 1996). Internal dispositions and states seem to be strictly connected with the attitude we show during our everyday life interactions. Therefore, we can speculate that as it happens for mentalizing abilities (Platek et al. 2003; Nichols, Stich 2003), the intentional stance might be influenced also by self-awareness and self-reflective abilities. Few studies demonstrated that self-processing abilities affect mental state attribution to others, and, specifically, that psychiatric population with deficits in self-awareness show impairments in the attribution of mental states to other agents (Williams 2010; Moriguchi et al. 2006). Studying whether such influences affect interactions with robots as well might help in the development of new explanatory models about human social abilities. Although human-computer interaction (HCI) and human-robot interaction (HRI) studies are focused on designing artificial interfaces that can satisfy the needs of humans, understanding the way humans interact with artificial agents can also improve our knowledge about human cognition in general. Theories and models of human cognition can be tested using such technologies, offering new challenges (Dautenhahn 2007). Cognitive architecture models, for example, are largely used to develop artificial intelligence (Kelley 2006; Metta et al. 2010). A case of particular interest might be the humanoid robot iCub, mentioned in the studies that we reported in previous sections. Its developers and designers took inspiration from models of cognitive architectures and implemented similar functions in the robot (Vernon et al. 2007). Specifically, the synthetic architecture that is fitted to the iCub is composed of components such as a distributed multi-functional perceptual-motor network, a system of inhibitory modulation and a system of action simulation. These systems originated from research and models proposed by neuroscientists, such as the presence in the brain of specific networks responsible for action selection (Chevalier, Deniau 1990, Deniau, Chevalier 1985) or models about action simulation (Shanahan 2005). Implementing a robot with such functions helps researchers to clarify the ecological validity of their models, providing new insights about the functioning of the human brain.

It would be crucial to include also social cognitive mechanisms in the artificial models implemented in robots. It is in this context that our theoretical questions about adopting the intentional stance, characteristics of robot behavior or inter-individual differences need to be addressed. One of the most interesting questions is whether endowing robots with intentionality is crucial for evoking adoption of the intentional stance, or rather whether emulation of human-like behavior is sufficient. This question is analogous to the issue of the Turing test (Turing 1950): Is it possible to produce responses in the Turing test that would make participants believe they are interacting with an intelligent agent, even though it is just an algorithm? Or rather, is it indispensable to have a genuine intelligence in order to pass the Turing test. In simple words, is it possible to "fake" intelligence or intentionality thanks to very well-designed behavioral characteristics? At present, researchers are still far from developing artificial agents that can be perceived as human-like. Although some technologies provided with artificial intelligence might be able to pass Turing test, humans are still fully aware of the boundaries between artificial agents and living beings (Kahn et al. 2006). The open question is whether it is only a matter of technological advances to "fake" intentionality, or is it in principle impossible.

4. Societal Impact

Evoking the adoption of the intentional stance might be desirable in certain applications. For example, pathological traits, such as autistic traits, severely compromise social interaction and a multitude of studies demonstrated that individuals diagnosed with autism spectrum disorder (ASD) are often unable to deal with the complexity of human social interaction (Wing, Gould 1979; Frith 2003; McConnell 2002). Specifically, children diagnosed with ASD often display less interest towards behaviors aimed at generating and maintaining social interaction (Vivanti, Nuske 2017). Severity, onset time and specificity of such deficits captured the attention of clinicians and neuroscientists, stimulating the development of innovative evidence-based therapeutic approaches (Yates, Couteur 2016; Fernandez et al. 2018). Recent literature suggests that robots might be efficiently used as clinical tools in order to enhance social competences of children affected with social impairments (Diehl et al. 2012; Feil-Seifer, Mataric 2010; Kajopoulos et al. 2015; Carlson et al. 2018). Several studies demonstrated that the advantages of using such artificial agents rely on their physical appearance and on the extensive control that clinicians and scientists can exert on their behaviors (Scassellati et al. 2012; Robins et al. 2005; Liu et al. 2008). Robots represent a safe, predictable and coherent environment in which training social interaction skills in people that display difficulties in interaction with other humans (Dautenhahn, Werry 2004) should be beneficial. It seems that artificial agents can significantly improve therapeutic efficiency in patients affected with social impairments. For example, in 2013, Zheng et al. demonstrated that after only four sessions with a specific rehabilitation protocol carried out with a robot, children with ASD improve their social skills (Zheng et al. 2013). Concerns might be raised about the development of a certain social dependence that the patient can develop toward assistive technology. However, some studies demonstrated that learning acquired during the activities with robots are automatically transferred to everyday-life activities and interaction with other humans (Francois et al. 2009; Robins et al. 2005).

We speculate that robot behaviors that evoke adoption of the intentional stance shall be even more efficient in training social skills that can be then transferred to human interactions. This is because adopting the intentional stance towards other humans is a natural process in social interactions of typically developing individuals. It is very likely that difficulty in the effortless adopting of the intentional stance towards others is part of the problem in social skill deficiencies (Griffin & Dennett 2008). Therefore, evoking the intentional stance in therapies with robots should have a beneficial effect for transferring the skills trained with robots to interactions with humans.

Apart from the application of social robots to healthcare, social robots are postulated to have a potential application also in elderly care. Robots that are designed for elderly care provide support in everyday-life activities (eating, bathing, etc.), in mobility, in housekeeping and in monitoring health condition of the user. Several researchers focused their attention on the consequences caused by the introduction of assistive robots (Pollak et al. 2002; Graf et al. 2004) and an assistive smart environment (Bahadori et al. 2003) in elderly care. Results showed that the introduction of robots in elderly care does not only improve the quality of life of people in terms of perceived autonomy, but also in terms of psychological wellbeing (Broekens et al. 2009; Sharkey, Sharkey 2012). This is because such robots are able to provide social feedback to the user, and can be treated as "companions". Companion robots seem to increase positive mood, while reducing the perception of loneliness and stress (for a review, see Broekens et al. 2009 and Bemelmans et al. 2012). Results emphasize the usefulness of robots that are able to induce some degree of social attunement. These applications will be needed in the near future, considering that the elderly population is constantly increasing, while the availability of healthcare professionals is decreasing (WHO, Investing in the health workforce enables stronger health systems, in Fact sheet. 2007: Belgrade, Copenhagen). In order to maximize the efficiency of healthcare professionals, repetitive and frustrating activities can be assigned to assistive robots. In parallel, this would allow lower institutionalization of medical care, and higher autonomy for individuals. Also in the case of elderly care, robots that evoke adoption of the intentional stance might prove more beneficial than those that not, due to potential higher degree of bonding.

However, some studies pointed out the risks that such social devices might entail. In the last decades, the main application of robots was industry. Recently, however, domestic robots started entering our homes. Some authors question the safety granted by new technologies, in terms of producing reliable and non-harmful behavior (Denning et al. 2009; Sharkey & Sharkey 2010), others point out the issues related to psychological risks, such as social isolation (Sparrow 2006; Hampton et al. 2009) or addiction to technology (Veruggio 2005). Finally, one of the most discussed controversial aspects related to robots is the issue of privacy. Privacy is an ancient concept, originated by ancient Greeks as a distinction between public and political activities (*polis*) and the domestic life (*oikos*) (Guthrie, 1962). Nowadays, privacy is conceived as the possibility of controlling one's own personal data. Recently, privacy management has caught the attention of media and legislative authorities. although no robot or artificial intelligence was involved in the 2018 scandal of Facebook, Facebook admitted that information of eighty-seven millions of users were given to Cambridge Analytica without their permission. If we think about assistive and domestic robots, one of the necessary requirements for them to work is collecting, processing and potentially recording data from users. Some of the commercial house assistants, like google home, need to save information collected during the interaction with the user in the cloud. In the context of the intentional stance, perhaps also in this area, if robots induce adoption of the intentional stance, they might be more likely to collect more personal or intimate information from the user, relative to robots that would be treated only as mechanistic artifacts. This is because humans might be more inclined to reveal their intimate information towards agents that are perceived as intentional (to make it more evident: we are certainly more likely to confide in another human than in an artifact such as coffee machine). Thus, designing robots that evoke the adoption of the intentional stance in users might have consequences for the type of data that is recorded and potentially stored. Companies producing robots for daily home use must guarantee privacy protection of such information, in order to avoid third parties to access those data without consent of the user.

In this context, it is crucial to inform users about all the aspects involved in interacting with technology. The more we aim to introduce technology in everyday life, the more we need to inform users about the potential risks, as well as benefits provided by having access to such tools. Giving as much information as possible to users, and educating them on how to use technology would on the one hand circumvent certain fears produced often by popular culture (e.g., the fear of robots "taking over the world") and would reduce the unnecessary hype related to AI and robotics that is based on misconceptions about what AI is capable of. Some authors, for instance, claimed that assistive robots are unethical since their effectiveness depends on creating an illusion of companionship (Sparrow 2002; Turkle et al. 2006; Wallach 2009). However, it is crucial to avoid creating illusions and misconceptions related to new technologies. As demonstrated by the results of our experiment, education plays a key role in the attribution of intentional stance towards artificial agents. This could imply that highly educated people might benefit less than others might from the interaction with a social robot, in terms of perceived social companionship. However, this is still an empirical question, and other studies are necessary to clarify the relation between education and social cognition mechanisms spontaneously adopted during human-robot interaction. We can argue that informing people about the nature, the functions and the possible applications of robots might solve part of the ethical issues related to their usage.

Taken together, we argue that adopting intentional stance towards artificial agents might be beneficial in some contexts, and is not necessarily something to be afraid of, given how automatic and default this mechanism is (Heider, Simmel 1944; Gray et al. 2007; Wiese, Metta, Wykowska 2017). However, it is generally important to provide sufficient information about robot technologies to all users, in order to avoid potential risks of misconceptions, misuse of private data as well as unnecessary fears related to fantasies created by popular culture.

5. Final Remarks

In conclusion, studying the conditions and consequences of implementing human-like behaviors on artificial agents that can potentially induce adoption of the intentional stance is fascinating from a theoretical perspective, and extremely important for the future of our societies. From the theoretical perspective, it can be informative with respect to the mechanisms of the human social cognition – showing how flexible our socio-cognitive system is. How far can the human brain go in extending mechanisms of social cognition from natural to artificial agents? What consequences does adopting the intentional stance have for other (perhaps more implicit) mechanisms of social cognition? Can we develop models of intentionality for artificial systems? Is endowing artificial systems with intentionality necessary for evoking adoption of intentional stance or is it sufficient to emulate human-like behaviors?

From the perspective of applications of robots and future societal impact, it is important to discuss when treating artificial agents as intentional systems is desirable. It is also important to keep in mind that proper information for the end users regarding the capabilities of the artificial systems can avoid misconceptions and hype on the one hand, and fears and risks on the other. All these issues need to be discussed not only among the scientific community, but also with the general audience. This would allow a strengthening of people's awareness about the technology, avoiding misuse and misjudgment.

6. Acknowledgment

This work has received support from the European Research Council under the European Union's Horizon 2020 research and innovation programme, ERC Starting grant ERC-2016-StG-715058, awarded to Agnieszka Wykowska.

The content of this paper is the sole responsibility of the authors. The European Commission or its services cannot be held responsible for any use that may be made of the information it contains.

7. Bibliography

Bahadori, Shahram , Cesta, Amedeo, Grisetti, Giorgio, Iocchi, Luca, Leone, Riccardo, Nardi, Daniele, Oddi, Angelo, Pecora, Federico, & Rasconi, Riccardo (2003, September) "Robocare: an integrated robotic system for the domestic care of the elderly". *In Proceedings of Workshop on Ambient Intelligence AI*IA-03*, Pisa, Italy.

Baron-Cohen, Simon, Campbell, Ruth, Karmiloff-Smith, Annette, Grant, Julia, & Walker, Jane (1995) "Are children with autism blind to the mentalistic significance of the eyes?" *British Journal of Developmental Psychology*, 13(4), 379–398.

Bartneck, Christoph, Croft, Elizabeth, Kulic, Dana & Zoghbi, Susana (2009). "Measurement instruments for the anthropomorphism, animacy, likeability, perceived intelligence, and perceived safety of robots". *International Journal of Social Robotics*, 1(1), 71–81.

Bemelmans, Roger, Gelderblom, Gert Jan, Jonker, Pieter, & De Witte, Luc (2012) "Socially assistive robots in elderly care: A systematic review into effects and effectiveness." *Journal of the American Medical Directors Association*, 13(2), 114–120.

Bérubé, Michael (2013) "A Theory of Theory of Mind." *American Scientist*, 101(2), 148.

Brentano, Franz (1874/1911/1973) *Psychology from an Empirical Standpoint.* London: Routledge and Kegan Paul.

Broekens, Joost, Heerink, Marcel, & Rosendal, Henk (2009) "Assistive social robots in elderly care: a review." *Gerontechnology*, 8(2), 94–103.

Burgos, Josè Enrique (2007) "About aboutness: Thoughts on intentional behaviorism." *Behavior and Philosophy*, 65–76.

Carlson, Kelsey, Wong, H. Y. Alvin, Dung, Tran Anh, Wong, C. Y. Anthony, Tan, Yeow Kee, & Wykowska, Agnieszka (2018, November) "Training Autistic Children on Joint Attention Skills with a Robot." In *International Conference on Social Robotics*. Cham: Springer, 86–92.

Chevalier, Gilles, & Deniau, Jean Michel (1990) "Disinhibition as a basic process in the expression of striatal functions." *Trends in neurosciences*, 13(7), 277–280.

Churchland, Paul Montgomery (1981) "Eliminative materialism and propositional attitudes." *the Journal of Philosophy*, 78(2), 67–90.

Cozolino, Louis (2014) *The neuroscience of human relationships: Attachment and the developing social brain.* WW Norton & Company.

Dautenhahn, Kerstin (2007) "Methodology & themes of human-robot interaction: A growing research field. International" *Journal of Advanced Robotic Systems,* 4(1), 15.

Dautenhahn, Kerstin, & Werry, P. Ian (2004) "Towards interactive robots in autism therapy: Background, motivation and challenges." *Pragmatics & Cognition,* 12(1), 1–35.

Davidson, Donald (1980). *Essays on Actions and Events,* Oxford University Press.

Deniau, Jean Michael, & Chevalier, Gillian (1985) "Disinhibition as a basic process in the expression of striatal functions. II. The striato-nigral influence on thalamocortical cells of the ventromedial thalamic nucleus." *Brain research,* 334(2), 227–233.

Dennett, C. Daniel. (1971) "Intentional systems." *The Journal of Philosophy,* 68(4), 87–106.

Dennett, C. Daniel (1981) "True believers: The intentional strategy and why it works."

Dennett, Daniel (2009). "Intentional systems theory". *The Oxford handbook of philosophy of mind,* 339–350.

Denning, Tamara, Matuszek, Cynthia, Koscher, Karl, Smith, R. Joshua, & Kohno, Tadayoshi (2009, September) "A spotlight on security and privacy risks with future household robots: attacks and lessons." In *Proceedings of the 11th international conference on Ubiquitous computing.* ACM, 105–114.

Diehl, J. Joshua, Schmitt, M. Lauren, Villano, Michael, & Crowell, R. Charles (2012) "The clinical use of robots for individuals with autism spectrum disorders: A critical review." *Research in autism spectrum disorders,* 6(1), 249–262.

Driver, Jon, Davis, Greg, Ricciardelli, Paola, Kidd, Polly, Maxwell, Emma, & Baron-Cohen, Simon (1999) "Gaze perception triggers reflexive visuospatial orienting." *Visual cognition,* 6(5), 509–540.

Erber, Ralph, Wegner, M. Daniel, & Therriault, Nicole (1996) "On being cool and collected: Mood regulation in anticipation of social interaction." *Journal of personality and social psychology,* 70(4), 757.

Feil-Seifer, David, & Mataric, J. Maja (2010) "Dry your eyes: examining the roles of robots for childcare applications." *Interaction Studies,* 11(2), 208.

Fernández, Marta, Mollinedo-Gajate, Irene, & Peñagarikano, Olga (2018) "Neural circuits for social cognition: Implications for autism." *Neuroscience,* 370, 148–162.

Fodor, A. Jerry (1975) *The language of thought* (Vol. 5). Harvard University Press.

Fodor, A. Jerry (1983) *The Modularity of Mind*; Cambridge, a Bradford Book.

Fonagy, Peter (2018) *Affect regulation, mentalization and the development of the self.* Routledge.

François, Dorothee, Powell, Stuart, & Dautenhahn, Kerstin (2009) "A long-term study of children with autism playing with a robotic pet: Taking inspirations from non-directive play therapy to encourage children's proactivity and initiative-taking." *Interaction Studies,* 10(3), 324–373.

Friesen, K. Chris, & Kingstone, Alan (1998) "The eyes have it! Reflexive orienting is triggered by nonpredictive gaze." *Psychonomic bulletin & review*, 5(3), 490–495.
Frith, D. Chris, & Frith, Uta (2008) "Implicit and explicit processes in social cognition." *Neuron*, 60(3), 503–510.
Frith, D. Chris, & Frith, Uta (2012) "Mechanisms of social cognition." *Annual review of psychology*, 63, 287–313.
Frith, Uta (2003) *Autism: Explaining the enigma*. Blackwell Publishing.
Gergely, Csibra, & Gergely, György. (1998) "The teleological origins of mentalistic action explanations: A developmental hypothesis." *Developmental Science*, 1(2), 255–259.
Gergely, György, & Gergely, Csibra. (2003) "Teleological reasoning in infancy: The naıve theory of rational action." *Trends in cognitive sciences*, 7(7), 287–292.
German, C. Tamsin, & Johnson, C. Susan (2002) "Function and the origins of the design stance." *Journal of Cognition and Development*, 3(3), 279–300.
Ghiglino, Davide, De Tommaso, Davide, & Wykowska, Agnieszka (2018, November) "Attributing human-likeness to an avatar: the role of time and space in the perception of biological motion." In *International Conference on Social Robotics*. Cham: Springer, 400–409.
Graf, Birgit, Hans, Matthias, & Schraft, Rolf Dieter (2004) "Care-O-bot II – Development of a next generation robotic home assistant." *Autonomous robots*, 16(2), 193–205.
Gray, M. Heather, Gray, Kurt, & Wegner, M. Daniel (2007) "Dimensions of mind perception." *science*, 315(5812), 619–619.
Green, F. Michael, & Horan, P. William (2010) "Social cognition in schizophrenia." *Current Directions in Psychological Science*, 19(4), 243–248.
Griffin, Richard & Dennett, C. Daniel (2008) "What does the study of autism tell us about the craft of folk psychology?" In T. Striano & V. Reid (Eds.) *Social Cognition: Development, Neuroscience, and Autism*. Wiley-Blackwell, 254–280.
Guthrie, K. C. William (1962). A History of Greek Philosophy: Volume 2, The Presocratic Tradition from Parmenides to Democritus (Vol. 2). Cambridge University Press.
Hampton, N. Keith, Sessions, F. Lauren, Her, J. Eun, & Rainie, Lee (2009) "Social isolation and new technology." In: Pew Research Center (Internet and Technology). Available at: https://www.pewinternet.org/2009/11/04/social-isolation-and-new-technology/.
Heider, Fritz, & Simmel, Marianne (1944) "An experimental study of apparent behavior." *The American journal of psychology*, 57(2), 243–259.
Husserl, Edmund (1900/01). Logical Investigations, trans. J. N. Findlay, London: Routledge [1900/01; 2nd, revised edition 1913], 1973.
Jacquette, Dale (1991) *The origins of Gegenstandstheorie: Immanent and transcendent intentional objects in Brentano, Twardowski, and Meinong*, Brentano Studien, 3, 277–302.

Kahn, H. Peter, Ishiguro, Hiroshi, Friedman, Batya, & Kanda, Takayuki (2006, September) "What is a human?-toward psychological benchmarks in the field of human-robot interaction." In: ROMAN 2006 – The 15th IEEE International Symposium on Robot and Human Interactive Communication, pp. 364–371.

Kajopoulos, Jasmin, Wong, H. J. Alvin, Yuen, W. C. Anthony, Dung, A. Tran , Kee, Y. Tan, & Wykowska, Agnieszka (2015, October) "Robot-assisted training of joint attention skills in children diagnosed with autism." In *International Conference on Social Robotics*. Cham: Springer, 296–305

Kamide, Hiroko, Kawabe, Koji, Shigemi, Satoshi, & Arai, Tatsuo (2015) "Anshin as a concept of subjective well-being between humans and robots in Japan." *Advanced Robotics*, 29(24), 1624–1636.

Karmiloff-Smith, Annette, Klima, Edward, Bellugi, Ursula, Grant, Julia, & Baron-Cohen, Simon (1995) "Is there a social module? Language, face processing, and theory of mind in individuals with Williams syndrome." *Journal of cognitive Neuroscience*, 7(2), 196–208.

Kelemen, Deborah (1999) "Function, goals and intention: Children's teleological reasoning about objects." *Trends in Cognitive Sciences*, 3(12), 461–468.

Kelley, D. Troy (2006) "Developing a psychologically inspired cognitive architecture for robotic control: The Symbolic and Subsymbolic Robotic Intelligence Control System (SS-RICS)." *International Journal of Advanced Robotic Systems*, 3(3), 32.

Korkmaz, Baris (2011) "Theory of mind and neurodevelopmental disorders of childhood." *Pediatr Res*, 69(5 Pt 2), 101R–8R.

Lillard, Angeline S., & Kavanaugh, D. Robert (2014) "The contribution of symbolic skills to the development of an explicit theory of mind." *Child development*, 85(4), 1535–1551.

Liu, Changchun, Conn, Karla, Sarkar, Nilanjan, & Stone, L. Wendy (2008) "Online affect detection and robot behavior adaptation for intervention of children with autism." *IEEE transactions on robotics*, 24(4), 883–896.

Lopes, N. Paulo, Salovey, Peter, Côté, Stéphane, Beers, Michael (2005) "Emotion regulation abilities and the quality of social interaction." *Emotion*, 5(1), 113.

Marchesi, Serena, Ghiglino, Davide, Ciardo, Francesca, Baykara, Ebru, & Wykowska, Agnieszka (2019) *Do we adopt the Intentional Stance towards humanoid robots?* Frontiers in psychology, 10, 450.

Matan, Adee, & Carey, Susan (2001) "Developmental changes within the core of artifact concepts." *Cognition*, 78(1), 1–26.

Mayer-Hillebrand, Franziska (1951) *Verzeichnis der Manuskripte Franz Brentanos*, Harvard: Typoscript.

McConnell, R. Scott (2002) "Interventions to facilitate social interaction for young children with autism: Review of available research and recommendations for educational intervention and future research." *Journal of autism and developmental disorders*, 32(5), 351–372.

Metta, Giorgio, Natale, Lorenzo, Nori, Francesco, Sandini, Giulio, Vernon, David, Fadiga, Luciano, Hofsten, Claes, Rosander, Kerstin, Lopes, Manuel, Santos-Victor José & Bernardino, Aalessandro (2010) "The iCub humanoid robot: An open-systems platform for research in cognitive development." *Neural Networks*, 23(8–9), 1125–1134.

Metta, Giorgio, Sandini, Giulio, Vernon, David, Natale, Lorenzo, & Nori, Francesco (2008, August) "The iCub humanoid robot: an open platform for research in embodied cognition." In *Proceedings of the 8th workshop on performance metrics for intelligent systems*. ACM, 50–56.

Millikan, G. Ruth (1984) *Language, Thought and Other Biological Objects*, Cambridge, Mass.: MIT Press.

Moriguchi, Yoshiya, Ohnishi, Takashi, Lane, D. Richard, Maeda, Motonari, Mori, Takeyuki, Nemoto, Kiyotaka, Matsuda, Hiroshi, & Komaki, Gen (2006) "Impaired self-awareness and theory of mind: an fMRI study of mentalizing in alexithymia." *Neuroimage*, 32(3), 1472–1482.

Natale, Lorenzo, Bartolozzi, Chiara, Pucci, Daniele, Wykowska, Agnieszka, Metta, Giorgio (2017) "The not-yet-finished story of building a robot child." *Science Robotics*, 2 (13), DOI: 10.1126/scirobotics.aaq1026.

Nichols, Shaun, & Stich, P. Stephen (2003) *Mindreading: an integrated account of pretence, self-awareness, and understanding other minds*. Clarendon Press/Oxford University Press.

Özdem, Ceylan, Wiese, Eva, Wykowska, Agnieszka, Müller, J. Hermann, Brass, Marcel, & Van Overwalle, Frank (2017) "Believing androids – fMRI activation in the right temporo-parietal junction is modulated by ascribing intentions to non-human agents." *Social neuroscience*, 12(5), 582–593.

Perner, Josef (1991) *Understanding the representational mind*. The MIT Press.

Platek, M. Steven, Critton, R. Samuel, Myers, E. Thomas, & Gallup Jr, G. Gordon (2003) "Contagious yawning: the role of self-awareness and mental state attribution." *Cognitive Brain Research*, 17(2), 223–227.

Pollack, E. Martha, Brown, Laura, Colbry, Dirk, Orosz, Cheryl, Peintner, Bart, Ramakrishnan, Sailesh, Engbers, Sandra, Mattews, T. Judith, Dunbar-Jacobs, Jaqueline, Mccarthy, E. Colleen, Montemerlo, Michael, Pineau, Joelle, & Roy, Nicholas. (2002, August) "Pearl: A mobile robotic assistant for the elderly." In *AAAI workshop on automation as eldercare* Vol. 2002, 85–91.

Putnam, Hilary (1967) "Psychological predicates." *Art, mind, and religion*, 1, 37–48.

Robins, Ben, Dautenhahn, Kerstin, Te Boekhorst, Rene, & Billard, Aude (2005) "Robotic assistants in therapy and education of children with autism: can a small humanoid robot help encourage social interaction skills?" *Universal Access in the Information Society*, 4(2), 105–120.

Russell, Bertrand (1905) "On denoting", *Mind*, 14: 479–93; reprinted in A. P. Martinich (ed.) (1996) *The Philosophy of Language*, Oxford: Oxford University Press.

Sabbagh, A. Mark (2004) "Understanding orbitofrontal contributions to theory-of-mind reasoning: Implications for autism." *Brain and cognition*, 55(1), 209–219.

Sallinen-Kuparinen, Aino, McCroskey, C. James, & Richmond, P. Virginia (1991) "Willingness to communicate, communication apprehension, introversion, and self-reported communication competence: Finnish and American comparisons." *Communication Research Reports*, 8(1), 55–64.

Scassellati, Brian, Admoni, Henny, & Matarić, J. Maja (2012) "Robots for use in autism research." *Annual review of biomedical engineering*, 14, 275–294.

Schaafsma, M. Sara, Pfaff, W. Donald, Spunt, Bob, & Adolphs, Ralph (2015) "Deconstructing and reconstructing theory of mind." *Trends in cognitive sciences*, 19(2), 65–72.

Schoemaker, M. Marina, & Kalverboer, F. Alex (1994) "Social and affective problems of children who are clumsy: How early do they begin?" *Adapted physical activity quarterly*, 11(2), 130–140.

Scholl, J. Brian, & Leslie, Alan (1999) "Modularity, development and 'theory of mind'." *Mind & Language*, 14(1), 131–153.

Searle, John (1979) "What is an intentional state?" *Mind*, 88(349), 74–92.

Searle, John (1980) "Minds, Brains and Programs." *Behavioral and Brain Sciences*, 3: 417–457

Shanahan, Murray (2006) "A cognitive architecture that combines internal simulation with a global workspace." *Consciousness and cognition*, 15(2), 433–449.

Sharkey, J. Amanda, & Sharkey, E. Noel (2012) "Granny and the robots: ethical issues in robot care for the elderly." *Ethics and information technology*, 14(1), 27–40.

Sparrow, Robert (2002) "The march of the robot dogs." *Ethics and information Technology*, 4(4), 305–318.

Taylor, Marjorie, & Carlson, M. Stephanie (1997) "The relation between individual differences in fantasy and theory of mind." *Child development*, 68(3), 436–455.

Terada, Kazunori, Shamoto, Takashi, & Ito, Akira (2008, August) "Human goal attribution toward behavior of artifacts." In *Robot and Human Interactive Communication*, 2008. In: ROMAN 2007 – The 17th International Symposium on Robot and Human Interactive Communication, pp. 160–165.

Thellman, Sam, Silvervarg, Annika, & Ziemke, Tom (2017) "Folk-psychological interpretation of human vs. humanoid robot behavior: exploring the intentional stance toward robots." *Frontiers in psychology*, 8, 1962.

Turing, M. Alan (1950) "Mind." *Mind*, 59(236), 433–460.

Turkle, Sherry, Taggart, Will, Kidd, D. Cory, & Dasté, Olivia (2006) "Relational artifacts with children and elders: the complexities of cybercompanionship." *Connection Science*, 18(4), 347–361.

Vernon, David, Metta, Giorgio, & Sandini, Giulio (2007, July) "The icub cognitive architecture: Interactive development in a humanoid robot." In: ICDL 2007 – The 6th International Conference on Development and Learning, 122–127.

Veruggio, Gianmarco (2005) *The birth of roboethics*. In: ICRA 2005 – International Conference on Robotics and Automation, Workshop on Roboethics, 1–4.

Vivanti, Gianmarco, & Nuske, J. Heather (2017) "Autism, attachment, and social learning: Three challenges and a way forward." *Behavioural brain research*, 325, 251–259.

Wallach, Wendell (2010) "Robot minds and human ethics: the need for a comprehensive model of moral decision making." *Ethics and Information Technology*, 12(3), 243–250.

Wiese, Eva, Metta, Giorgio, & Wykowska, Agnieszka (2017) "Robots as intentional agents: using neuroscientific methods to make robots appear more social." *Frontiers in psychology*, 8, 1663.

Wiese, Eva, Wykowska, Agnieszka, Zwickel, Jan, & Müller, J. Hermann (2012) "I see what you mean: how attentional selection is shaped by ascribing intentions to others." *PloS one*, 7(9), e45391.

Williams, David (2010) "Theory of own mind in autism: Evidence of a specific deficit in self-awareness?" *Autism*, 14(5), 474–494.

Wing, Lorna, & Gould, Judith (1979) "Severe impairments of social interaction and associated abnormalities in children: Epidemiology and classification." *Journal of autism and developmental disorders*, 9(1), 11–29.

Wykowska, Agnieszka, Kajopoulos, Jasmin, Obando-Leiton, Miguel, Chauhan, S. Sushil, Cabibihan, John-John, & Cheng, Gordon (2015) "Humans are well tuned to detecting agents among non-agents: examining the sensitivity of human perception to behavioral characteristics of intentional systems. International" *Journal of Social Robotics*, 7(5), 767–781.

Wykowska, Agnieszka, Wiese, Eva, Prosser, Aaron, & Müller, J. Hermann (2014) "Beliefs about the minds of others influence how we process sensory information." *PLoS One*, 9(4), e94339.

Yates, Kirsty, & Le Couteur, Ann (2016) "Diagnosing autism/autism spectrum disorders." *Paediatrics and Child Health*, 26(12), 513–518.

Young, N. Simon, & Leyton, Marco (2002) "The role of serotonin in human mood and social interaction: insight from altered tryptophan levels." *Pharmacology Biochemistry and Behavior*, 71(4), 857–865.

Zheng, Zhi, Zhang, Lian, Bekele, Esube, Swanson, Amy, Crittendon, A. Julie, Warren, Zachary, & Sarkar, Nilanjan(2013, June) "Impact of robot-mediated interaction system on joint attention skills for children with autism." In: ICORR 2013 – International Conference on Rehabilitation Robotics, 1–8.

Revisiting the Dancing-Qualia Argument for Computationalism

Stefan Reining

Abstract

According to computationalist theories of the mind, consciousness does not depend on any specific physical substrate, such as carbon-based biological material, but automatically arises out of the right kind of computational structure. Even though this thesis has become an unquestioned assumption in most of current AI literature, there exist only few direct arguments in favor of it. One, and probably the most prominent, argument for computationalism from the philosophy of mind is David Chalmers' dancing-qualia argument. The aim of this paper is to challenge this argument from a hitherto neglected angle, by arguing that it is undermined by some experimental results in neurobiology regarding the workings of phenomenal memory. However, I will argue that Chalmers' overall case for the possibility of conscious AI can still be vindicated.

1. Computationalism and Its Role in AI Research

Defendants of the possibility of strong AI, that is, the possibility of AI with human-like consciousness, very often ground this position on a computationalist account of the mind. According to computationalism, the mind is basically a computing system, generating certain outputs from certain inputs on the basis of some computational structure. In the case of humans, this computational structure is realized by the structure of the human brain. The crucial aspect for the possibility of strong AI is that, according to computationalism, anything replicating the structure of a conscious being's brain will itself be conscious, and will have mental states of the same quality (given the same inputs), regardless of the actual physical material on the basis of which a conscious being's brain structure has been replicated. Neurons in the human brain are carbon-based, but, according to computationalism, artificial neurons that are, say, silicon-based[1], will do the same job of contributing to conscious mental states, as long as the brain's computational structure remains unaltered.

1 Actually, the reputation of silicon as a promising candidate for a material of artificial neurons has diminished, due to the capacity of carbon to form more stable chemical bonds. In

 | DOI:10.30965/9783957437488_007

In order to avoid circularity, one should not define 'computationalism' using 'computational structure'. In line with Chalmers (Chalmers 2011: 337–338), I will instead use the notion of a causal topology. Roughly speaking, two systems have the same causal topology if and only if they are disposed to produce the same outputs under the same input conditions. This notion now allows for a broad definition of 'computationalism':

(C) Conscious beings are conscious in virtue of their brains' causal topology.

A question that arises about this definition concerns the fineness of grain with which the causal topology of actual brains has to be replicated in order to produce mental states of the same kind. As Chalmers notes, "[i]t may turn out that the relevant computational description of these processes is very fine-grained, reflecting extremely complex causal dynamics among neurons" (Chalmers 2011: 350). Chalmers approaches the difficulty of determining the necessary fineness of grain by choosing a criterion that addresses the system's overall behavior, stating that "the relevant level of causal topology will be a level fine enough to determine the causation of behavior" (Chalmers 2011: 337), and that "[f]or the brain, this is probably the neural level or higher, depending on just how the brain's cognitive mechanisms function" (Chalmers 2011: 337). However, this way of determining the necessary level of fineness is in tension with Chalmers' own critique of behaviorism (i.e., the theory that mental states are constituted by the respective system's behavioral dispositions). As Chalmers himself notes, "two mentally distinct systems can have the same behavioral dispositions" (Chalmers 2011: 336), which is inconsistent with his claim that two computational structures that produce the same behavioral dispositions will also be disposed to experience the same mental states. But apart from this point of criticism, from a computationalist perspective, it is certainly intuitive to assume that the necessary fineness of grain indeed lies at the neural level or higher, and most computationalists will probably subscribe to this assumption.

What makes computationalism a particularly relevant topic at a conference like the one at which the present paper was held is the fact that it has become an unquestioned assumption in most of today's popular AI literature.

---

fact, the whole question whether consciousness depends on a certain biological substrate would turn out trivial if only carbon-based neuronal cells can provide the desired input/output-relations. Nowadays, a more promising candidate material of artificial neurons are themselves carbon-based – namely, carbon nanotubes. However, the makeup of these nanotubes is very different from human neurons, so they may still be characterized as a different biological substrate.

Futuristic scenarios in which conscious beings are realized by the execution of computer programs that emulate human brains all depend on a rather strong computationalist view (especially since physical realizations of such programs will not even remotely resemble the shape of human brains). One example particularly well-suited to illustrate this role of computationalism in contemporary AI literature can be found in Nick Bostrom's 2003 paper "Are We Living in a Computer Simulation?". In the following passage, Bostrom is quite explicit regarding the role of computationalism (or, as he calls it here, the thesis of substrate independence) in his argument:

> Provided a system implements the right sort of computational structures and processes, it can be associated with conscious experiences. It is not an essential property of consciousness that it is implemented on carbon-based biological neural networks inside a cranium: silicon-based processors in a computer could in principle do the trick too. Arguments for this thesis have been given in the literature, and although it is not entirely uncontroversial, I shall here take it as given. (Bostrom 2003: 244)

Bostrom himself does not provide a reference to any of the arguments given in the literature in favor of substrate independence. And in fact, hardly any arguments for this thesis can be found at all. Plenty of arguments have been given for preferring certain computationalist theories over computationalist or non-computationalist competitors. However, the raw thesis of substrate independence, which lies at the core of computationalism in general, has, maybe due to its fundamental nature, not been subject to many direct arguments (be they pro or con). One notable exception is the work of David Chalmers. In the 1990s, Chalmers first presented his so-called dancing-qualia argument to defend computational sufficiency; the thesis that the right kind of computational structure is sufficient for consciousness. Chalmers' argument is well-known in the philosophy of mind, and accessible to a broader audience.

In what follows, I will first reiterate Chalmers' argument and then provide a novel kind of criticism of it.

2. The Dancing-Qualia Argument for Computationalism

The argument is based on a thought-experiment, actually, on a scenario that has to be regarded as futuristic, regardless of the time that has passed since the publication of the argument. Chalmers asks the reader to imagine a person whose brain can be accessed in such a way as to abruptly replace certain neurons by input/output-equivalent silicon chips (or chips made of any other material different from that of our actual biological makeup). In particular, the neurons whose replacement Chalmers asks the reader to imagine are those whose firing correlates with certain color sensations. Moreover, the reader is to imagine that the neuronal replacement in question does not alter the causal topology of the person's brain, or, at least, does not alter any level of causal topology "fine enough to determine the causation of behavior" (Chalmers 2011: 337). Thus, the surgery will not affect the person's behavior. (In order to imagine this, it is probably best to imagine that the surgery happens unbeknownst to the person.) This includes, for instance, that the person's use of color terms will be unaltered, that is, the person will continue to use certain color terms for certain objects in the same way as before the surgery.

The question at hand is now the following: Will the replacement affect the person's phenomenal color perception? The dancing-qualia argument is basically a *reductio ad absurdum* of the claim that the replacement would indeed affect the person's phenomenal color perception. If it did affect the person's phenomenal color perception, the person would notice. However, since the causal topology of the brain has not been altered by the surgery, the subject would show no observable sign of noticing any phenomenal change, even if the switch between the old and the new brain circuit is being continuously flipped, "so that [for instance] red and blue experiences 'dance' before the system's inner eye" (Chalmers 2011: 340). And assuming that noticing something always involves some change in behavioral dispositions, which stay the same in this case, "there is no room for any noticing to take place" (ibid.). According to Chalmers' argument, the only tenable option for avoiding the kind of absurdity in question, namely the absurdity of assuming that the subject notices a change in her phenomenal color perception without being in any way disposed to show signs of noticing this change, is to drop the assumption that the neuronal replacement in question affects the person's phenomenal color perception. Accordingly, the computationalist conclusion of the dancing-qualia argument is that as long as the brain's causal topology, and thereby its computational structure, remains unaltered, artificial replacements will produce exactly the same phenomenal states as those the person would have had without the surgery.

3. The Dancing-Qualia Argument in the Face of Neurobiological Findings

I will now go on to identify an assumption made in the course of the dancing-qualia argument, an assumption necessary for the argument to be sound, and

I will argue that this assumption is incompatible with some experimental findings from neurobiology.

According to the assumption in question, it is possible to alter the way in which neurons process perceptual input while at the same time preserving the subject's ability to properly compare the new perceptual experiences with those perceptual experiences the subject had in the past when processing perceptual input of the relevant sort. Put differently, the assumption in question, which will hereafter be referred to as the Unimpaired Memory Assumption (UMA), runs as follows:

> (UMA) If some neuronal replacement changed a subject's phenomenal color perceptions, the subject would (*ceteris paribus*) be able to compare the new color perceptions with an unimpaired memory of prior phenomenal experiences of this color.

To illustrate what UMA amounts to, imagine again the neuron-replacement scenario described in the previous section. More specifically, imagine a version of this scenario in which those neurons whose firing correlates with the subject's phenomenal experience of the color red are replaced by silicon chips without altering the brain's causal topology. This kind of replacement will not change the way the subject phenomenally experiences the color red, because even though the subject would notice such a change, the subject would, since the brain's causal topology has not been altered, not express any kind of wondering, which would be an absurd combination of consequences. Applied to this version of the scenario, UMA implies that if the replacement in question altered the way the subject experiences the color red, the subject would still remember her previous experiences of this color, compare these old experiences with her new ones, and come to the conclusion that something has changed in the way she experiences this color.

UMA may initially appear quite plausible. However, it is undermined by relatively recent experimental results in neurobiology supporting what may be called the Reactivation View (RV) of phenomenal memory:

> (RV) Recalling a past perceptual experience results from a reactivation of the very same neurons whose activation produced the original experience during the perceptual episode (or at least from a reactivation of that subset whose activation was responsible for those features of the original experience that are being recalled).

RV has been experientially confirmed both in the case of smaller animals (Reijmers et al. 2007, Liu et al. 2012, Ramirez et al. 2013, and Tayler et al. 2013) and humans (Gelbard-Sagiv et al. 2008)[2]. It has the following implications regarding the dancing-qualia argument. The possibility that neuronal replacement causes changes in the subject's phenomenal experiences can no longer be denied by arguing that this possibility would have the absurd consequence that the subject would notice the change despite being unable to express any corresponding wonder or amazement. Instead, a neuronal replacement that alters a subject's color experience would have corresponding effects on the subject's memory of the color's phenomenal character.

Applied to the specific scenario above, RV implies that if the replacement in question altered the way in which the subject experiences the color red, it would in the same way alter the subject's memories of her past experiences of this color. For according to RV, both the subject's present experiences of red things and her memories of such experiences depend on the firing of the very same set of neurons. Consequently, if replacing this set of neurons affects the subject's present experiences of the color red in a certain way, it will also affect the subject's memories of such experiences in this way. Thus, no phenomenal changes would be noticed by the subject, explaining why the subject is not disposed to show any signs of wonder or amazement. RV thereby undermines the dancing-qualia argument.

To be sure, accepting RV only commits one to admitting that UMA is *actually* false in the case of humans and some smaller animals. For all we know, the existence of beings for which UMA is true still remains a theoretical possibility. However, the actual falsity of UMA suffices for undermining the dancing-qualia argument, because once it has turned out that there are cases in which the kind of neuronal replacement in question can, without absurdity, lead to changes in the subject's phenomenal experiences, the dancing-qualia argument no longer provides a good reason for believing in computational sufficiency, since computational sufficiency was supposed to be necessarily true, or at least true for actual humans, and not just true for some imaginary beings.

My argument in this section is not a positive argument for the anti-computationalist claim that the kind of neuronal replacement in question would indeed result in some kind of phenomenal modification. Instead, it undermines Chalmers' positive argument for the computationalist claim that the

2 In the case of humans, the experiment involved epilepsy patients implanted with depth electrodes that enabled the experimenters to track single neurons spatiotemporally: "[W]e report here that a subset of neurons in the human hippocampus and entorhinal cortex exhibited highly reliable and specific responses during viewing of video episodes. […] The same neurons showed an increased firing rate again with free conscious recall, before the verbal report, when the sequence of physical sensory stimuli was absent and no external cues were provided." (Gelbard-Sagiv et al. 2008: 100)

kind of neuronal replacement in question would *not* result in some kind of phenomenal modification.[3]

4. Resorting to Fading Qualia

I have argued above that, given the Reactivation View of phenomenal memory, the dancing-qualia argument does not establish computational sufficiency regarding the quality of phenomenal experiences. However, it is still worth investigating whether a different version of the argument, also proposed by Chalmers, namely, the so-called *fading*-qualia argument, is undermined by RV as well. The fading-qualia argument uses the same reasoning as the dancing-qualia version, but instead of the conclusion that a neuronal replacement of the kind described above does not lead to *different* phenomenal experiences, it involves the conclusion that the replacement in question does not result in consciousness slowly fading away. Let's suppose the anti-computationalist were to claim that the kind of replacement in question results in the subject's experiencing the color in question in a less clear way than before, and that she will eventually, after a sufficient number of neurons have been replaced, not be able to have any phenomenal experiences of this color. By hypothesis, the subject's behavioral dispositions stay the same (since, just as in the case of the dancing-qualia thought experiment, the brain's causal topology is unaffected by the replacement), and the subject's use of the respective color term will thus not be altered.[4] Analogously to the above remarks concerning the dancing-qualia argument, the subject cannot, given RV, simply introspectively compare her new phenomenal experiences with memories of more vivid past experiences of the respective quality. (The reason is, once again, that the subject's memories of the phenomenal quality in question suffer from the same

3 Defendants of the dancing-qualia argument might still try to resort to the mere possibility of stumbling upon beings for which UMA is true, by embracing the following line of argument: "Once a being appears for which UMA turns out to be true, an application of the dancing-qualia argument will show that regarding such a being, it is absurd to assume that the kind of neuronal replacement in question would result in any kind of phenomenal modification. And what could explain this absurdity other than computational sufficiency?" Proponents of this possible line of reasoning would still owe us a story about why computational sufficiency, understood as a metaphysical claim about all conscious beings, should be regarded as a good explanation of the kind of absurdity in question. This is a rather complicated issue, and I take it that the burden of proof lies with those who wish to embrace this kind of reasoning.

4 In order to ensure that it makes sense to stipulate that the subject's behavioral dispositions do not change in any relevant way, it is probably best to imagine that the neuronal replacement happens unbeknownst to the subject.

degree of desaturation as her current experiences of the respective color.) Thus, analogously to my remarks concerning the dancing-qualia argument, the anti-computationalist could claim that the kind of neuronal replacement in question will result in the subject's phenomenal color experiences becoming less clear than before, without being committed to the absurd consequence that the subject would register the change but could, by hypothesis, not express even the slightest degree of wonder or amazement. Instead, as in the case of dancing qualia, the phenomenal change would simply remain unnoticed by the subject.

However, a closer look at Chalmers' scenario involves trouble for the anti-computationalist. Imagine the subject in the scenario to possess the kind of conception of phenomenal experience whose possession leads one, in the face of the description of the thought experiment, to describe this scenario as one in which the subject is no longer able to experience a certain phenomenal color quality that she had experienced before the neuronal replacement. Even if the subject is no longer able to remember the phenomenal color quality in question, and therefore does not directly notice any phenomenal modification in the course of the neuronal replacement, there is no reason to assume that her possession of the kind of conception of phenomenal experience just mentioned could not lead her to judge, or even explicitly state (maybe in the context of some philosophical discussion), that her ability to phenomenally experience colors is deficient regarding some specific color. This, however, contradicts the stipulation that the subject's behavioral dispositions have not been altered. That is, by hypothesis, the neuronal replacement does not alter the brain's causal topology and therefore does not cause the subject to react differently to certain color experiences than before. However, it is absurd to assume that the subject does not know (or at least believe) that she used to have vivid experiences of the color in question in the past, and is no longer able to have these experiences now, and it is equally absurd to assume that her behavioral dispositions (in particular, her dispositions regarding the way in which she describes her own ability to perceive colors) have not changed accordingly.

An analogous line of reasoning can arguably be applied if one were to suppose that the neuronal replacement in question does not only cause the subject's experiences of a single color to fade, but instead her color experiences in general, so that the subject would eventually no longer be able to phenomenally differentiate between any two colors. Here again, even though the subject would, given RV, no longer be able to directly remember the phenomenal quality of any color, she would of course nonetheless be able to remember, on a more abstract level, *that* she used to have different phenomenal experiences when seeing objects of different colors. And she would therefore detect, on this more abstract level, the difference from her present state. But since, by hypothesis, the brain's causal topology has not been altered by the neuronal replacement, thus leaving no room for the expression of any wonder or amazement, the fading-qualia argument can still be used to derive the desired kind of absurdity. That is, even in the face of RV, the fading-qualia argument is still a sound *reductio ad absurdum* of the claim that the kind of neuronal replacement in question will affect the subject's phenomenal experiences.

5. Summary

The Reactivation View (RV) of phenomenal memory undermines Chalmers' argument for one version of the computational sufficiency claim, namely, the version according to which the quality of a subject's phenomenal experiences remains completely *unaltered* by neuronal replacement, as long as the latter does not affect the brain's causal topology regarding the subject's behavioral dispositions. However, RV does not undermine Chalmers' argument for the claim that consciousness does not *fade* and eventually disappear due to the kind of neuronal replacement in question. Chalmers' argument can therefore still be regarded as a promising case for the possibility of strong AI, since the latter does not depend on the assumption that causally isomorphic brains produce phenomenal experiences of exactly the same quality. In conclusion, beings with artificial brains causally isomorphic to ours might have phenomenal experiences different from ours, but they will probably still have phenomenal experiences of some sort.

6. Bibliography

Bostrom, Nick (2003) "Are We Living in a Computer Simulation?" *Philosophical Quarterly*, 55, 243–255.

Chalmers, David John (1996) *The conscious mind. In search of a fundamental theory.* New York & Oxford: Oxford University Press.

Chalmers, David John (2011) "A computational foundation for the study of cognition." *Journal of Cognitive Science*, 12(4), 323–357.

Gelbard-Sagiv, Hagar, Roy Mukamel, Michal Harel, Rafael Malach, and Fried Itzhak (2008) "Internally generated reactivation of single neurons in human hippocampus during free recall." *Science*, 322, 96–101. https://doi.org/10.1126/science.1164685.

Liu, Xu, Steve Ramirez, Petti T. Pang, Corey B. Puryear, Arvind Govindarajan, Karl Deisseroth, and Susumu Tonegawa (2012) "Optogenetic stimulation of a hippocampal engram activates fear memory recall." *Nature*, 484, 381–385. https://doi.org/10.1038/nature11028.

Ramirez, Steve, Xu Liu, Pei-Ann Lin, Junghyup Suh, Michele Pignatelli, Roger L. Redondo, Tomás J. Ryan, and Susumu Tonegawa (2013) "Creating a false memory in the hippocampus." *Science*, 341, 387–391. https://doi.org/10.1126/science.1239073.

Reijmers, Leon G., Brian L. Perkins, Naoki Matsuo, and Mark Mayford (2007) "Localization of a stable neural correlate of associative memory." *Science*, 317, 1230–1233. https://doi.org/10.1126/science.1143839.

Tayler, Kaycie Kuss, Kazumasa Z. Tanaka, Leon G. Reijmers, and Brian J. Wiltgen (2013) "Reactivation of neural ensembles during the retrieval of recent and remote memory." *Current Biology*, 23, 99–106. https://doi.org/10.1016/j.cub.2012.11.019.

PART II

Superintelligence

Will Superintelligence Take Over?

A Critical Reflection on the Apparent and Actual Dangers of Artificial Intelligence

Kilian Karger

Abstract

The successes of self-learning AI systems lead to highly speculative but very influential theories involving the concept of superintelligence. The following text will illustrate what is meant by the concept of superintelligence, which prerequisites and conditions are connected with it and to what extent these are realizable. On this basis, it will be critically reflected to what extent such an understanding of artificial intelligence challenges our human self-conception and whether the dangers in the use of AI predicted by advocates of the superintelligence hypothesis are actually the most pressing questions in dealing with this versatile and promising technology.

1. The Expanding Influence of Artificial Intelligence

Modern technology has found its way into almost all areas of our lives, and we have become accustomed to the fact that machines relieve us of many tasks that used to require a lot of time and effort. Starting with the textile industry in Great Britain in the 18th century, a worldwide change took place that transformed the agrarian nations of the modern era into industrial nations dominated by the widespread use of machines. This development, which continues with a steady increase in the efficiency of automation, is the greatest factor influencing all aspects of life in modern society, and has a deep impact on its social, cultural and political structures.[1] The scale in the production of technical devices achieved by automated mass production has led to lower costs for automobiles, televisions and smartphones, while digitalization has enabled ever greater networking of the globalized world economy.[2]

A new dimension of technical progress was achieved in the development of the computer. According to the mathematician Alan Turing, the computer is a machine that can manipulate formal symbols according to syntactic rules

1 Cf. Kurzweil 1990: 3.
2 Cf. Ramge 2018: 50.

 | DOI:10.30965/9783957437488_008

and thus algorithmically process mathematical problems. The next stage of this development is now seen in artificial intelligence, i.e. in computers whose algorithmic processing through "machine learning" is plastic and can adapt to certain tasks "automatically", and automatically – after a certain period of training. The buzzwords associated with this dominate the economic and feature sections and "not a week goes by without a new robot, a self-learning program, a new smart gadget coming onto the market, hardly a day without headlines conjuring up an upcoming revolution or one that is already in full swing: through artificial intelligence".[3] A prominent example is the victory of the AlphaGo computer system, produced by Google, over a South Korean master player in the Go board game. This millennia-old Asian game represents strategic thinking, human creativity and intelligence. It is considered the most complex board game and has more game variations than there are atoms in the universe.[4] Artificial intelligence also shines in games like Chess and Checkers and is now able to bluff better than the best poker champions in the world.[5] But the success of artificial intelligence is not limited to mastering games; AI systems are also being used successfully in the medical field for the diagnosis of certain cancer cells and the analysis of MRI images.[6]

Computer systems today provide legal advice[7], regulate the temperature and energy consumption of so-called smart homes and, under good external conditions, control automobiles and aircraft more safely and precisely than human drivers and pilots.[8]

2. Artificial Intelligence as a Tool – Saviour or Threat?

These examples illustrate the great benefits that the use of Artificial Intelligence, as an almost universal tool, has for us in a variety of areas of our lives. This universality is also the special feature of this tool, because where other devices or machines have been used since the beginning of human history to support humans in a certain task, a self-learning AI system is capable of apparently adapting to any task on the basis of the same basic learning algorithms. Where previously specialized expert systems were tailored to a specific task, universally applicable systems will soon dominate. As the examples show, the

3 Lenzen 2018: 10.
4 Cf. Wewetzer 2016.
5 Cf. Ramge 2018: 8.
6 Deutscher Ethikrat 2018: 48f
7 Meusers 2016.
8 Cf. Ramge 2018: 16.

fields of application of such systems now extend to tasks in which it was until only recently thought that they could only be processed by a human being who had to make use of his cognitive abilities in some way.[9] Although machines "have always been there to do something better than humans […], the relationship between humans and machines is now reminiscent of a battle of retreat"[10]. This development has effects both on how we regard ourselves as human beings and on our assessment of the machines, which we have hitherto regarded as helpful tools. The special tool available to us in the form of artificial intelligence has become the subject of highly speculative, but nevertheless extremely influential theories in recent years. The role of the AI in these theories alternates between exaggeration, to be the saviour of mankind, and portrayal as a spectre that threatens the existence of our entire species. Both visions of the future have in common that they assume that man loses his significance as the most important factor in the shaping of reality and gives it to AI systems of all kinds. Thus, both theories are variants of transhumanism based on the premise that the human species is subject to continuous evolution, and must evolve in order not to be pushed from the throne or even exterminated by other, competing, life forms.[11] According to Ray Kurzweil, inventor, author and head of technical development at Google, "Man will have lost his position as the most intelligent and powerful being on earth"[12] before the end of the next century. Self-learning artificial intelligence will then lead to incredible progress in the sciences, which can neither be understood nor controlled by man, and to which Kurzweil refers to as singularity.[13] Man will merge with the machine intelligence he originally created and coexist with intelligent machines either as a cyborg or as bodiless consciousness uploaded into a virtual world.[14] For Kurzweil, however, this development, with all its imponderables, has a very positive connotation, as the human species evolves in interaction with artificial intelligence. Man's cognitive abilities are increased to immeasurable proportions, his life span and especially his health span can be extended by advanced medicine and technical implants, and even his dream of *immortality* becomes reality. This is to become possible thanks to Kurzweil's vision of the so-called mind-upload, in which a detailed scan of the brain captures a person's entire personality, including all his memories and abilities. This

9 Cf. Ramge 2018: 19.
10 Lenzen 2018: 12.
11 Sorgner 2016: 10.
12 Kurzweil 2000: 18.
13 Cf. Lenzen 2018: 14.
14 Cf. Kurzweil 2000: 359.

digital person can be transferred as "mental software" to artificial bodies like files from one computer to another.[15]

The philosopher and Oxford professor Nick Bostrom, on the other hand, has a much gloomier vision of the future. He warns of the risks of machine super intelligence, which he characterizes as "an intelligence that far exceeds human cognitive performance in almost all areas."[16] Bostrom assumes that man-made artificial intelligence that is capable of learning can produce such a superintelligence in the course of a continuous further development of itself. He constructs scenarios in which he tries to make plausible that a "superintelligence decides the fate of life on Earth, that it could very easily have non-human ultimate goals, and that it probably has instrumental reasons for appropriating more and more resources."[17] From Bostrom's point of view, these developments pose dangers that can lead to the rapid annihilation of mankind, and with his prognoses he wants to sensitize people to greater caution in the development of artificial intelligence. Similarly dark visions of the future have recently been expressed by central representatives of the large Silicon Valley companies, such as Elon Musk, Bill Gates and Steven Wozniak.[18] Such extreme positions make the headlines and are helpful in the struggle for attention in the highly competitive market for high-tech products. However, with their highly speculative and more distant forecasts, they are also paralysing the debate about the real opportunities and risks of artificial intelligence, which already affect our society to a large extent. Due to their sensational linguistic style and the sharp focus on drastic consequences, certain background assumptions on which these theories are based often fall out of sight, which affect both the cognitive abilities of the AI and our view of man, which is challenged in particular by transhumanism. In order to be able to provide a critical classification of such prognoses, these aspects will be briefly examined below.

3. The Challenge of the Human Image through Artificial Intelligence

At this point, the relevance of AI research for the human image in the 21st century should be mentioned. This makes a philosophical examination of the latest technical developments unavoidable.

---

15 Cf. Kurzweil 2014: 330.
16 Bostrom 2016.
17 Bostrom 2016.
18 Cf. Lenzen 2018: 14.

For the possibility of realizing cognitive achievements, which were previously reserved for the intellectually and consciously gifted human being, through self-learning machines based on purely algorithmic calculability, now calls into question to an unprecedented extent the special position of the human being, according to which he was regarded as the only living being to whom attributes such as autonomy, rationality, normativity, creativity and emotionality can be fully ascribed as a conscious subject. The human being was conceived as a thinking subject that "is at least in principle capable of orienting its actions towards reasons and thus of determining its interaction with other subjects and its dealings with its environment rationally, i.e. through goals, values and purposes"[19]. Such a subject can orient itself on norms, by which it critically questions itself and to which it can once again relate itself.

In many cases, the current dominance of artificial intelligence, which is apparently capable of performing superhuman cognitive tasks, is included in the series of offences that, according to Sigmund Freud, have been inflicted on man by new scientific discoveries. "Homo Sapiens had to learn that he is neither the center of the universe nor the crown of creation. He had to learn that his ego is dominated by drives. And now, apparently, even the mind is no longer a bastion of man. [...] That is the fourth offence."[20]

Thus the progress of the new technology of artificial intelligence has not only great consequences for everyday life, the economy, politics and the way we live together as a society, but it confronts us like hardly any other phenomenon in a special way with our self-interpretation. According to the interpretations of many AI researchers, man can only be understood as a machine. Conversely, must machines with certain cognitive abilities be granted certain rights?

Such questions are also reflected within AI research, whose goal from the beginning went beyond constructing intelligent machines. Already at a symposium in 1956 in Dartmouth, which is regarded as the starting signal of the efforts for artificial intelligence, the research goal was proclaimed "to learn and to describe all other characteristics of intelligence so precisely that with these findings a machine can be built which simulates these processes"[21]. A major part of the research that gave rise to today's cognitive sciences has therefore, since its beginnings, been devoted to trying to find out more about human beings and the way their abilities function. The artificial systems that are designed serve, besides their practical use in computer science, industry or the

19 Müller 2015: 31.
20 Wewetzer 2016.
21 Ramge 2018: 33.

facilitation of our everyday life, to "test hypotheses about the functioning of human thought"[22]. The decisive assumption of this research approach is the so-called computer metaphor. According to this model, man is to be understood as a computer, with his body corresponding to the hardware and his mind to the software of the computer.[23] In the opposite dynamic, on the one hand, the human being is regarded exclusively as a neurocybernetic machine and described with information technology vocabulary. On the other hand, the abilities and characteristics that were previously regarded as genuinely human, such as rational thinking, autonomous decision-making or creativity, are now also attributed to the artificial computer system.

One side of the dynamic is based on a reductionist understanding of man and reality, which is now seen only as a pure sequence of actual events that no longer permits freedom of action or normativity.[24] Man, like all reality, is viewed purely mechanistically, for example as a natural information processing system that has arisen in the course of biological evolution and whose cognitive abilities give him an advantage in the evolutionary arms race.[25] "With the advent of computers, they became the model for the brain: the brain is therefore a kind of computer, the mind its program."[26] This model is proving to be extremely popular today, especially in the neurosciences, and some of its most prominent representatives, such as Gerhard Roth[27], Wolf Singer[28] and Thomas Metzinger, represent it partly openly, partly implicitly.

The second side of the dynamics of the computer metaphor, which is intended to legitimize the attribution of human abilities to the computer, lies in the computational functionalist theory of the mind. This theory regards cognitive human abilities, especially thinking, as equivalent to data processing in a computer. Both the brain and the computer are regarded as physical information processing systems, pure input-output machines that, in the case of humans, receive sensory stimuli as inputs and output behavioural forms.[29] In both cases, processing is carried out according to a program that syntactically determines the manipulation of symbols that are entered, i.e. according to formal rules. Since "the same program can in principle run on different machines,

22 Lenzen 2018: 30.
23 Cf. Mutschler 2002/2003: 108.
24 Cf. Cramm 2008: 44.
25 Cf. Metzinger 2014: 299.
26 Lenzen 2018: 38.
27 Cf. Roth 2003: 16
28 Cf. Singer 2002: 36.
29 Cf. Kutschera 1993: 76.

the character of the machine cannot be decisive for the cognitive process"[30]. Because it is therefore irrelevant on which material basis the program is realized, subjective abilities such as perception, thinking, acting, etc. must also be attributed to the computer – according to the representatives of computer functionalism.[31]

4. AI Systems as Power-addicted Agents? On the Difference between Man and Machine

In order to examine whether the visions of thinkers such as Bostrom or Kurzweil are plausible and whether the fear of developments in the field of artificial intelligence is justified, it is necessary to take a look at what systems today's research has created so far, how they work and which prerequisites of transhumanist theories they can fulfil. At this point, it should be noted that such a reconstruction of the abilities of computer systems is made considerably more difficult by the imprecise and anthropomorphic way of speaking, which is used both in popular-scientific and relevant scientific texts. It is quite natural to speak of systems that think, learn, plan, decide, that are autonomous or even self-confident. These terms are used partly according to their meaning in everyday language, but partly also as technical terms with different content. Since the use of the terms is usually not marked, a confusing gibberish is created from upscaling everyday and technical language, which transports countless misunderstandings. It would therefore be necessary, as a first step, to reconstruct the language with which the object of research is discussed.[32] Such a project is reasonably and methodically possible by reconstructing the basic technical prerequisites and trying to determine their basic scope in scientific theory. This would create a basis for methodically critically examining the visions of the future, which focus on artificial intelligence.

If one considers the term artificial intelligence, "these two words [...] today stand for a broad, interdisciplinary and unclear field in which researchers with quite different questions and methods are working to develop systems that can [...]: use language, form concepts, solve problems."[33] Some of these tasks can be done by existing or at least foreseeable self-learning AI systems. Today, these systems are mostly artificial neural networks (KNN). These networks are

30 Fischer 2003: 34.
31 Cf. Lenzen 2018: 19.
32 Cf. Janich 2009: 13f.
33 Lenzen 2018: 23.

based on a statistical procedure. Nerve cells are simulated according to the model of the human brain. These differently weighted formal models of neurons are interconnected and layered to form a hierarchical network.[34] Data is fed into this network in which the algorithms of the system are to find patterns or regularities. Such a KNN can learn in different ways: Either the programmer already tells the system what to learn and gives hints, or the network tries to recognize patterns in the entered data. A combination of supervised and unsupervised learning, the so-called reinforcement learning, has proven to be particularly promising. Here "a system tries out possible actions and receives feedback again and again, also called rewards, and then changes its behaviour a little in order to get as much positive feedback as possible"[35]. With this process, KNNs are able to improve their performance in a given task and have already achieved a superhuman level of precision and speed in many fields when evaluating huge amounts of data. The plastic networks created in the course of training can now become so complex that it is no longer possible for the human user or programmer to understand, through all layers and neurons of the system, how the network came to a certain conclusion.[36] The systems presented here are called forms of weak AI in research and already prove to be extremely helpful assistants to human beings in many areas.

The horror scenarios in which Bostrom identifies artificial intelligence as a realistic threat to humanity, however, require a system that is referred to as strong AI and which must be capable of much more than just recognizing patterns in data according to a learning algorithm and, on the basis of this, selecting an option for further action from a limited selection of technical implementations. The artificial system is attributed with a will to self-preservation and a striving for power, which is characterized as a striving for self-improvement and maximization of the available resources. Why a strong AI that pursues these goals must be able to actually make autonomous decisions is discussed below.

Although the term 'autonomy' is frequently used in the AI debate, which becomes apparent, for example, in the discussion about self-propelled cars, which are often referred to as autonomous, on closer inspection it refers to a form of gradual autonomy, which consists, for example, in the ability of a robot to "carry out an intended task on the basis of its current state and its perception without human intervention"[37]. Autonomy in the full human sense would

34 Cf. Ramge 2018: 46.

35 Lenzen 2018: 51.

36 Cf. Ramge 2018: 27.

37 Lenzen 2018: 124.

only come to an artificial system when it is able not only to behave according to predetermined rules, but to reflect on them and, if necessary, to disregard them. However, this is not possible in principle for artificial intelligence for a number of reasons, no matter how autonomous the system may appear to the outside world.

One reason for the impossibility of true autonomy in artificial systems is that the attempt to attribute an essential prerequisite of autonomy to a machine has already failed. This condition is a will to survive, i.e. an urge to secure one's own survival by one's own drive. Living beings are not rigid stimulus-reaction machines, but are located within a context of life or a useful purpose.[38] This means that if there are several options for action in their environment, they have a very strong disposition to choose the alternative that best serves their survival, for example the intake of food or the escape from predators. Such a context of self-preservation, which seems to be fundamentally inherent in all life, is, however, sought in vain in machines, unless a human programmer has programmed the concept of such a context into the machine.[39]

Such basic dispositions can also be found in humans and are expressed in the fact that we eat when we are hungry, sleep when we are tired and try to avoid situations that threaten our physical integrity.

Furthermore, in addition to these dispositions for survival, people also behave according to learned norms[40], such as rules of social coexistence, or according to the norms of certain practices that they have acquired through socialization or education. However, this is not just a matter of blindly following rules to fulfill them. If it is a matter of rational action practice, it must also be possible to critically question the respective rules and norms themselves, so that they remain controllable in principle through reflection processes. An action practice in which these abilities become particularly clear is, for example, the conduct of scientific research and, in particular, experimentation as a necessary component of this research. Every scientist commits himself to certain norms that define what is considered a scientific experiment and guarantee that its results are reproducible, controllable and intersubjectively verifiable. Scientific research is therefore only possible if the people who carry it out are capable of orienting themselves towards standards. The scientist is not a blind machine that spits out scientific results as output from the input it receives by observing events in the world. Rather, the scientist is – at least in principle – an

38 Cf. Müller 2018: 104.

39 Cf. Lenzen 2018 127.

40 Norms are conglomerates of implicitly traditional and explicitly learned rules and guidelines that structure a particular practice.

autonomous subject who, for rational reasons, freely decides to orient him or herself to the norms of his or her practice. This enables him to intervene in the world within the experiment, to create an idealized observation situation and to share and discuss the results of the experiment with the scientific community. However, an autonomous subject, through its orientation towards the norms of scientific practice, is by no means bound to let itself be guided by them in every situation of its life. What is special about a human subject is that it is able to critically question these norms, and thus possesses the ability to once again behave in accordance with its own dispositions.[41] Thus, when faced with a decision, man can step out of the context of life and the norms by which he orients himself, shed light on these from various perspectives, and in a spontaneous act decide on the option for action that seems sensible in the current situation. He is thus in a position "to orient his actions to reasons and thus to determine his interaction with other subjects and his dealings with his environment rationally, i.e. through goals, values and purposes"[42]. The human being is therefore not a blind input-output machine that gives out a certain form of behaviour when inputting sensory stimuli. With an artificial system, however, one has to deal with exactly such a system, because also AI systems are ultimately syntactic machines, whose semantic framework is given by their manufacturer, programmer and user. The great benefit of artificial intelligence lies in its ability to recognize patterns in data, "in machine learning, a system uses examples to identify patterns or regularities in order to classify new data"[43]. As a result, a system can better deal with raw data that has not been formalized for the computer and discover patterns in that data. However, the system always proceeds algorithmically according to the same rules in order to arrive at a result. It is completely irrelevant to the machine what kind of data is entered and what is to be done with the out-put it produces. This is also the greatest strength of computers, they are "optimization machines that have to be blind in their construction, because only then they are universal tools"[44]. Therefore, a self-learning artificial intelligence system that has been trained to triumph in the Go game can beat masters in chess or checkers after a short training phase. The AI, which recognizes patterns in MRI images and thus assists the oncologist in diagnosing cancer, is based on the same learning algorithm as the assistance system, which finds patterns in the stock market figures for an investment banker that he interprets as opportunities to maximize

41 Cf. Müller 2018: 101 ff.

42 Müller 2015: 31.

43 Lenzen 2018 50.

44 Mutschler 2002/2003: 106.

profits. The artificial system only recognises factual patterns in a lot of data, it does not know what it is doing, why it is doing this or that it knows something. In other words, it lacks the freedom to determine as the capacity to orient itself to norms, to question them, and thus to determine a judgment or an action in the light of reasons.[45] The artificial system is therefore not itself part of a normative action practice which gives orientation as to what its goal is, what means are to be chosen in what order for an attempt to achieve this goal, and when this attempt was unsuccessful and right or wrong.[46] The normative action practice is determined by the human user and programmer, he determines the purpose and the goal of the action, the machine has no relation to meaningful and reasonable embeddings in a large whole, but optimizes means or partial actions, which are necessary to reach the goal set by man. The universal tool character of the machine is also reflected in its structural design. The basis of every computer is first of all a virtual machine, i.e. a mathematical model in which it is fixed how the system calculates the input data for certain outputs. This fixation is done by algorithms that prescribe how a target is to be calculated step by step. The most important and widespread mathematical model of this kind is the so-called Turing machine. This virtual machine is now realized on a real machine, which for example consists of transistors made of silicon. Every state that such a system goes through is the causally necessary consequence of the previous state. But from a physical description of the system's state, nothing can be said about which task it is currently performing. A first attempt at interpreting the system-internal states is to translate them into a syntactically structured physical symbol system. "But physical symbol systems [...] do not carry their meaning in themselves. Independent of a code that is able to decode them, they are not more than a sum of electrical processes [...]. The mentioned structures only become a symbol system for a perceiving individual who is able to map the physical order [...] to a mental order of meaningful entities"[47]. From this it follows that even the level of pure syntax is read into the machine and exists only for the programmer who has an intentional semantics, i.e. the ability to recognize signs and to see this meaning. The level of semantics is not accessible to the computer and the ability achieved by today's self-learning AI systems "to calculate arbitrary inputs by no means means that it can now [...] connect signs with imaginary content relating to real or invented objects, processes, classes, etc."[48]. For the artificial system nothing

45 Cf. Fischer 2003: 39.
46 Cf. Müller 2018: 103.
47 Cf. Stekeler-Weithofer 2012: 104.
48 Fischer 2003: 39.

matters, it plays chess without knowing what it does and translates languages it does not understand, as John Searle illustrated in his famous thought experiment "the Chinese room". "It is only for the human observer that a string of characters handled by a computer becomes a symbol of something in the world."[49] This meaning is not determined by the individual, but the entire common normative practice of action is included in this process.[50] Semantics is of the greatest significance for the question of whether a system, such as Bostrom's super intelligence, can have autonomy, since only against its background can it be meaningfully stated that something or someone is striving for something or wants to set goals for himself. These goals lie in reality, which, however, can only be accessed by semantics. An artificial system that receives its goals and purposes from a human user does not need meaningful access to reality for its performance as a tool for optimizing resources and sub-processes. But if the system is to be able to set its own goal autonomously or wants to break out of the task assigned to it, it must have a relation to reality that is not only in the descriptive recognition of patterns. It must be able to recognize what these patterns mean, be able to judge on the basis of normative criteria how these recognized structures relate to its goal, and be able to decide accordingly for an action.

The fear of a super-intelligence that swings up to world ruler and wants to extinguish mankind is ultimately unfounded. This is not only due to the fact that it is not yet possible on the part of AI research to technically realize a strong artificial intelligence, but in particular to the impossibility in principle of a superintelligence which sets its own goals and which has a self-preservation will and a striving for power out of itself. For such strong AI has prerequisites that an artificial system is categorically incapable of providing, as long as it is based on a Turing machine that calculates program steps algorithmically and can be characterized purely functionally. Nevertheless, we are fascinated by dark science fiction visions in which evil supercomputers threaten humanity. These are all based on an old myth: "Man creates an artificial assistant to serve him. But the assistant trumps his creator by learning to learn and finally developing his own interests and setting his own goals."[51]

49 Fischer 2003: 42.
50 Cf. Stekeler-Weithofer 2012: 114.
51 Ramge 2018: 83.

5. Risks in Dealing with Artificial Intelligence

But even if we cannot regard artificial intelligence as an independent and power-hungry agent, but as a promising tool that is already revolutionizing many areas of our lives today, there are dangers that can easily get out of sight in the light of the horror scenarios that are more present in the media. The real danger does not emanate from artificial intelligence itself, but from people who use or even abuse it.

A major and often overlooked problem in dealing with self-learning systems is that the results of their machine learning processes are strongly dependent on the data with which they are trained. If the system learns with training sentences that are too small or poorly selected, it can take over structures from these that the human trainer has not noticed in them. Since the learning system depicts patterns that are actually present in the data in a contextless and meaningless manner, it is possible that it forms structures that were neither foreseen nor desired by the programmer to this extent. The result of such a process can be artificial intelligence, which is strongly prejudiced in its field of application, because it bases its decision making on patterns, which were present in the training data, but have no correspondence in reality.[52] If one considers how complex and intransparent the decision making process in a neural network is for the user, it becomes clear how difficult it is to locate such prejudices; because the machine merely spits out a result without being able to explain or show how it came about. If, for example, such a system is instructed by a bank to decide which customers are to receive a loan, it can happen that "on the basis of the training data it thinks it can recognize that an ethnic minority or men over 53.8 years of age or cyclists with yellow helmets and 8-speed gears pay back loans less reliably [...] – regardless of whether this is illegal or completely nonsense."[53] Nevertheless, many companies are enthusiastic about the incredible speed and efficiency of artificial intelligence judgments and point to the advantage of the machine's apparent objectivity, which cannot be dazzled by subjective preferences, rhetorical tricks, or lies. An AI system can be a useful classification tool, but one must always be aware that it does not judge as objectively and neutrally as it may appear on the surface. It is therefore important to critically question any judgement made by an AI system.

Another problem is how artificial intelligence changes our relationship to machines. While machines used to serve a certain purpose, which was usually to relieve us of physical work, and were otherwise of little importance for

52 Cf. Lenzen 2018: 61.

53 Ramge 2018: 26.

our lives, the role of technology has changed significantly, especially with the introduction of AI into our tools. The example of the smartphone, which has become the constant companion of many in recent years, shows how much influence such devices have on us. The "intelligent telephone" has become a universal tool in our everyday lives, it ensures that we receive information tailored to our interests via social networks with the help of algorithms, it helps us to orientate ourselves everywhere, and via countless apps, mostly free of charge, it offers us the possibility of controlling almost all areas of our lives comfortably on the move; be it our own diary, our fitness program, or the temperature in our homes. The smartphone has become the ultimate personal assistant and, thanks to self-learning voice control, is increasingly intuitive to use. We ourselves are giving up more and more responsibility without asking ourselves in whose interest the virtual assistant is actually acting. However, many accept this paternalism and lack of transparency because this so-called "nanny tech" makes our lives so comfortable and easy.[54]

People are also inclined to give machines, with which they can apparently talk, and simulate compassion, a different status than that of a mere tool. We are thus tempted to succumb to their functional simulation and attribute feelings and empathy to them. In the world of work and in the care and education sector, machines are already in use which, thanks to artificial intelligence, are capable of such simulations and thus enjoy a higher degree of emotional acceptance among their human employees or those under protection than machines which perform their tasks with cool efficiency. Nevertheless, simulated empathy does not replace real human attention, and there is a danger that artificial systems will be used as gap fillers in task areas where there is a shortage of human labor due to unattractive activity or insufficient funding. It is not possible to predict what long-term effects will be associated with the fact that adolescents in particular will increasingly have more contact with artificial companions than with humans. Our society is in the process of handing over a multitude of competences to artificial intelligence, and thus to a relatively new technology, without being able to assess these with regard to future developments. Even if it cannot be assumed that artificial systems will take over power, they are becoming more and more indispensable and we are slipping into immaturity because we seem to believe that intelligent technology is more capable of dealing with the problems of our increasingly complex world than ourselves.[55]

---

54 Cf. Ramge 2018: 89f.

55 Cf. Lenzen 2018: 246.

It is not taken into account that neither our private everyday tasks nor more global problems are transferred to a neutral agent when we delegate them to artificial systems. For "not the machines take control, but those who own and control the machines influence and control more and more areas of society"[56]. These actors acting in the background are global technology groups such as Google, Facebook, Microsoft, Amazon, and Apple, some of which have already secured quasi monopolies on the much sought-after artificial intelligence. These corporations have their own agenda and the artificial systems that advise us, do so in their interest. These monopoly structures threaten to become even more entrenched due to the so-called feed-back effect, leaving competitors only a small chance of establishing themselves in the technology sector. This mechanism is of paramount importance in the AI sector, where data are the most important raw material. Self-learning systems can improve themselves on the basis of feedback data and "the more often they are used, the more market shares they conquer, the more difficult it will be to catch up on their lead"[57]. Some companies are already exploiting this advantage and are already exerting their influence outside the market by taking on tasks that society or a democratically legitimized government should actually regulate. Here, for example, questions of data protection, the influence of social media on democratic elections or the regulation of communication in such networks should be considered.

But even in the hands of the state, the use of artificial intelligence can entail great risks. Self-learning systems, whose pattern recognition competence is used for facial recognition on cameras, spying on telephone conversations and collecting and evaluating data from digital communication, provide the police and intelligence services with powerful tools for monitoring the population. AI systems, which use data analysis to predict where crimes will occur or which individuals are most likely to commit crimes, are also in the testing phase in many countries.[58] These developments are worrying, as they partly conflict with personal rights such as data protection and legal principles such as the presumption of innocence. An actual horror scenario arises when these technologies are used in the hands of unscrupulous dictators against their own population. If artificial intelligence is misused for tyrannical purposes, it can provide the ruler with an almost complete surveillance of the citizens and provide him with subtle but extremely powerful means of manipulating the masses.

56 Lenzen 2018: 247.

57 Ramge 2018: 88.

58 Cf. Lenzen 2018: 169f.

With all these dangers, it should be emphasized that artificial intelligence is an extremely powerful tool with the potential to bring great progress to mankind. We are already seeing how the use of self-learning systems in medicine combats disease, how driver assistance systems and, in the near future, self-propelled cars will make road traffic safer, and how robots can save lives in disaster areas that are inaccessible to humans. However, we as autonomous and critical subjects, who have to ensure that this new promising technology is in the right hands, are decisively challenged. We must not allow ourselves to be paralysed by the horror stories of power-hungry super intelligence, nor by the convenience-promising offers of the technology companies. It is our task to educate and critically challenge the use of artificial intelligence as a tool in order to identify where it can be used meaningfully and where its application must be stopped.

6. Bibliography

Bostrom, Nick (2016) *Superintelligenz. Szenarien einer kommenden Revolution*. Berlin.

Cramm, Wolf-Jürgen (2008) „Zur kategorialen Differenz von Vernunft und Natur." In: W.-J. Cramm / G. Keil (eds.) *Der Ort der Vernunft in einer natürlichen Welt*, Göttingen.

Deutscher Ethikrat (2017) "Big Data und Gesundheit – Datensouveränität als informationelle Freiheitsgestaltung." Available at: https://www.ethikrat.org/fileadmin/Publikationen/Stellungnahmen/deutsch/stellungnahme-big-data-und-gesundheit.pdf last checked: July 23rd 2019.

Fischer, Klaus (2003) „Drei Grundirrtümer der Maschinentheorie des Bewusstseins." In: Wolfgang R. Köhler / Hans-Dieter Mutschler (eds.): *Ist der Geist berechenbar?* Darmstadt.

Janich, Peter (2009) *Kein neues Menschenbild. Zur Sprache der Hirnforschung*. Frankfurt/M.

Kurzweil, Ray (1990) *The Age of Intelligent Machines*. Cambridge 1990.

Kurzweil, Ray (2000) *Homo S@piens*. München.

Kurzweil, Ray (2014) *Menschheit 2.0*. Berlin.

Kutschera, Franz von (1993) *Die Falsche Objektivität*. Berlin.

Lenzen, Manuela (2018) *Künstliche Intelligenz. Was sie kann & was uns erwartet*. München.

Metzinger, Thomas (2014) *Der Ego-Tunnel. Eine neue Philosophie des Selbst: Von der Hirnforschung zur Bewusstseinsethik*. München 2014.

Meusers, Richard (2016) „Website DoNotPay. Mit dem Roboanwalt gegen Knöllchen." Availiable at: http://www.spiegel.de/netzwelt/gadgets/donotpay-robo-anwalt-hilft-bei-einspruechen-gegen-knoellchen-a-1078887.html last checked: Mai 8th 2018.

Müller, Tobias (2015) „Naturwissenschaftliche Perspektive und menschliches Selbstverständnis, Eine wissenschaftsphilosophische Analyse zur Unverzichtbarkeit lebensweltlicher Qualitäten." In: T. Müller / T. M. Schmidt (eds.) *Abschied von der Lebenswelt*, Freiburg.

Müller, Tobias (2018) „Der moderne Naturbegriff, unser lebensweltliches Freiheitsverständnis und die Grenzen der Freiheit." In: M. Reder/D. Finkelde/ J. Wallacher (eds.) *Jahrbuch praktische Philosophie in globaler Perspektive*, München.

Mutschler, Hans-Dieter (2002/2003) „Zur Herrschaft der Computermetapher." In: *Scheidewege* 32, Stuttgart.

Ramge, Thomas (2018) *Mensch und Maschine. Wie künstliche Intelligenz und Roboter unser Leben verändern*. Ditzingen.

Roth, Gehrhardt (2003) *Aus der Sicht des Gehirns*. Frankfurt/M.

Singer, Wolf (2002) *Der Beobachter im Gehirn. Essays zur Hirnforschung*. Frankfurt/M.

Sorgner, Stefan Lorenz (2016) *Transhumanismus – »Die gefährlichste Idee der Welt« !?*. Freiburg.

Stekeler-Weithofer, Pirmin (2012) *Denken. Wege und Abwege in der Philosophie des Geistes*. Tübingen.

Wewetzer, Hartmut (2016) „Die vierte Kränkung." Availiable at: http://www.tagesspiegel.de/wissen/menschliche-computer-die-vierte-kraenkung/13312632.html last checked: Mai 8th 2018.

Why Artificial Intelligence is a Matter of Design

Andreas Theodorou

Abstract

Unlike other human-made objects, the ability for intelligent systems to exhibit agency, and even appear anthropomorphic, leads to *moral confusion* about their status in society. As Himma states "If something walks, talks, and behaves enough like me, I might not be justified in thinking that it has a mind, but I surely have an obligation, if our ordinary reactions regarding other people are correct, to treat them as if they are moral agents." Here, I present an evaluation of the requirements for *moral agency* and *moral patiency*. I examine human morality through a presentation of a high-level ontology of the human action-selection system. Then, drawing parallels between natural and artificial intelligence, I discuss the limitations and bottlenecks of intelligence, demonstrating how an 'all-powerful' Artificial General Intelligence would not only entail omniscience, but also be impossible. I demonstrate throughout this Chapter how culture determines the moral status of all entities, as morality and law are human-made 'fictions' that help us guide our actions. This means that our moral spectrum can be altered to include machines. However, there are both descriptive and normative arguments for why such a move is not only avoidable, but also should be avoided.

1. Introduction

The words *ethics* and *morals* are often regarded as interchangeable, when they are not. The word ethics derives from the Greek word ήθος (ethos), meaning *moral character*. However, *moral* comes from the Latin word *mos*, meaning *custom* or *manner*. Ethics is the *system* of moral values, providing the framework for examining these values. When a society agrees to follow an ethical system, it produces *policy* to ensure the enforcement of those values.

Since the Ancient Greek philosophers, morality has been recognised as unique to humans – with the exception of anthropomorphised versions of our gods and other deities. We are deceived by our own genes into believing that there is a pro-social objective morality that binds us all (Ruse and Wilson 1986). We can still maintain that morality is rooted in our unique action-selection

 | DOI:10.30965/9783957437488_009

system; like all animals, we are not born as blank slates without goals, biases, and means to trigger behaviours. However, our uniqueness is in our ability to produce cultural tools to increase our list of possible behaviours and dramatically enhance the performance of our existing ones. Central to our culture, as we will discuss further, is our language. It allows us to not only signal intentions, but also facilitate complex social hierarchies and cooperation (Bryson 2007).

Now, we are at an unprecedented point in human history, where we create artefacts, *intelligent systems* like robots, *which* can perform action selection based on the decision-making systems found in nature. The same artefacts exploit our language and influence our culture. Our understanding of intelligent agents is rooted to popular science and science, which leads to the creation of inaccurate mental models and the consequential attribution of anthropomorphic elements to them (Gunkel, 2017; Rotsidis et al. 2019).

This anthropomorphism leads to believing in an *Asimovian* future, where all-knowing human-like machines will be part of our societies and be able to follow and be owed legal rights and obligations (Carsten Stahl 2004; Gunkel 2012). This overidentification with intelligent systems creates a *moral confusion* about the moral status of these human-made objects, construing them as moral subjects (Bryson and Kime 2011; Wortham 2018).

In this Chapter, first I provide clarification of terms, such as *Artificial Intelligence* (AI), and then define the various 'labels' we place on entities to denote their moral status and subsequently any rights and obligations they have or are owed. I discuss what *moral agents* and *moral patients* are and the requirements for an entity to gain either of the two moral statuses. Next, I provide a high-level overview of human morality, by examining our own action-selection mechanism from an evolutionary perspective. I explain how our ability for cultural accumulation, which language both exemplifies and further enables, is the key behind human uniqueness in the animal kingdom. I exploit this discussion to draw parallels to Artificial Intelligence and how we prescribe agency to our robots. Further, I debunk the 'myth' of the possibility of creating an 'all-powerful' Shelleyan[1] machine that will turn against its creators and instead discuss *what is possible*. Finally, I make descriptive and normative arguments against the elevation of intelligent systems into moral subjects.

1 From Mary Shelly author of the classic Frankestein book.

2. Terminology

The greatest challenges of appropriately regulating AI are social rather than technical. Especially as we cannot even agree on a definition of the term, even though there are perfectly well established definitions of both *artificial* and *intelligence.* The primary problem is that we as humans identify ourselves as intelligent, which certainly is one of our characteristics, but that does not imply that *intelligent* means exclusively 'human-like'. Hence, to ensure a common vocabulary with the reader, widely-used terms are defined for at least the scope of this chapter.

Agency refers to the ability of an entity to effect change; through the usage of sensors, an agent can perceive its environment and internal state and through the use of actuators to change them. Throwing rocks off a cliff in order to satisfy hunger is not optimal – or rational – but it is a sign of agency as the rocks move and the environment changes. We should not confuse agency with *intelligence.* The later implies the selection of the *right action* at the right time; e.g. eating hallumi to satisfy hunger. Thus, *intelligent agents* are a special category of entities characterised by their ability to select rationally (i.e. perform action selection) and as optimally as possible their actions. If the agent can perform a real-time search between all its available actions to select how to react to a situation, then following from Bryson and Winfield (2017) it may also be termed *cognitive.* Hence, *artificial intelligence* refers to the human-made action-selection systems that exhibit signs of intelligence.

Physical agents, such as humans and robots, inhabit, populate, and alter the material world. Within this chapter, the term *robot* denotes the combination of both the software and hardware of an autonomous system and always implies as such agency. I do not discuss less intelligent robotics here. Virtual agents can be either an independent piece of software, running autonomously, or part of a larger system. Such agents, as is case of simulations and video games, may inhabit dynamic, virtual worlds, while another virtual agent may be a simple agent running in the background, performing tasks various tasks without any user interaction.

Agents that act individually, instead of as a dependent of a Multi-Agent System (MAS), are called *complete* agents. Such an agent that has multiple – and often conflicting – goals and mutually-exclusive methods of fulfilling those goals, is also a *complex* agent (Bryson 2001). These different goals may also be of different priorities. For example, we, humans, prioritise satisfying hunger (and ultimately survival) over entertainment. If we are both very hungry and bored, we will first try to satisfy the former by finding food and eating it.

Agents, such as humans, animals, plants, and micro-organisms, are *natural agents*; their existence is the consequence of biological reproduction. Their traits, abilities, and limitations are due to *biological evolution*. While all-natural agents may behave intelligently, not all of them are cognitive. Manufactured agents are *artificial* agents. They are designed and produced to serve a particular purpose, they are in other words, intentionally made. Their functionalities are set by their creators and their limitations are either set due to resources constraints or again, intentionally. In this chapter, the terms *agent* and *intelligent agent* are used interchangeably to refer specifically to artificial agents. Unless explicitly stated, these agents can be either *virtual* or *physical*.

3. Our Morality Spectrum

A special case of agency is *moral agency*. A moral agent is an entity that could be sensibly held accountable for its action; that is, it can be held *morally responsible*. Over the last three millennia, there have been arguments on what constitutes an action carrying a *moral value*. Various schools of ethics have been debating on what is the 'correct way' to judge an action. For example, for a utilitarian an action must maximise the utility (i.e. a criterion of what is morally right for the utilitarian) gained, with the ends justifying the means.

While there are various schools of normative ethics, most refer back to Aristotelian conditions of moral responsibility and in turn moral agency. Hence, in this Section, I discuss the Aristotelian requirements for moral agency and compare them with the ones set by other philosophers. Further, I discuss another category of moral subjects: *moral patients*. I conclude by discussing the relationship between morality and law.

3.1 *Moral Agency*

Aristotle in *Nicomachean Ethics* links *autonomy* with moral responsibility (Aristotle). He distinguishes actions whose origins were 'inside' a person from those whose origins were from the 'outside' (i.e. due to external influence). In his model of autonomy, what defines an action as autonomous is seen as its point of origin; it must have an 'immaculate conception'. Actions performed without external influences (i.e. of 'inside' origins), i.e. pressure and compulsion, are to be considered voluntary and the only ones worthy of a moral value.

Ultimately, the Aristotelian version of *autonomy* depends on how we define external influences. Our actions, as I will discuss in detail in the next section, can often be in response to an environmental change and external stimuli but still considered 'our own.' Aristotle took this into account in his discussion on the problems that arise from defining 'compulsion' and by estimating the degree of severity of pressure that could make an action not voluntary. Aristotle considers not only an agent's autonomy as an essential requirement, but also the actor having knowledge of the action.

Aristotle, building upon Plato's work, is the Western founder of *Virtue Ethics*, one of the major frameworks of normative ethics. For a virtue ethicist, a moral action should be performed *only if* it will help its actor to gain a *virtue*, a positive trait of character. Only agents which can accept a distinctive range of considerations as reasons for an action that lead to the acquisition of a virtue can possess it. In layman's terms, the agent should not just perform the morally-correct action, but also be able to understand why this is the right action and what the wrong could be.

Similar to Aristotle, Kant associates morality to the idea of freedom (Johnson and Cureton 2019). Kant argues that full autonomy implies that an agent is the lawmaker of its own rules. Kantian Ethicists refer to such rules as *duties*. A free agent acts with no foreign forces dictating its actions and obligations. Yet, a moral agent can still act contrary to its duty – it can act immorally. In order to assess whether an action is morally permissible or not, an agent must consider if it is consistently universal; all moral agents, given the same context, would act in the same manner.

Kant argues that the *moral law* stems from reason alone, acting as a moral compass for all rational beings. Yet, rational beings can still act irrationally. Their actions might be influenced by basic animalistic needs for survival, the temptations of pleasure, and other emotions. Thus, for Kant the moral law acts as a constraint to natural desires. A moral agent is acting on 'good will' only if its actions are wholly determined by their moral demands. Kant's moral framework is *Deontological*; it is founded on the idea that doing what is right is nothing else other than doing one's duty. These duties are self-determined and exclusive for rational beings, as an agent needs to be able to rationalise and understand moral concepts; what is right and what is wrong. The ability to reason before selecting an action is essential for a Kantian moral agent.

While Aristotle, Kant, and other philosophers *disagree on what constitutes an action as morally right, they overlap on which are the principal requirements an agent needs to fulfil to be granted moral agency*. Himma (Himma 2009) summarises and generalises these requirements as the following two conditions:

1. The agent has a *capacity to freely choose its actions*; they are not forced by an external agent, e.g. held hostage or under the influence of a demon. This is widely understood as having autonomy or self-control.
2. The agent has the *ability to apply moral concepts and principles*; it has an understanding of what is *right* and *wrong*.

The second requirement is what the different schools of ethics disagree upon; *how to judge* the moral worth of an action. Despite the disagreement, there is still an agreement on having *the ability to do so*. This is why children – and even adults with severe cognitive disabilities, such as schizophrenia – are not considered moral agents. We consider the cognitive abilities to be underdeveloped or insufficient and the agent unable to make clear rational choices. Societies across the world have different age requirements for adulthood (usually post secondary education) and thus for moral agency.

Himma (2009) argues that there is a precondition implied in these two conditions; the agent has the capacity for *consciousness or at least self-consciousness*. As a result, he discards doings such as 'breathing' and 'waking up' from being considered as (moral) actions. Himma argues that the lack of an intention – of a mental state or desire – causing the doings is what discounts them from being considered as actions. Yet, when we breath the air around us changes. Plants, in the process of photosynthesis, consume carbon dioxide and produce oxygen. These are signs of agency. Implicit awareness is not a requirement for a *doing* to be consider as an *action* or even for an action to have a moral value. I clarify this further, in the next Section, where I provide an ontological view of our action-selection system and how our culture and language remains a *sine qua non* for our concept of morality.

3.2 *Moral Patiency*

It is important to make a distinction between a moral agent and a *moral patient*. Moral patients are entities, which are owed at least one right or obligation by moral agents. On this definition, all moral agents are also moral patients, but not all moral patients are moral agents. Unlike moral agency, there are no widely-accepted requirements on which entities we can attribute moral patiency. Humans who suffer from a cognitive disability are not expelled from our moral spectrum, but granted moral patiency. Similarly, animals are not considered moral agents, but are widely accepted as moral patients.

Animals are a good example of how moral patiency is spread over a wide area of our moral spectrum. Different animals are treated differently (Franco and Olsson 2015). Our legal system reflects how we find it harder to kill cats, dogs, and non-human primates than 'game animals', e.g. pheasants, rodents, and rabbits. For example, EU legislation regulating animal welfare explicitly states: "animals such as dogs and cats should be allowed to be re-homed in families as there is a high level of public concern as to the fate of such animals" (European Commission, 2010). In Western countries, we currently consider it moral to raise in captivity specific breeds of animals and to 'domesticate' others. While an extended discussion of animals' mortality is far beyond the scope of this chapter, the message here is how the utility of an animal, alongside with the culture of the moral agent interacting with it, determine its exact position in the moral spectrum of a society.

In the Greek Orthodox world, icons (paintings of Saints and other deities) and the Evangelion are to be treated with respect. In Islamic countries, a similar respect is provided to copies of the Qur'an. Likewise, flags are to be treated with specific rules – they must never touch the ground and uniformed personnel, e.g. military, have to formally salute them. These objects, without agency, have been granted moral patiency due to the entities they symbolise. An Orthodox believer will not kiss an icon to honour the wood and paint pigments that it is made of. Instead, she will perform this 'ritual' to request the blessing of the deity drawn on the wooden plank. A non-orthodox person may admire such an icon as a piece of art, but will not follow the same social norms as an Orthodox believer.

In short, artefacts do not receive any obligations, but rather indirectly 'benefit' from the duties and protection we show towards the entities they represent. Therefore, objects receive protection and participate in 'rituals' based on their utility and moral values of the agents' interacting with them.

3.3 *Morality and Law*

Each society has different moral values they want to enforce, making high-level keywords found in AI ethics guidelines, such as *fairness*, too abstract and impossible to generalize across the globe. Instead, there is a need for context-specific interpretations and implementations of such values (Aler Tubella et al. 2019). The disparities between societies extend beyond moral values into the acceptable behaviour of the human actors that make up those societies. For example, there is a significant global variation in humans' willingness to cooperate, and treatment of those who engage in cooperative behaviour. In low-GDP countries, especially where there is evidence of corruption and lack of rule of law, individuals punish those who behave prosocially (Sylwester, Herrmann, and Bryson 2013). When a society decides to protect and promote specific moral values, it does so through its legal framework.

Moral agents' rights and obligations towards a society are enforced through its *legal framework*, by assigning a *legal personhood* to them (Smith 1928). Legal persons have been granted rights, such as owning property and conducting financial transactions, but they are also bound by the laws of the societies in which they operate. Such laws include taxation to ensure contributions to public goods. If a legal person performs an action forbidden or did not perform one mandated by the law, they can also be held *legally liable* for these actions or inactions. Legal liability can be enforced through the placement of sanctions, such as momentary fines or time spent in prison, when there is a breach of the law.

The law is itself an artefact. It is made by humans, a tool aimed to enable human agents protect themselves and their societies. Yet, when the term 'person' comes to legal concepts, it is not necessarily confined to and synonymous with humans. Any entity *or even specific objects* can be granted a legal personhood. Lawyers call this a 'legal fiction'. Bodiless entities such as corporations, which are not physical objects, are considered as legal persons (Johnson 2012; Solaiman 2017).

The law treats organisations, corporations, and even countries as legal persons. Such legal entities have rights, obligations, and can be held legally liable. Still, the law recognises that an organisation can't act on its own. Each action performed by it is the result of the collective actions and decisions made by the individual agents that are affiliated with it. Individuals, such as stockholders and employees, are the residual claimants of corporate assets and the ones benefited by profits in the form of dividends, capital gains, or residual payments. Similarly, if a corporate acts unlawfully, depending on the country and the severity of the act, individuals may also be held responsibly in addition to the company. Even if individuals are not held directly responsible, any sanctions against a company will result in damages and an indirect punishment of its key stakeholders.

It is not unheard of to provide legal personhoods to non-cognitive objects, such as rare religious idols and even rivers (Solaiman 2017). There is a lack of uniformity across legal systems – a potential by-product of moral values variation between societies – to dictate which entities can (or should) be considered legal persons. Thus, the extent of what is 'allowed' to be granted by a legal personhood depends upon a given jurisdiction of each country's independent legal system. This also brings an interesting case of misalignment between moral philosophy and the law: for thousands of years the concept of moral agency has been universally accepted as only applicable to humans. We created law to enforce it and now we are developing legal fiction to allow non-natural agents to be granted rights and obligations towards our societies. If we are giving legal personhood to non-human entities, should we give a 'higher' moral status and perhaps even moral responsibility to artificial agents?

In the next Section, I will discuss morality from an evolutionary perspective and why we are the only species considered as moral agents. I exploit this discussion in the section following after to answer the above question, by arguing why artificial agents are neither eligible nor should ever be.

4. Natural Intelligence

We breathe air to generate energy and stay alive, drink coffee while writing our dissertations to stay awake, and so on. Throughout the day we perform doings without what Kant calls *reasoning*. 'Deciding' to use our legs to walk to the coffee machine is chosen from a pool of *possible acts*. Emotions, past experiences, and even 'gut feelings' often trigger what we consider *spontaneous* behaviours.

We are complex agents with multiple goals and an even larger pool of possible actions to select from. There could have been numerous combinations of sensory inputs and actuator output, but natural selection trims down this search space. Contrary to Skinner (1935), Gallistel et al. (1991) demonstrates that pigeons can learn to peck for food and flap their wings to escape shock, but not to flap their wings for food or to peck to avoid shock (Bryson 2001).

Biological evolution provides the architecture to scaffold viable working systems that withstand natural selection. This Section discusses this scaffolding from a high-level ontological view, by exploring the relationship between *conscious* and *unconscious* actions. It elaborates on the costs and bottlenecks associated with real-time search. It concludes by arguing why humans are the only moral agents.

4.1 *Kinds of Minds*

Dennett (1996) suggests a high-level ontology of four 'different minds' that make up our own and to some extent other biological creatures'. We are products of biological evolution, making us what Dennett calls 'Darwinian creatures'. When we are born our brains are not a *tabula rasa*, without any drives, behaviours, unable to use our sensors and actuators. Instead, as thanks to our Darwinian mind, we know how to breath, how to 'eat' food, and how to cry. We have inherited hardwired goals and behaviours that are executed – often in parallel to other actions – to ensure our survival and eventually reproduction.

The Skinnerian Mind allows *ABC learning* (Associationism, Behaviourism, and Connectionism) by testing actions in the external environment. This reinforcementlearning system allows us to make associations and generalisations through trial and error. When we first received building toys, in our early childhood, we probably spent hours and hours learning what to do with them. Now, as adults, we are able to build objects with the same blocks in a matter of minutes. That joy we felt on our first successful construction helped us learn how to properly use these blocks. This positive reinforcement happened so early in our lives, that it makes us assume that this ability was always there. Such prior-learned skills are considered as common sense.

Common-sense knowledge is the collection of behaviours that we can learn, most at such an early enough age that we can't recall their origins (Minsky 2006). Our Skinnerian mind helps us build this collection, allowing us to create plans, sequences of actions needed to achieve a specific goal, reducing the need to re-formulate such plans, when we encounter similar scenarios.

Popperian creatures run an internal environment, a simulation, after importing sensory input to preview and select amongst possible actions. Unlike Skinnerian creatures, the hypothesis testing is done internally, allowing hypothesises to die without being executed. Our Popperian Mind is what helps us deliberate options and select sequences of actions to form behaviours. It is one of our *conscious* minds; it allows an online modelling of the expected outcomes across a range of candidate behaviours.

Finally, unique to humans, is the Gregorian mind. Our inner environment is informed by the designed parts of the outer environment. We import cultural tools to facilitate this self-reflection. These imported tools expand our pool of available behaviours, enhance our decision making, and give us an unprecedented power over our environment (Wheeler 2010).

These 'different minds' allow us to compute which is the right action. We can loosely group Dennet's Kinds of minds into two groups: an implicit minds group, consisting of the Darwinian mind, and an explicit mind group consisting of the Popperian and Gregorian minds. The Skinnerian belongs to both groups; model training is done explicitly, but action repetition can be done implicitly.

4.2 *Dual-process Theory*

The argument that our decision-making mechanism actually consists of two distinctively separate systems, a 'System 1' capable of implicit actions and a 'System 2' to deal with explicit processes, is not new (Evans 1984; Epstein 1994; Stanovich and West 2000; Kahneman and Frederick 2005). This dual-process theory states that our brain works in parallel, triggering both conscious and unconscious actions at the same time. System 1 makes decisions fast, based on intuitions, impressions, impulses, and feelings. In contrast, System 2 follows a slow serialised deliberative conscious process.

Haidt (2006) uses 'the elephant and the rider' metaphor to describe this dual-process theorem. The elephant is our System 1, performing implicit actions at the background. In Dennett's ontology, this will be our Darwinian mind and even part of our Skinnerian mind. Haidt's rider represents explicit control; the actions performed by System 2 after deliberation. The rider, being weaker, can only steer the elephant by providing appropriate stimulus, making its control precarious. Spontaneous decisions and emotional triggers by the elephant can make the rider lose control and the elephant to change its route.

4.3 *Conciousness and Action Selection*

Whenever the rider makes a decision on which path to follow or tries to retain control of the elephant, it consumes energy and a time delay is introduced. Our brain represents about 2% of our body weight, yet accounts for over 20% of our daily consumption of calories (Raichle and Gusnard 2002). Not every agent can accommodate the energy and time requirements of System 2. Consciousness is, by necessity, adaptive in nature. Unless a new or surprising situation arises, a conscious creature will keep using its fast inexpensive System 1 to perform automatic habitual actions (Isoda and Hikosaka 2007).

For 'consciousness', I use the definition placed by Bryson (Bryson 2012) to exclusively refer to the mechanism that generates awareness of the moment and episodic memory, enabling the learning of new behaviours and explicit decision making. I do not consider any of the other psychological, ethical, or even religion-related phenomena that consciousness has been associated with over the years (Dennett 2001).

Any cognitive processing can delay taking action, as consciousness introduces a lag and noise into action selection (Norman and Shallice 1986; Cooper, Shallice, and Farringdon 1995; Schneider and Chein 2003). This is shown by Libet (Libet 1985), who ran a study where participants had to flex their hand at the wrist while noting the position of a revolving spot. Albeit a simple task, when the experimental subjects were making conscious decisions, there was a delay between 350 and 400 msec behind the onset readiness potentials. This delay may reflect an allocation of time to real-time search for a better solution. However, if the agent acts too slow, another agent may take advantage of a situation before it (Bryson 2010).

If consciousness is so ineffective, why and how do conscious creatures survive natural selection? Intelligent species can only survive natural selection due to the actions they perform (West, Griffin, and Gardner 2007). At the same time natural selection tunes these actions over generations to maximise their effectiveness (Dawkins 1996). At the cost of performance speed, we can perform real-time search (cognition) to solve problems and take advantage of opportunities that change more rapidly than other ways of performing action selection (Bryson 2012). Consciousness allows individuals to flexibly adjust their behaviour in previously-unseen dynamic environments (Wortham and Bryson 2016).

4.4 *The Power of Language*

While biological evolution plays a key role in the capacities of all species, including of *Homo sapiens*, it is our cultural evolution – particularly in its extent – that sets us apart. We have an unsurpassed ability to participate in a collective cognition, allowing us to domesticate other animals or even build intelligent artefacts, to extend our physical and cognitive capabilities beyond our originated evolutionary scaffold.

Our unsurpassed competence for cultural accumulation is enabled by our – both written and spoken – language (Bryson 2007). The ability to communicate is not unique to us; many animals have vocal and other means of signalling each other (Wortham and Bryson 2018). However, language is unique to humans. We can communicate not only information about the surrounding world, like animals do, but also gossip. Gossip leads to information exchange about others and eventually the development of social hierarchies and organisation (McAndrew and Milenkovic 2002). Even if that information comes with a high degree of uncertainty, cooperation can be sustained (Mitchell, et al. 2016).

Language also facilitates the creation of fictional constructs, such as names for our tribes and deities to defend them. No monkey will name its forest and then call "its spirits" to protect it. These constructs provide us a system of guiding beliefs and symbols of in-group identities (Ysseldyk, Matheson, and Anisman 2010). Migration between groups results in spatial structuring of cultural skill accumulation, as specialised skills and tools become available to both the migrants and their receiving group (Powell, Shennan, and Thomas 2009). Due to the acquisition and ascription of such memes, alongside with gossip, we are able to cooperate and organise ourselves in far larger and more complex societies than any other animal. Cultural accumulation at scale is the incontrovertible aspect of human uniqueness compare to other biological creatures. This accumulation significantly enhances and extends our consciousness and general cognition capabilities (Wheeler 2010).

4.5 *Goals and Dithering*

Explicit decision making at all times has both a cost and is subject to *dithering*: switching from one goal to the other so rapidly that little or no progress is made in achieving any goals (Humphrys 1996; Rohlfshagen and Bryson 2010). Nonetheless, an agent *should be able to* devote sufficient resources to at least achieve survival-related goals. An appropriate balance between the responsiveness and persistence at pursuing the currently selected goal is needed to avert dithering. In order to achieve this balance, we have both built-in individual and external regulation mechanisms to modulate how the agent invests its resources in trying to complete different goals.

In nature, drives and emotions may be seen as a chemically-based latching system, evolved to provide persistence and coherence to the otherwise electrically based action selection, provided by the central nervous system (Bryson and Tanguy 2010). The hormone and endocrine systems, which underlie drives and emotions, are evolved to provide smooth regulation of behaviours in all animals. They determine the current focus of attention by managing the trade-off of losing resources, such as energy and time, when switching between the different goals.

However, such regulation mechanisms focus primarily on helping the agent achieve some of its *own goals*. If multiple complete, complex agents work together, they form a community with both common and individual objectives – unlike a multi-agent system, where there is a common objective for all agents. A common objective can be the acquisition or development and maintenance of public goods, resources shared within a single social group. For a society to remain sustainable its agents must spend a sufficient amount of resources at both their own private goals and the common public ones. When group members take advantage of a good shared among the community members, without contributing towards it, they are *free riders*. Communities do not passively accept the free riding of others. Instead, when they have the opportunity to punish free riders, they do so.

Punishment takes the form of an actor paying a cost to reduce a fitness value of the punished target. However, for this to be consider as punishment, the punished agent needs to experience an unpleasant mental state (Himma 2009). We ordinarily aim to avoid receiving punishment, and in so doing allow the law to influence our decision making. The claim is not that the law 'limits our free will' and that actions taken under the threat of sanctions have no moral value. Taxes are meant to ensure sufficient contributions to public goods and the wellbeing of the community at large. We still have the option to break the law. When people break the law, e.g. tax evaders, they have an understanding that they will be sanctioned if caught and then chose to damage their societies by acting anti-socially. The law, like all cultural tools, is imported by our Gregorian mind. It helps us focus our decision-making process, by adding biases that influence our action selection to avoid pursuing behaviours that may inflict damage to our societies.

4.6 *Morality for Humanity*

The legal system is a reflection of the moral values a society wants to protect. As a society, we have complete control over it. Creating a new law is not different from software development. Bugs (loopholes) and flaws (exclusion clauses) are bound to happen during the implementation, which is why we try to find them and fix them during internal testing – when policy-makers debate for ratifying the law. Arguably, in the process we may introduce further bugs and flaws. Once the system (law) is made available, some users may discover new bugs. In Common Law, judges do not just interpret the law, but often fix it through their rulings. As a society grows and evolves, we may update, retract, or replace completely a piece of legislation by voting new lawmakers to represent our views and interests. In short, the law and its foundation – morality – are artefacts.

There is no universal objective morality. We may all have the same kinds of minds, but our Skimmerian mind received different training, our Gregorian imported different tools, and so on. This does not mean that we cannot evaluate the optimality of an action, but that our evaluations will be different if an action holds any moral gravity. At the end, morality is not only subject to cultural variation, but it also provides group-specific moral norms for group members to follow and enhance their in-group identity (Ellemers, Pagliaro, and Barreto 2013). Like language, it influences culture and is being influenced by it.

Animals, which are biological agents like us, are not considered able to satisfy the requirements for moral agents (Gruen 2017). If morality is a matter of benevolent inclinations, accepting as good that which is agreeable or useful to ourselves for others, then some animals can be moral agents. For example, a lioness providing her hunt to the whole pride is an act of altruism. Yet, due to human exceptionalism and speciesism, we discard such actions as 'mere instinct' (Johnson 1983; Gruen 2017). I argue – as an exceptionalist myself – that as morality is part of culture, therefore, access to the unique-to-humans Gregorian mind is needed to understand moral gravity of an action. This does not exclude animals from performing acts of moral value, but exonerates them from being held morally responsible.

5. Artificial Intelligence

Artificial Intelligence is not a newly evolved life form, but a field of research aimed at producing products which provide some utility to us (Brundage and Bryson 2016). We have the ability to decide their shape, action-selection system, and so on. Sometimes even the act of developing an agent is by itself the purpose of act, e.g. for educational purposes. The same can be said with any other tool, from a scythe made to help us mow to nuclear weapons that provide us a competitive advantage over others. In short, we are not morally or otherwise obliged to develop any artificial agents, but we do so to improve our own performance.

Miller (2015) argues that their human-made nature is the exact reason why there is *no question what moral status intelligent artefacts deserve*; the only question is what *we may want assign them*. Yet, not everyone considers AI as *just* an artefact; purposely designed and developed to work as extensions of our own agency. Instead, there have been advocates of granting a moral status to (some) intelligent artefacts (Carsten Stahl 2004; Gunkel 2012). Others argue that an elevation of artefacts' moral status does not apply to current specialist systems, but only to future powerful and potentially conscious system (Himma 2009; Coeckelbergh 2009).

In this Section, I discuss how the bottlenecks of real-time search in Natural Intelligence exist also in Artificial Intelligence. I argue that the development of an 'all-powerful' system is unlikely and even practically unnecessary. Furthermore, I discuss why even a hypothetical system does not fulfil the requirements for moral agency set earlier in this Chapter. Finally, I discuss through normative and descriptive arguments that making AI moral agents or patients is not only an intentional and avoidable decision, but also an undesirable one.

5.1 *The Omniscience of Artificial General Intelligence (AGI)*

In Section 4 I discussed how real-time search costs time and energy, which is why biological agents limit or even avoid it. These two requirements impose limits not only to natural intelligence, but also to Artificial Intelligence (Bryson 2012; Bryson 2018). No agent is able to compute all possible solutions to all problems. The recent progress is largely due to a combination of substantial corporate investments in AI research, improvements in the design of computer hardware, and the availability of larger datasets. Yet, 'there is no such thing as a free lunch'; regardless of how energy efficient and whichever fabrication process we use, our processors still require *some* time to run an algorithm, consume energy, and take space – not only for the processor itself, but also for its cooling system, as heat is a by-product of energy consumption.

Moravec (1988) argues that: 'It is comparatively easy to make computers exhibit adult level performance on intelligence tests or playing checkers, and difficult or impossible to give them the skills of a one-year-old when it comes to perception and mobility'. We already have systems that can outperform humans in 'hard' problems, such as complex games like go and DOTA2 (OpenAI 2018). While we are making steps at object identification and classification (He, et al. 2015), due to the combinatorial complexity of explicit action selection, we are nowhere near *Artificial General intelligence* (AGI). Even if we can build and power a high-performing computing centre at the size of Sweden to run an 'all-powerful' AGI, what will be the actual gain? There is no practical benefit to use the same agent someone develops to play DOTA2 against to also drive our cars, manage our calendars, and do our taxes.

This does not stop AGI advocates from claiming that machine learning techniques, such as reinforcement learning, can lead to 'human-like' AI – some go as far as claiming that such a discovery can happen 'accidentally'! In reinforcement learning, we specify a reward function giving us an imperfect control over what the agent should consider as good behaviour and what not to. Our designs can include fail-safe mechanisms and other means to focus learning and action selection in general. In Bryson and Theodorou (Bryson, Joanna, and Theodorou 2019) we present such a design methodology and ontology to maximise our control and hence safety. There can be no 'accidental' development of a machine that will 'decide to turn us all into paperclips'. However, bad code may still lead to accidents costing human lives. Artificial agents are coded and in the process we, intelligent-systems developers, prescribe our own moral agency to them.

Consider a pet, such as a dog, trained to respond to specific commands. If you shout *sit* and it sits, you reward this behaviour by either petting it or handing a treat to it. At each re-occurrence of the *sit* command followed by a treat, the actionreward association is enforced. On the sound of *sit* a dog can also be trained to attack humans, roll on the floor, and so on. The behaviours that can be associated to *sit* are only limited by the physical and cognitive constraints of the dog. A dog *does not understand the context behind our words*; it lacks a Gregorian mind. It will never understand it – let alone communicate back using – our language. This does not reduce a dog's ability to react to the specific sounds that make up words. Similarly, the words *sit*, *eat*, and so on have no real context to an intelligent system. If there is a sensory trigger associated with the detection of that sound or input of text, an action will take place.

An example of this is the chatbot *Tay*, which tweeted pro-Nazi messages as a group of people exploited the lack of appropriate filters to feed it with malicious data (Vincent 2016). The bot did not actively support Nazism or far-right political views. Instead, as its creators failed to develop the necessary filters. It then acted in the way it was programmed, to maximise interaction by using – what it perceived due to the attack as – commonly-used words and popular phrases. This also demonstrates that there is no need for intelligent agents to understand our culture to exploit it. In fact, search tools, such as Google, already exploit our language by looking at the distribution of keywords in documents. Dating apps don't need to understand relationships or romance to suggest potential matches. Yet, it needs to be clear to all that any training data used to train AI models inevitably contains our biases (Caliskan, Bryson, and Narayanan 2017). This not only weakens the argument for granting moral agency to intelligent systems, but instead it also makes an argument for *transparency*, i.e. the ability to get an accurate interpretation of the agent's status (Theodorou, Wortham, and Bryson 2017). Transparency can ensure efficient auditing of the system, its debugging, and, therefore, help attribute accountability to the responsible actors when necessary (Bryson and Winfield 2017; Bryson, Joanna, and Theodorou 2019).

5.2 *Extending of Our Agency*

Let's consider the game *catch* for a second. Its roots go back to when dogs were trained to be used by hunters to bring their shot-down prey. Dogs are faster and better at tracking than humans are, thus, their use in hunting. We use pets as an extension of ourselves to improve our abilities, expand our pool of available actions, and increase our performance. Dogs are autonomous intelligent agents, but through reinforcement training we domesticate them.

A bird eats worms to survive, flaps its wings to fly, and procreates for its genes to carry over through its offspring. These are behaviours refined by natural selection. Its Darwinian Brain drives their execution. There is an unconscious intention behind them; the short-term survival of the individual and long-term survival of species. I argue that an artificial agent does not have this intention; such agents are not bound by natural selection with a need to perform any action. Their deployment and actions *should* serve – ultimately – their owners.

Artificial agents are developed to be used as enhancement of our agency – like a domesticated dog is. During the development and training of artificial agents, we ascribe agency to them. Intelligent agents, as their name suggests, perform action selection; *they decide amongst a human-defined list of possible actions* which one to follow. Artificial agents are programmed with access to sensors/actuators based on a number of criteria, e.g. the utility of the agent, costs and availability of the sensor/actuator, and so on. We have the same control over deciding whether to add an ultrasonic sensor as we have on which bathtub to install in our new bathroom. In both cases, one or more human stakeholders decide to increase the cost and add a certain utility.

Nonetheless, the emerging behaviour of our systems can be surprising, as the world is dynamic and complex. Simulators and testing procedures can only cover what the developers of a system thought of. As Dignum (2017) describes, using the infamous trolley problem as example, even if we implement systems that adhere to well-studied ethical theories, the resulting behaviour can possibly lead to different possible outcome. Where incidents occur, they must be addressed, in some cases redressed, and in all cases used to reduce future mishaps. This is important in principle, but probably impossible to implement without having an understanding of what lead to the error. However, we can, through the use of system designs that promote transparency and traceability, mitigate the problem of unexpected emerging behaviour. Access to transparency-related information can help designers and developers design and debug their systems, investigators to make better-educated judgements on what happened, and end users to understand when the system is functioning the way it should be and when it is not (Theodorou 2019).

In the end, it is our design choices that define their pool of available actions in the first place. We prescribe any 'freedom' of choice they have and even our own biases to them. Moreover, like animals, there is no understanding of the moral gravity of its actions. An artificial agent *is intelligent and can act autonomously*, but its actions are context agnostic, as its agency is prescribed. Still, as morality is a human-defined concept, the moral law can be updated to accommodate artificial agents. Next, I will discuss why we should not do that.

5.3 *Incidents Happen*

Even the best trained dog may still ignore its owner and instead start barking at a passing car. Our control over all domestic animals is precarious. When they ignore us, we shout at them, we pull them by the collars, and so on. Our response to their misbehaviour does not aim to punish the animal by attributing responsibility to them. Instead, it aims at teaching them that their actions were wrong – as always, based on our own understanding of what is right and wrong. We do not hold the dog responsible, but rather we try to educate it in order to *fix the bugs in its training*. Dogs, like all animals, are not subject to moral governance and hence not morally accountable for their actions (Himma 2009; Bryson 2018). Any moral responsibility is passed to their owner and trainer.

From a functional perspective, the attribution of responsibility is as effective as the punishment that it carries. For an act to be considered as *punishing*, there must be a successful introduction of an unpleasant mental state and a reduction of fitness value. Himma (Himma 2009) writes on punishment "You cannot punish someone who loves marshmallows, as conceptual matter, by giving them marshmallows; if it doesn't hurt, it is not punishment, as a matter of definition – and hurt is something only a conscious being can experience". Our legal systems map different sanctions with different behaviours we, as a society, consider worthy of punishing. We only live a finite amount of time, any time spent in prison reduces our ability to acquire wealth or competitive advantages. Potentially, it leads to long-term disadvantages. Our backup mechanism, written language, takes a significant amount of time and effort. It is imperfect; we can't just read and 'restore' someone's memories. Two biological agents can never be exactly the same – even monozygotic twins raised in a shared environment (Freund, et al 2013). There is an element of uniqueness that affects both ourselves directly when we lose time under punishment, but also others who benefit from our actions.

Autonomous systems perform various actions on our behalf. Defects, operational mistakes, and even malicious interference with their operation can result in enormous harm to humans, animals, and property. When such incidents happen any errors (e.g. bugs that allowed the malicious hack to take place) need to be addressed – and often to be re-addressed. We need to not only deter occurrences, but also distribute responsibility. In such the correct implementation of post-incident transparency can help with the accurate discovery of a fault and attribution of responsibility to the right parties (Winfield and Jirotka 2017).

The idea of punishing an intelligent agent that caused an incident is similar to punishing a gun instead of the shooter in a murder. For the agent the 'trigger' is pulled indirectly by the system's stakeholders upon its design, development, and sequential deployment. However, both the agent and the gun are artefacts; they are items developed and owned.

Providing robots with a moral agency requires us to not only re-examine the concept of morality, but also its legal implications. A precedent in favour of granting moral agency non-human entities is the legal personhood granted to organisations. However, there are two distinct differences between granting personhood to a company and to a robot: 1. the effects of punishing a company, and 2. the motivation and benefits from such an act.

In a company, there is a collective responsibility for the actions taken. Any sanctions placed on a company affect directly and indirectly all personnel. If a regulator fines a company, its stakeholders lose money and individual employees may be found also responsible. In heavily regulated industries, companies are required to have compliance and legal officers. These officers are require to audit and consult other executives. They discourage any wrongdoing, as they have to stand to answer for any violations of the law found in their organisations. Over the years, we have been enforcing these and other mechanisms to discourage – and deal with – any wrongdoings for the protection of our societies.

If we were to hold a robot responsible, similarly to a company, how can we punish it? A robot can be made to *simulate pain* (Kuehn and Haddadin 2017). Similar to the language processing, the agent does not have a context, but rather simulates a response as if it is in pain. AI affords the capacity to fake similarity (Wortham and Theodorou 2017). Emotions simulation can be useful as latching mechanism to avoid dithering (Tanguy, Willis, and Bryson 2003) or even facilitate communication with end users (Collins, Prescott, and Mitchinson 2015). I am not arguing that robots should be made to suffer, but that we have control over the ability to make them simulate such emotions. Like with all other attributes we describe on them, simulating pain is a deliberate design decision. Likewise, the pain-simulation algorithm can be turned off at any moment with the robot remaining operational. Furthermore, putting a machine in jail is inefficient at best. Why do we want to keep a machine 'switched on' to experience time spent in a confined space, wasting electricity and taking space? A machine can be repaired and its memories restored prior to entering jail. At the same time, its owners – who are the main benefactors of its actions – can make replicas of the machine and receive the same benefits. The impracticalities of punishing machines do not stop here. If an infringement is found a company can hire lawyers to challenge the prosecutors and mitigate damages. A robot granted moral agency would either hold assets itself or be given legal representation by the state. This raises another question: will robots – which are property themselves – be granted the right to own assets to pay their lawyers? Such a right, alongside with the general notion of granting them rights, could lead to societal disruption and to an eventually increase of (human) inequality (Bryson, Diamantis, and Grant 2017; Solaiman 2017).

Corporate personhood allows individuals to take higher risks and participate in larger ventures (Johnson 2012). It increases the stability, size, and complexity of corporations. Our legal fiction aims at contributing to our socio-economic growth. Granting legal personhood to companies is not a decision taken overnight. It took years to establish how to treat corporations as separate persons and up to this day, we still update our corporate law to better organise and protect our societies. Allowing machines to be granted personhood has arguably no possible socio-economics benefits. On the contrary, it may seriously disrupt our ability to govern, as well as our economy (Bryson 2018).

5.4 *Patiency not Agency*

So far, I have focused the discussion on robots as moral agents. However, our spectrum of morality is far wider. Entities that are not eligible for moral agency may be given moral patiency. Gunkel (Gunkel 2012) states "The question of moral patiency is, to put it rather schematically, whether and to what extent robots, machines, non-human animals, extraterrestrials, and so on might constitute an other to which or to whom one would have appropriate moral duties and responsibilities."

Coeckelbergh (Coeckelbergh 2010) argues that even if we regard robots as 'just' property, we still have *indirect* obligation towards them. He bases his argument on the fact that robots, like all objects, have a value. Coeckelbergh argues that out of respect of their owners, we have indirect obligations towards robots. Moreover, due to their value, robot owners and users will protect robots – even with violence – from others. I agree with Coeckelbergh that robots have the value that we place on them. Their value is not only the costs associated with their development, marketing, and selling, but also of the utility they provide. However, I argue that we do not protect items because we feel obliged to do that. Instead, we protect items – and all public goods at large – because we aim to avoid a reduction of our fitness value and the loss of any competitive advantage we had from the ownership of those items. No one enters an unpleasant mental state when she drops her phone due to feeling bad for 'making the phone suffer' but rather to its repairs cost or property loss. This is why the law specifies how any compensation for property damage goes to the owner and not to the property. The owner does not have to repair or replace with a replica the damaged object. Even artefacts such as religious icons, flags, and so on, as discussed in Section 3.2, are not directly owed any obligations. Any signs of respect, salute, and so on are due to their status as in-group identifiers – in other words, is directed to the in group.

Ultimately, morality is an artefact; it varies 'in shape' across different social groups. Like all cultural tools, it is the product of our social evolution. It is used by our Gregorian mind to regulate our decision making. Assigning a moral status to any entity is an intentional action. In this Chapter I have shown that moral patiency is not a binary *all of them or none of them* choice, but rather a spectrum. I retain my earlier stated position that robots – like all artefacts – are purposely developed and can only be considered as property. In this sense, they are no different than a computer, a calculator, or even a hammer. They are designed, implemented and used by us. Hence, we are responsible for when incidents happen and that chain of responsibility includes all of us; from the researchers to the policymakers, developers, and even end users. We all need to actively work towards the next steps for producing concrete ethical and socio-legal governance to ensure the societal-beneficial development, deployment, and usage of AI technologies (Theodorou 2019; Theodorou and Dignum 2020).

6. Conclusions

Morality is itself an artefact, a fictitious concept developed by us, varied across societies. Like all other social constructs, we developed it to help our own action-selection system and ensure cooperation within a society. Like all fiction, it provides a common identity. As it is stated in the introduction of this Chapter, there are serious concerns regarding the anthropomorphism and misunderstanding of the mechanical context-agnostic nature of robots, which leads to a confusion about the moral status of the robots (Bryson and Kime 2011; Coeckelbergh 2010; Gunkel 2012).

In this Chapter, I presented an evaluation of the generalised requirements for moral agency, by tracing them back into Aristotelian and Kantian ethics. Then, I discussed another type of moral entity: moral patients, demonstrating how not only the utility of an entity, but also the cultural background of the moral agents interacting with it determines its moral status.

I provided a high-level ontology of the human action-selection system, arguing that morality and law are human-made 'fiction' to help us guide our actions. They are the consequence of our cultural evolution that is enabled and is enhanced by the source of our uniqueness, language. I discussed the limitations and bottlenecks of natural intelligence; dithering and the costs associated with cognition. Then, I explained that AI is actually subject to the same limitations, thus, the idea of AGI or omniscience is impractical. Instead, we should purposely limit the application domain of our systems to ensure their performance, similar to how cognition – and consciousness to an extent – is adaptive in nature.

I argued that even an Asimovian robot does not fit the requirements for moral agency. However, I acknowledged the fact that morality, as it is a human-made construct, can be altered to include machines. Hence, I made further descriptive and normative arguments why such a move is not only avoidable, but also disruptive to our societies. Finally, I discussed moral patiency and demonstrated why *objects are not moral patients*. Instead, in rare cases, such as when they are used by as embodied symbols of our fiction, the object 'benefits' from the obligations and rights attributed toward the fiction it represents.

It is important to specify that I do not claim that an artificial agent cannot take actions of moral worth (e.g. actions that could lead to injury or death of humans). Rather, I claim that all actions performed by an intelligent system are executed as an extension to their developers and/or owners moral agency. Hence, the study of Ethics in AI (or AI Ethics as the field is also called) deals exclusively with the moral values of the *human agents* involved in the design, development, usage, and even governance of AI regardless of the tools or techniques used to the develop a system.

This includes the tools used and actions taken by stakeholders, such as the researchers, designers, developers, end users, and policy makers of AI-related products and services. Any policy produced are made by and meant to be followed by humans.[2]

2 Acknowledgements: I would like to acknowledge Joanna J. Bryson and Andrea Aler Tubella for their valuable insights and proofreading this article. Moreover, I would like to acknowledge Robert H. Wortham for suggesting to me to read Daniel Dennett's work and for our informative discussions on theory of mind and evolutionary biology. I thank the anonymous

7. Bibliography

Aler Tubella, A., Theodorou, A., Dignum F., and Dignum V. (2019) "Governance by Glass-Box: Implementing Transparent Moral Bounds for AI Behaviour" *Proceedings of the Twenty-Eighth International Joint Conference on Artificial Intelligence (IJCAI 2019)*, 5787–5793.

Aristotle. *The Nicomachean Ethics*.

Brundage, M. and Bryson, J. J. (2016) *Smart Policies for Artificial Intelligence*. Retrieved from https://arxiv.org:https://arxiv.org/abs/1608.08196

Bryson, J. J. (2001) *Intelligence by Design: Principles of Modularity and Coordination for Engineering Complex Adaptive Agents*. PhD thesis, MIT, Department of EECS, Cambridge, MA. AI Technical Report 2001–003.

Bryson, J. J. (2007) "Embodiment versus memetics." *Mind & Society*, 7(1):77–94.

Bryson, J. J. (2010) "Crude, cheesy, second-rate consciousness." In: Hernandez, C. and Sanz, J. G. R. (eds.) *From Brains to Systems: Brain-Inspired Cognitive Systems*. Madrid, 454–467.

Bryson, J. J. (2018) "Patiency is not a virtue: the design of intelligent systems and systems of ethics." *Ethics and Information Technology*, 20(1), 15–26.

Bryson, J. J., Diamantis, M. E., and Grant, T. D. (2017) "Of, for, and by the people: the legal lacuna of synthetic persons." *Artificial Intelligence and Law*, 25(3):273–291.

Bryson, J. J. and Kime, P. P. (2011) "Just an artifact: Why machines are perceived as moral agents." In: *Proceedings of the 22nd International Joint Conference on Artificial Intelligence*. Barcelona. Morgan Kaufmann, 1641–1646.

Bryson, J. J. and Tanguy, E. A. R. (2010) "Simplifying the design of human-like behaviour: Emotions as durative dynamic state for action selection." *International Journal of Synthetic Emotions*, 1(1), 30–50.

Bryson, J. J. and Theodorou, A. (2019) "How Society Can Maintain Human-Centric Artificial Intelligence." In: Toivonen-Noro, M., Saari, E., Melkas, H., and Hasu, M. (eds) *Human-centered digitalization and services*.

Bryson, J. J. and Winfield, A. (2017) "Standardizing Ethical Design for Artificial Intelligence and Autonomous Systems." *Computer*, 50(5), 116–119.

Caliskan, A., Bryson, J. J., and Narayanan, A. (2017) "Semantics derived automatically from language corpora contain human-like biases." *Science*, 356(6334),183–186.

reviewers for their insightful comments. I would like to express my gratitude to the EPSRC grant [EP/L016540/1] for funding this research. An earlier version of this article has also appeared to Theodorou (2019) "AI Governance Through a Transparency Lens." PhD Thesis. Finally, I will like to thank Sabine Hauert and Marina De Vos for their feedback on the earlier version of this manuscript.

Carsten Stahl, B. (2004) "Information, Ethics, and Computers: The Problem of Autonomous Moral Agents." *Minds and Machines*, 14(1):67–83.

Coeckelbergh, M. (2009) "Virtual moral agency, virtual moral responsibility: on the moral significance of the appearance, perception, and performance of artificial agents." *AI & Society*, 24(2), 181–189.

Coeckelbergh, M. (2010) "Moral appearances: Emotions, robots, and human morality." *Ethics and Information Technology*, 12(3), 235–241.

Collins, E. C., Prescott, T. J., and Mitchinson, B. (2015) "Saying it with light: A pilot study of affective communication using the MIRO robot." In: *Lecture Notes in Computer Science (including subseries Lecture Notes in Artificial Intelligence and Lecture Notes in Bioinformatics)*, volume 9222, Cham: Springer, 243–255.

European Commission (2010) *Directive 2010/63/EU of the European Parliament and of the Council of 22 September 2010 on the protection of animals used for scientific purposes Text with EEA relevance.*

Cooper, R., Shallice, T., and Farringdon, J. (1995) Symbolic and continuous processes in the automatic selection of actions. In Hallam, J. (eds) *Hybrid Problems, Hybrid Solutions, Frontiers in Artificial Intelligence and Applications*, pages 27–37. IOS Press, Amsterdam.

Dawkins, R. (1996) *The Blind Wathcmaker. Why the Evidence of Evolution Reveals a Universe without Design.* New York: Norton.

Dennett, D. C. (1996) *Kinds of minds: Towards an understanding of consciousness*. Weidenfeld and Nicolson.

Dennett, D. C. (2001) "Are we explaining consciousness yet?" *Cognition*, 79, 221–237.

Dignum V. (2017) "Responsible autonomy". *Proceedings of the 26th International Joint Conference on Artificial Intelligence (IJCAI 2017)*. 4698–4704.

Ellemers, N., Pagliaro, S., and Barreto, M. (2013) "Morality and behavioural regulation in groups: A social identity approach." *European Review of Social Psychology*, 24(1), 160–193.

Epstein, S. (1994) "Integration of the cognitive and the psychodynamic unconscious." *American Psychologist*, 49(8), 709–724.

Evans, J. S. B. T. (1984) "Heuristic and analytic processes in reasoning." *British Journal of Psychology*, 75(4), 451–468.

Franco, N. H. and Olsson, I. A. S. (2015) "Killing animals as a necessary evil? The case of animal research." In: *The end of animal life: a start for ethical debate*, 12–187.

Freund, J., Brandmaier, A. M., Lewejohann, L., Kirste, I., Kritzler, M., Kruger, A., Sachser, N., Lindenberger, U., and Kempermann, G. (2013) "Emergence of individuality in genetically identical mice." *Science*, 340(6133), 756–759.

Gallistel, C. R., Brown, A. L., Carey, S., Gelman, R., and Keil, F. C. (1991) "Lessons From Animal Learning for the Study of Cognitive Development." In: Carey, S. and Gelman, R. (eds.) *The Epigenesis of Mind*. Hillsdate, NJ: Lawrence Erlbaum, 3–36.

Gruen, L. (2017) "The Moral Status of Animals." In: Zalta, E. N. (ed.) *The Stanford Encyclopedia of Philosophy*. Metaphysics Research Lab, Stanford University, fall 2017 edition.

Gunkel, D. J. (2012) *Critical Perspectives on AI, Robots, and Ethics*. Cambridge, Mass: MIT Press.

Gunkel, D. J. (2017) "The Other Question: Can and Should Robots Have Rights?" *Ethics and Information Technology*.

Haidt, J. (2006) *The Happiness Hypothesis; Putting Ancient Wisdom to the Test of Modern Science*. Basic Books.

He, K., Zhang, X., Ren, S., and Sun, J. (2015) Delving Deep into Rectifiers: Surpassing Human-Level Performance on ImageNet Classification. *ICCV '15 Proceedings of the 2015 IEEE International Conference on Computer Vision (ICCV)*, 1026–1034.

Himma, K. E. (2009) "Artificial agency, consciousness, and the criteria for moral agency: what properties must an artificial agent have to be a moral agent?" *Ethics and Information Technology*, 11(1), 19–29.

Humphrys, M. (1996) "Action selection methods using reinforcement learning." In: Maes, P., Mataric, M. J., Meyer, J.-A., Pollack, J., and Wilson, S. W. (eds.) *From Animals to Animats 4 (SAB '96)*, Cambridge, MA. MIT Press.

Isoda, M. and Hikosaka, O. (2007) "Switching from automatic to controlled action by monkey medial frontal cortex." *Nature Neuroscience*, 10(2), 240–248.

Johnson, L. E. (1983) "Can animals be moral agents?" *Ethics & Animals*, 4(2), 50–61.

Johnson, L. P. (2012) "Law and Legal Theory in the History of Corporate Responsibility: Corporate Personhood.." *35 Seattle University Law Review 1521 U of St. Thomas Legal Studies Research Paper*, 12–19.

Johnson, R. and Cureton, A., "Kant's Moral Philosophy." *The Stanford Encyclopedia of Philosophy*.

Kahneman, D. and Frederick, S. (2005) "A model of heuristic judgment." *The Cambridge Handbook of Thinking and Reasoning*, 267–294.

Kuehn, J. and Haddadin, S. (2017) "An Artificial Robot Nervous System To Teach Robots How To Feel Pain And Reflexively React To Potentially Damaging Contacts." *IEEE Robotics and Automation Letters*, 2(1), 72–79.

Libet, B. (1985) "Unconscious cerebral initiative and the role of conscious will in voluntary action." *Behavioural and Brain Sciences*, 8(4), 529–539.

McAndrew, F. T. and Milenkovic, M. A. (2002) "Of Tabloids and Family Secrets: The Evolutionary Psychology of Gossip." *Journal of Applied Social Psychology*, 32(5), 1064–1082.

Miller, L. F. (2015) "Granting Automata Human Rights: Challenge to a Basis of Full-Rights Privilege." *Human Rights Review*, 16(4), 369–391.

Minsky, M. L. (2006) *The Emotion Machine: Commensense Thinking, Artificial Intelligence, and the Future of the Human Mind*. Simon & Schuster.

Mitchell, D., Bryson, J. J., Rauwolf, P., and Ingram, G. P. (2016) "On the reliability of unreliable information." *Interaction Studies*, 17(1), 1–25.

Moravec, H. (1988). *Mind children: the future of robot and human intelligence*, Harvard University Press.

Norman, D. A. and Shallice, T. (1986) Attention to action: Willed and automatic control of behaviour. In Davidson, R., Schwartz, G., and Shapiro, D.(eds) *Consciousness and Self Regulation: Advances in Research and Theory*, volume 4, pages 1–18. Plenum, New York.

OpenAI (2018). "OpenAI Five" Retrieved from: https://openai.com/blog/openai-five/.

Powell, A., Shennan, S., and Thomas, M. G. (2009) "Late Pleistocene demography and the appearance of modern human behaviour." *Science*, 324(5932), 1298–1301.

Raichle, M. E. and Gusnard, D. A. (2002) "Appraising the brain's energy budget." *Proceedings of the National Academy of Sciences*, 99(16), 10237–10239.

Rohlfshagen, P. and Bryson, J. J. (2010) "Flexible Latching: A Biologically-Inspired Mechanism for Improving the Management of Homeostatic Goals." *Cognitive Computation*, 2(3), 230–241.

Ruse, M. and Wilson, E. O. (1986) "Moral Philosophy as Applied Science." *Philosophy*, 61(236), 173–192.

Schneider, W. and Chein, J. M. (2003) "Controlled & automatic processing: Behaviour, theory, and biological mechanisms." *Cognitive Science*, 27, 525–559.

Skinner, B. F. (1935) "The generic nature of the concepts of stimulus and response." *The Journal of General Psychology*, 12(1), 40–65.

Smith, B. (1928) "Legal Personality." *The Yale Law Journal*, 37(3), 283.

Solaiman, S. M. (2017) "Legal personality of robots, corporations, idols and chimpanzees: a quest for legitimacy." *Artificial Intelligence and Law*, 25(2), 155–179.

Stanovich, K. E. and West, R. F. (2000) "Individual differences in reasoning: Implicaitons for the rationality debate?" *Behavioural and Brain Sciences*, 23(5), 645–65.

Sylwester, K., Herrmann, B., and Bryson, J. J. (2013) "*Homo homini lupus?* Explaining antisocial punishment." *Journal of Neuroscience, Psychology, and Economics*, 6(3), 167–188.

Tanguy, E. A. R., Willis, P. J., and Bryson, J. J. (2003) "A layered Dynamic Emotion Representation for the creation of complex facial animation." In Rist, T., Aylett, R., Ballin, D., and Rickel, J. (eds.) *Proceedings of the Fourth International Workshop on Intelligent Virtual Agents (IVA '03)*. Springer, 101–105.

Theodorou, A. (2019) *AI Governance Through a Transparency Lens*. PhD Thesis. University of Bath.

Theodorou, A. and Dignum V. (2020) "Towards ethical and socio-legal governance in AI" *Nature Machine Intelligence* 2, 10–12.

Theodorou, A., Wortham, R. H., and Bryson, J. J. (2017) "Designing and implementing transparency for real time inspection of autonomous robots." *Connection Science*, 29(3), 230–241.

Vincent, J. (2016) "Twitter taught Microsofts AI chatbot to be a racist asshole in less than a day." Retrieved from: https://www.theverge.com/2016/3/24/11297050/tay-microsoft-chatbot-racist.

West, S. A., Griffin, A. S., and Gardner, A. (2007) "Evolutionary Explanations for Cooperation." *Current Biology*, 17(16), R661–R672.

Wheeler, M. (2010) "Minds, things, and materiality." In: Malafouris, L. and Renfrew, C., (eds.) *The cognitive life of things recasting the boundaries of the mind*, London: McDonald Institute for Archaeological Research, 29–37.

Winfield, A. F. and Jirotka, M. (2017) "The case for an ethical black box." In: *Lecture Notes in Computer Science (including subseries Lecture Notes in Artificial Intelligence and Lecture Notes in Bioinformatics)*, volume 10454 LNAI, 262–273.

Wortham, R. H. (2018) *Using Other Minds: Transparency as a Fundamental Design Consideration for Artificial Intelligent Systems*. PhD Thesis, University of Bath.

Wortham, R. H. and Bryson, J. J. (2016) "A role for action selection in consciousness." In: *CEUR Workshop Proceedings*, pages 25–30.

Wortham, R. H. and Bryson, J. J. (2018) "Communication." In Prescott, T. J., Lepora, N., and Verschure, P. F. (eds.), *Living machines: A handbook of research in biomimetics and biohybrid systems*. Oxford University Press, ch. 33, pp. 312–326.

Wortham, R. H. and Theodorou, A. (2017) "Robot transparency, trust and utility." *Connection Science*, 29(3), 242–248.

Ysseldyk, R., Matheson, K., and Anisman, H. (2010) "Religiosity as Identity: Toward an Understanding of Religion From a Social Identity Perspective." *Personality and Social Psychology Review*, 14(1), 60–71.

Between Epistemic Progress and Existential Risk Minimization

Julia Alessandra Harzheim

Abstract

Rapid developments in the field of Artificial Intelligence prompt many questions and dilemmas; amongst others, they might force us to formulate definitive answers to controversial ethical questions. Future AGI systems will need to possess a certain degree of autonomy to yield the results we aim for, yet their decisions should preferably not be made arbitrarily – especially when it comes to moral dilemma scenarios. Instead, their actions should be guided by moral maxims we consider valid for epistemically justified reasons. The responsibility to equip AGI systems with an appropriate normative framework and certain premises that prevent them from being a threat to humankind is incredibly high. At the same time, our epistemic access to normative questions remains extremely limited. Since the very early days of philosophy, human beings have tried to figure out what they ought to do, and hence what the morally good that we should strive for could possibly be. Although philosophers have kept coming up with new moral theories, claiming to provide an answer or at least a generalizable approach, it appears that a broader consensus on the matter which is not highly culture-bound has never been reached. In this chapter, the compartmentalization of different branches of AI applications is examined as a possible approach to the aforementioned epistemic and normative issues. The core idea is to tackle epistemic and ethical questions before developing technologically powerful tools. Making use of the allegedly higher cognitive capabilities of AI systems in order to expand our epistemic access to the normative, we could eventually figure out how to guide future AGI systems safely, based on a normative position we carefully worked out.

1. Introduction: Concepts and Definitions

Being aware that this chapter will have a very diverse readership that may not be familiar with all concepts appearing in the following work, some basic terms and concepts are explained below. Concepts or ideas that are more complex

 | DOI:10.30965/9783957437488_010

and part of the core idea will be explained in detail within the development of this chapter.

AGI and Superintelligence: *General Artificial Intelligence*, also referred to as *Strong/Full AI* as opposed to *weak/narrow/applied AI*, describes an Artificial Intelligence System that is not only able to perform certain specific tasks but that could basically perform general intelligent behavior at the same level as any human being, not bound to a specific area or field. *Super Artificial Intelligence* or simply *Superintelligence* is a very advanced form of Artificial Intelligence that is supposed to even exceed the abilities of human beings. Nick Bostrom, whose work will play an important role within the proceedings of this work, defines it as "any intellect that greatly exceeds the cognitive performance of humans in virtually all domains of interest." (Müller 2016) How likely it actually is that the development of AI systems will ever reach this state, whether this scenario constitutes a desirable goal (taking into consideration the concerns and risks that inevitably accompany the undertaking) and what other philosophical problems would be implied by such an AGI or Superintelligence are questions that will be discussed below.

Intelligence: Talking about any kind of Artificial Intelligence, it is worth mentioning that there is a valid debate about whether the term "intelligence" is even suitable within this context; there is no universally accepted definition of the concept of intelligence. Within the context of Artificial Intelligence, it is rather meant to describe a form of efficiency, i.e. the ability to effectively solve problems given limited knowledge and resources. Whether intelligence actually requires consciousness and self-reflection is one question that remains a matter for debate and cannot be elaborated within the framework of this chapter, as with many other justified discussions of the concept of intelligence.

The Normative: Generally, there are two ways to make a statement about something: First, there is the purely descriptive one, which does not involve any judgement about the content from a moral position but is supposed to convey it as neutral as possible. The descriptive statement stands in contrast with the normative, hence prescriptive one, which is also making a judgement about how things *ought to be* and can do so in an implicit or an explicit way. The normative does not only surface by means of language, but it is also nonverbally involved regarding conventions, morals, various kinds of rules, law and social behavior. In philosophy, the normative is supposed to describe how something is ought to be according to what is considered good and right. It is thus highly intertwined with morality, ethics and the concept of values. As it currently seems impossible to reach any general consensus on the matter, this lack of consensus or progress will hereinafter be referred to as *The Normative Problem.*

Epistemology: Epistemology is the branch of philosophy concerned with the question of what we can know. It consequently addresses all philosophical problems related to the concept of knowledge, including the bare definition of the term to begin with. Throughout history, the field of epistemology has largely been divided into two main movements: rationalism on the one hand (regarding reasoning as the source of all knowledge, not sensory input or similar) as opposed to empiricism (regarding sensory input as main source of knowledge, emphasizing the importance of empirical evidence, e.g. through experiments), although more moderate or even completely new positions have emerged since. Epistemology overlaps with many other philosophical disciplines, e.g. metaphysics, ontology or philosophy of science.

2. Artificial Intelligence and The Bootstrapping Problem

2.1 *About AGI: Big Hopes, Big Fears and a Bunch of Hypotheses.*

Although reality has turned out to fall behind a lot of the predictions that were once made about the pace of progress regarding AGI, and although there is still a lively debate going on about whether developing a truly intelligent[1] AGI or Superintelligence is even possible, there is no denying the fact that some remarkable advance can be assigned to the field and its development during the last decades. In certain domains, AI systems have surpassed human abilities by far; the most prominent examples may be chess or the ancient Chinese board game Go.[2]

The approaches to creating Artificial Intelligence through machine learning[3] are as diverse as their results; to name just a few of them, scientists came up

1 For a brief comment on the usage of the term intelligence within the context of AI development, cf. Introduction: *Intelligence*.

2 For an example, cf. the game between Google's DeepMind AlphaGo AI that beat Go world champion Lee Sedol in March 2016 or IBM's Deep Blue defeating Garry Kasparov in 1997.

3 *Machine Learning* is the study of computer algorithms provided with huge data-sets and thus a strategy through which researchers hope to approach AI. Both terms, *AI* and *Machine Learning*, are often used interchangeably, although they differ semantically and should not be equated.

with neural networks, supervised vs. unsupervised learning, gradient descent learning, slow automatic adjustment, back-propagation of error, data mining, pattern recognition, clustering, reinforcement learning, generative vs. discriminative models, probability density distributions, recursive self-improvement and Bayesian networks – technically interesting strategies whose entire elaboration goes beyond the scope of this discussion.

However, the development of autonomously driving cars has entered a new stage just recently; market leaders such as BMW or Audi are working on more advanced autonomous systems and have constituted a special council, consisting of technical experts as well as philosophers and ethicists in particular. One of their most challenging tasks is to equip such cars with a framework accounting for the kind of moral decision making involved in interpersonal rescue dilemmas. A huge debate has evolved around the supposedly right approach in such scenarios: Is it morally justifiable to determine the odds of surviving of human beings from an ex ante position?

Even if we look at the two most prominent approaches to moral dilemma situations like the trolley dilemma[4], i.e. the consequentialist vs. the deontologist approach, we cannot say for sure which one would be morally right. Opinions on this question differ vastly, as shown in many experiments on the matter.[5]

Nonetheless, the development of technologically powerful tools seems to be advancing more pressingly than ever before. The uncertainty remaining in ethical regards appears to be considered negligible by many persons in charge. The aforementioned sequence of developments creates the flawed impression that ethical concerns could be taken care of later – when there might be more resources for research on those topics, and more general interest in finding an answer.

Meanwhile, in a kind of parallel universe of intellectual debate, particularly ethicists are concerned about this course of events. Many hypotheses have been arising about where and how far AI is going to take us. It is worth taking a quick look at some of them in order to see what could be at stake for us before returning to the epistemic and ethical issues.

4 The *Trolley Dilemma* is an interpersonal rescue dilemma which has been used to examine test subjects' ethical responses, trying to infer conclusions about their reasoning and lately even about the connection between ethical theories and certain parts of the brain; for a critical review of one of these examinations by Joshua D. Greene and Peter Singer, cf. *The Normative Insignificance of Neuroscience* by Selim Berker (Berker 2009).

5 For a discussion on the Trolley Dilemma, cf. e.g. S. Kagan, *The Limits of Morality* (Kagan 1989: 233).

A pivotal hypothesis about the future development of AI and its effects that evoked many debates and concerns in the first place is called the *Singularity Hypothesis*.

John von Neumann seems to be the first person who used this term in the context of technology (Neumann 2015). Afterwards, many authors picked up on it, speculating about possible outcomes and the pace of their development.

The core idea is that technological progress continues to grow exponentially, in principle allowing any (artificially) intelligent agent to reach the state of Superintelligence by means of an infinite cycle of self-optimization. This fits Irvin John Good's hypothesis of an *intelligence explosion* (Good 1966), stated as early as in 1965. The result would be a Superintelligence whose capabilities go far beyond those of the human race, and it is disputable whether the latter would still be able to exist under these circumstances.[6]

These concerning assumptions are often justified by referring to Moore's Law[7], which is thereby extended and applied to the area of technology in a special way. Ray Kurzweil, who also wrote a book on the Singularity Hypothesis[8], came up with the *Law of Acceleration Returns* (abbreviated: LOAR). He himself describes it as follows:

> The primary idea in my three previous books on technology […] is that an evolutionary process inherently accelerates (as a result of its increasing levels of abstraction) and that its products grow exponentially. I call this phenomenon the law of accelerating returns (LOAR), and it pertains to both biological and technological evolution. The most dramatic example of the LOAR is the remarkably predictable exponential growth in the capacity and price/performance of information technologies. (Kurzweil 2013: 4)

Kurzweil also argues on grounds of Moore's Law and gives a few impressive examples that emphasize his thesis.[9] Although there are many scientists who doubt that a generalization and application of Moore's Law in this manner is

6 Usually, the extinction of the human race is seen as one possible scenario as opposed to a transformation of the human race itself, c.f. e.g. Kurzweil: „The Singularity will allow us to transcend these limitations of our biological bodies and brains […]. There will be no distinction, Post-Singularity, between human and machine." (Kurzweil 2005: 9).

7 *Moore's Law* is an observation made by Gordon Moore in 1965 (Moore 1965) that describes the increase in the number of components of a dense integrated circuit within a certain period of time and has been proven accurate in the past. Since then, it has been applied in the field of digital electronics a lot to make predictions, e.g. about the pace of its advancements.

8 Cf. R. Kurzweil: *Singularity is near* (Kurzweil 2005: 9).

9 Cf. e.g. Hilbert, M. and López, P.: The World's Technological Capacity to Store, Communicate, and Compute Information (Hilbert and López 2011).

correct, or even doubt the hypothesis of Technological Singularity in general[10], the possibility that a future Superintelligence and the growth of technology will pose a threat for mankind is taken very seriously – especially by experts like Stephen Hawking and Elon Musk, who both expressed their concern about this so-called Existential Risk.[11]

A paper published by Nick Bostrom, one of the most reputable scientists working on these questions, has caught a lot of attention: In *The Superintelligent Will* (Bostrom 2012), his *Orthogonality thesis* is explained and put into context. Tracing back to Hume and his thesis about the independence of reason and morality (Hume 1793), Bostrom argues that the equivalent ought to be true when it comes to the relationship between any level of intelligence and any possible goal. More precisely, he states that "Intelligence and final goals are orthogonal axes along which possible agents can freely vary: more or less any level of intelligence could in principle be combined with more or less any final goal." (cf. Bostrom 2012). This thesis is indeed debatable[12] and, as we can already infer from the many relativizations it entails, can only be considered to be possibly true under certain circumstances. An extended examination of this thesis goes beyond the scope of this work.

However, it provides an interesting perspective and bears on the core questions: Assuming the Orthogonality thesis were correct and we could equip our future superintelligent agents with any possible "final goal" – how could we identify those goals we ought to be striving for (as morally and epistemically justified)?

It also leads back to the initial example of autonomously driving cars and their possibilities of reacting within ethical dilemma scenarios, and thereby illustrates that we are trying to circumvent one of the most important issues whilst we are making progress on the technical side: the lack of epistemic access to normative questions and the resulting ignorance of possible outcomes.

2.2 *The Bootstrapping Problem*

What has been alluded to so far is called, in short, the *bootstrapping problem.* As described by Jonathan Weisberg in 2012, "Bootstrapping is a suspicious form of reasoning that verifies a source's reliability by checking the source against

10 E.g. Jeff Hawkins and Gordon Moore himself, whose law is often cited as an argument in favor of the Singularity Hypothesis; furthermore, c.f. Paul Allen.: The Singularity isn't near (Allen and Greaves 2011) for reasons to dispute the plausibility of the aforementioned hypothesis.

11 E.g. Hawking in the BBC documentary *AI could end human race* (Hawking 2014).

12 For a useful review and interesting thoughts on Bostrom's Orthogonality thesis, cf. S. Armstrong: *General purpose intelligence: arguing the Orthogonality thesis* (Armstrong 2012).

itself. Theories that endorse such reasoning face the bootstrapping problem." (Weisberg 2012: 597).

This phenomenon is exactly what we end up with if we try to equip an AGI (which is supposed to be acting as a moral agent itself) with a basic moral framework, whose validity and hence legitimation we cannot be sure of today, and then expect this system to acquire a better moral framework by means of recursive self-improvement. The ideal scenario that many AI researchers have in mind is that these self-generating models would thereby become better and better, leading to an exponential increase of intelligence and knowledge not bypassing the field of morality and ethics.

But the bootstrapping problem does not only concern morality, it concerns every kind of knowledge. We want AGI systems to be superior to us human beings in terms of cognitive abilities, operating free from human biases and fallacies – and yet we have to create those systems ourselves and will, by doing so, probably not be able to avoid programming biases or fallacies into the system. One well-known, albeit extreme example is the so-called *existence bias*. It describes the fact that most human beings consider the mere existence of something (and especially of themselves) as both desirable and good eo ipso, without requiring evidence in the form of reasoning or performing a trade-off, e.g. between suffering and happiness.

There are many more illustrations of human biases that we cannot or only hardly avoid, even if we are aware of their existence. Some of them would be considered of minor importance, some of them we would want to get rid of, and some of them we might not even want to know about – because this knowledge would leave us in a very unpleasant or even dangerous position.[13] The two questions of interest in this case are:

(1) Will an AGI be able to correct those biases and fallacies, even though they might be entailed in its code?

(2) Do we want to enable an AGI to act according to what it will infer – even if its inferences seem counterintuitive to us or this autonomy includes the possibility that it could pose a threat to the existence of mankind?

Both questions will be discussed below.

2.3 *What Can We Know? Epistemology in Charge*

"What can I know? What should I do? What may I hope?" – these famous questions were posed by Immanuel Kant in his *Critique of Pure Reason* towards the end of the 18th century (Kant 1996). According to Kant, these are the most

13 Cf. the BAAN scenario described by Thomas Metzinger (Metzinger 2017) that will come to be discussed within this work.

essential questions philosophy should be able to answer. Epistemology, the discipline which is concerned with the first question[14], has come up with a lot of theories; however, many of those theories have been proven to be flawed.

Uncountable examples of how insecure our supposed knowledge generally is and how little we should actually trust our judgement regarding alleged facts are provided by history: The past has illustrated that even well-established methods of science are not beyond doubt, e.g. by means of a phenomenon commonly referred to as the *Gettier Problem*.[15]

For illustrations of how human bias affects our reasoning and our judgements without us being able to elude it, and how observation selection effects affect science and philosophy, making them prone to flaws eo ipso, cf. e.g. Nick Bostrom's book called *Anthropic bias* (Bostrom 2010).

Many issues mentioned above are largely owing to insufficient progress in the field of epistemology; and without the aid of AI systems, further progress within the near future seems very unlikely. Reasons for this estimation include the limited epistemic access of one single human being as well as the restricted and biased capabilities of the human mind, which are obviously reciprocally impacting. Consequently, the core task a suitable AI system ought to pursue in the first place could be defined as making progress in epistemology, and thereby gaining further epistemic access to relevant (philosophical, normative, …) questions.

It also means that we have to deal with a delicate dance between flexibility due to autonomy versus control: If we want AGI systems to surpass us in various areas, we have to grant them a certain power to change the underlying premises in case they turn out to be wrong. However, this concession is inevitably accompanied by the huge responsibility to allow for it in a way that minimizes the risk for our own lives to be at stake – the so-called Existential Risk.

3. Values: Towards a "Friendly AI"?

The expression "Value Alignment" seems to be omnipresent in discussions about AI and the Existential Risk that it may pose for humankind. The idea of a *Friendly AI*, an AI superior to us in all domains but still benevolent by

14 Cf. Introduction of this chapter: *Epistemology*.

15 The *Gettier Problem* illustrates that, even though we can establish certain criteria for an assumption to become knowledge per definition, there are still situations which will prove all these approaches to be flawed despite meeting all necessary criteria. For a critical review and an innovative approach to the Gettier Problem, s. e.g. W. G. Lycan: *On the Gettier Problem* (Lyan 2006).

preserving human values, is considered the ultimate goal by many researchers. "Abandoning inheritance from human value […] will yield futures worthless even from the standpoint of AGI researchers who consider themselves to have cosmopolitan values not tied to the exact forms or desires of humanity." (Yudkowsky 2011), says Eliezer Yudkowsky from the Machine Intelligence Research Institute (MIRI).

In the following section, three problems concerning the idea of striving towards a "friendly AI" or even towards "Value Alignment" will be elucidated and discussed, bearing in mind that the problems addressed make no claim to being exhaustive.

3.1 *Value is not only Fragile….*[16]

… but seems to be something extremely hard to grasp, and the distinction between values, goals and motivation appears diluted within the context of AI. Values, as one can understand the word very broadly, describe what we most appreciate in life; what is of uttermost importance for us and what we (ideally) consider to be intrinsically good and desirable. According to this already very broad and disputable definition, it seems dubious to even attribute this concept to a future AGI. We do not know what this AGI would consist of, whether it would possess an equivalent to consciousness or self-reflection and how we would need to re-define certain concepts and terms – possibly even the concept of a living being.

Leaving those linguistic and conceptual issues aside, we still do not know a lot about values, even with regard to the role they might be playing for human beings and their actions. There have been hypotheses suggesting that values determine how we act (i.e. that they play the pivotal role within decision making) if we do not want to get into the state of a cognitive dissonance. Others claim that values might not have such a striking impact on us and our behavior, but rather provide a non-causal system of justification – a narrative, so to say– for our more primitively originated desires and the pursuant actions. According to neuroscience, our decisions are mostly formed unconsciously and are already determined before we can even think about whether they are in line with our moral framework (and consistent with what we would define as our "values").[17] Moreover, it remains open to debate how steady said values are, whether they exist from birth on or are acquired later in life, whether we can

---

16 An allusion to E. Yudkowsky's paper named *Value is Fragile* (Yudkowsky 2009), which provides an interesting perspective on Value Theory within the context of AI development.

17 Cf. e.g. *Cognitive neuroscience: flutter discrimination: neural codes, perception, memory and decision making* (Romo and Salinas 2003).

distinguish intrinsic from extrinsic values, how culture, biography or other essential variables impact them – and lastly, how easily they can be altered once they have been (consciously or subconsciously) defined.

The distinction between *terminal* vs. *instrumental* values, which is often encountered within the debate, does not seem very helpful either. In fact, these expressions may even seem highly inappropriate in case they refer to hierarchical goal states programmed into a machine. Said goal states could, in turn, function as the attempt to *represent* certain values on a technical level, and thereby assure that an AGI cannot easily violate those (moral) principles we previously defined. Nonetheless, I do not think that the term is suitable, and argue that it should be replaced by more accurate descriptions.[18] In one of his papers about Value Theory, Eliezer Yudkowsky tries to describe the alleged distinction between instrumental and terminal values literally. As he fails to do so in an accurate way, he states: "So forget English. We can set up a mathematical description of a decision system in which terminal values and instrumental values are separate and incompatible types – like integers and floating – point numbers, in a programming language with no automatic conversion between them." (Yudkowsky 2007). He then proceeds to prove that the equivalent mathematical descriptions could make sense within a Bayesian decision system, which sounds valid and will be elucidated below. For this work, I will stick to the commonly used terminology despite my objections raised above, as Values and Value Alignment are established concepts within the field of AI development by now.

3.2 *Technical Issues: No Ghost in the Machine*[19]

Assuming that the normative problem had been solved and we knew exactly which values we wanted to implement into our AGI system, it is still not as easy as it sounds. In fact, it is extremely complex to program a system such that it accounts for every possible scenario in a way that we might have expected. As Yudkowsky describes it:

> As in all computer programming, the fundamental challenge and essential difficulty of Artificial General Intelligence is that if we write the wrong code, the AI will not automatically look over the code, mark off the mistakes, figure out what we really meant to say, and do that instead. [...] It is not that the AI is absolutely *obedient* to its code; rather the AI simply *is* the code. (Cf. Yudkowsky 2011: Abstract, 1)

18 E.g. *higher-level premises* vs. *lower-level premises*.

19 An allusion to E. Yudkowsky's abstract for his work named *Complex Value Systems in Friendly AI*, cf. (Yudkowsky 2011).

This cautionary finding can be considered an advantage and a disadvantage at the same time. On the one hand, it means that an AI system could very well make decisions that we did not wish for, which might even be opposed to our original preferences – because the programmers made a mistake, and this mistake is entailed within the AI system's code.

But on the other hand, it also means that terminal values or higher-level premises could be obtained despite the recurrent self-improvement of a possible AI; or, as Yudkowsky puts it: "This argues […] that under ordinary circumstances, rational agents will only self-modify in ways that preserve their utility function." (Cf. Yudkowsky 2011: Abstract, 2) And, furthermore: "Indeed, […], it seems exceedingly probable that most intelligences will not want to circumvent their own utility function. It is not as if there is a ghost-in-the-machine, with its own built-in goals and desires […] which can look over the code and find ways to circumvent the code […]." (Cf. Yudkowsky, 2011: Abstract, 2)

However, we should not underestimate the hidden complexity of our desires and values.[20] One scenario that, albeit extreme, is not impossible and illustrates this perfectly is developed by Thomas Metzinger in his essay called *Benevolent Artificial Anti-Natalism* (*BAAN*):

> The Superintelligence knows that one of our highest values consists in maximizing happiness and joy in all sentient beings, and it fully respects that value. However, it also empirically realizes that biological creatures are almost never able to achieve a positive or even neutral life balance. […] Through an analysis of our behavior and its empirical boundary conditions it reveals implicit hierarchical relations between our moral values of which we are subjectively unaware […]. Being the best analytic philosopher that has ever existed, it concludes that, given its current environment, it ought not to act as a maximizer of positive states or happiness, but that it should instead become an efficient minimizer of […] pain, unpleasant feelings and suffering. […] The Superintelligence concludes that non-existence is in the own best interest of all future self-conscious beings on this planet. (Cf. Metzinger 2017: 2f.)

3.3 *The Normative Problem*

Although many scientists argue on the basis of AI systems employing a utility function to maximize happiness and minimize suffering[21], it is, from an epistemic perspective, not at all obvious that happiness should be a terminal value or high-level goal and suffering should be completely obliterated. Those are

20 For a vivid illustration about this complexity and how difficult it is to program our wishes and preferences such that the outcome accounts for them, cf. (Yudkowsky 2011: 7).

21 Which seems understandable as they pose simple and illustrative examples.

intuitive answers a lot of people tend to give when asked about their preferences or values; and whilst there is no epistemic reason to object to the possibility that those are indeed morally justified standpoints, we do not have any evidence for the opposite being true either – unless the sheer fact that many people tend to have those intuitions is taken as evidence, or as indication that has to suffice. Furthermore, there are conflicts of interests, preferences and values that cannot be ignored, e.g. due to religion or culture. How can we cope with the resulting moral pluralism, being aware that there are no epistemically liable grounds on which one position could be deemed superior to another?

One approach called *reflective equilibrium* is supposed to create a balance between a set of beliefs or values by deliberative mutual adjustment whilst trying to obtain coherence among them.[22] Although this description does sound promising in theory, we are currently unable to outline how such a desirable state could be reached, and the concept does not provide any conclusive answer to the question of what we ought to do (for being known to be morally correct) either. Which values and moral guidelines are justified – not merely by intuition, but as a result of careful and eclectic examination? In an attempt to draw a conclusion, one cannot avoid to state that all these issues bear on problems of epistemology and epistemic access to the normative; an issue previously introduced as *The Normative Problem* (cf. Introduction of this work).

4. Compartmentalization of Branches of AI Application

4.1 *What Do We Want to Achieve? About Compartmentalization and* Oracle AIs.

When hoped-for improvements possibly brought about by AGI are addressed, the fact that most of those achievements could also be realized by a domain-specific, narrow AI system is frequently neglected.

In cases requiring indeed the capabilities of AGI or Superintelligence, moral judgement would be involved in some way or another, with the most obvious and recent example being autonomous cars. The possible application of AGI systems in medical treatment poses another controversial subject within a domain that challenges applied ethics like no other, and is thus in need of a

22 Nelson Goodman, John Rawls: *Reflective equilibrium;* for a helpful and general explanation of the concept, cf. (Rawls 1951). It was first implicitly used by Goodman to justify the principles of inductive logic, then named and explicitly introduced by Rawls in his work *A Theory of Justice*. For a review of the latter, cf. (Hare 1973).

strong ethical guideline to protect the affected (and often highly dependent) individuals from being harmed.

Assuming we were technically capable of developing such AGI systems already, we would likewise have to decide on the apt ethical framework (or the "values") we would equip it with today – from an ex ante position and without any idea how to solve the normative problem. As mentioned above, most scientists hope that those self-generative and self-improving systems would also expand their knowledge about ethics whilst they are making observations, learning more and more about mankind; but they could not account for the actual outcome of this process, and every statement about it has to remain mere speculation. Let us have a look at what is at stake vs. what tempts us regarding AGI by doing a short trade-off.

AGI is supposed to make our lives easier, more interesting and longer; moreover, it is supposed to bring about general progress and improvement – two expressions which are omnipresent in contemporary discourse and sound very promising, although it would be hard to find anyone who could actually define those seemingly desirable goals. The most concrete hope that can be encountered within the debate is that AGI could help us to maximize our happiness within the framework of a Utility Function.

Apart from the technical issues already raised above, which concern the implementation of any given goal in a sensible and responsible way, we fear the power that such a domain-independent and superintelligent agent may possess. As long as there is so little knowledge about possible outcomes, and as long as our epistemic access to normative questions remains that restricted and debatable, I argue that we should avoid AI systems making normative judgements and thereby choosing certain ethical theories as a maxim for their actions. This also applies to the ongoing development of autonomous cars, which has to appear rather irresponsible in that light.

Instead, we should strive for a differential development of AI applications, keeping them domain-specific at first. We would thereby end up without any AI system that gets to decide in real-life ethical dilemma situations, but with one branch of AI applications being devoted exclusively to making predictions and expanding our epistemic access.

This overall expansion would naturally comprise our knowledge about intelligence, the application of algorithms and all other AGI-related issues. Most importantly, it would also include the normative component, i.e. the Normative Problem. Questions such as the following illustrate some of the controversy we should approach first and foremost by means of a differential AI development:

(1) Is there any chance that an objective morality may exist?
(2) Alternatively: Is there a strategy to achieve a reflective equilibrium between conflicting values and beliefs in order to cope with moral pluralism?
(3) What kind of outcome would an AGI probably achieve, equipped with what kinds of values?

I do not argue that AGI systems should never be able to act as moral agents themselves; but I do argue that we need several specific applications first to ensure that our choices regarding any ethical framework or implemented values are justified from an epistemic and rational standpoint. Those choices would also take into consideration the minimization of any Existential Risk that an AGI system may pose for humankind.

Although it may sound a bit too figurative, the best metaphor for such an AGI system might be an Oracle. Below, we will come to see on which grounding such an Oracle AGI may be created. Judging from all we currently hope to know about it, a suitable AGI would most likely possess the following characteristics:

- It would have to be self-generative and self-optimizing.
- It would have to possess different levels of hierarchy to account for different levels of goals, motivation, beliefs, etc.
- It would have to possess different dimensions of temporal, spatial, visual, … abstractness, again preferably achieved through a hierarchical framework.
- There would be different degrees of likelihood or attributed trust for every new input, depending on variables like e.g. the source, the overall fit into the current set of beliefs, the simplicity of the resulting overall hypothesis, etc.
- It would thus need to possess the ability to correct or improve previous beliefs, depending on said trust that is attributed to the new input.
- It would constantly update its overall hypotheses.
- Predictions and input would have to be carefully adjusted to each other, just as the different hypotheses would have to be adjusted according to the overall coherence among them.
- It would need to be very efficient, as it might not be possible to test all hypotheses against each other and predict all possible outcomes; accordingly, it should (preferably) be able to work with a binary code.
- It should function based on approximations, taking into consideration the most established heuristic methods, a likelihood density distribution, a conditional probability function, a utility function or similar.
- It should, despite the efficiency, possess a high degree of preciseness.
- Ideally, all possible scenarios or hypotheses would have to be taken into account, but their significance would be weighted according to their relative likelihood.

5. Epistemic Progress

Over time, a lot of promising theories deemed able to improve our epistemic access or even epistemology itself have arisen. However, the core idea remains that an autonomously self-optimizing and self-generating system with a domain-independent efficiency and a high cognitive performance is being provided with all the factual knowledge we nowadays think to possess; with all those logical laws and rules we might have discovered over recent centuries, and with all our philosophical theories, including those that are considered to be flawed by now. To give some randomly chosen examples, it would contain theories like Hume's fork, all available data on the Trolley Problem, the Internalism vs. Externalism debate including its arguments, Syllogism and all other kinds of reasoning. The system would have access to a database that no thinkable human being could ever grasp. Epistemically, it would be superior by far.

Equipped with this epistemic and cognitive advantage, it should be able to find incoherence within or between hypotheses that we still believe to be true; it should also be able to find explanations or draw conclusions about observations that we cannot fully explain yet. Furthermore, it could hopefully get rid of any human bias that might have been influencing the programmers by means of recurrent self-optimization (s. below for some technical suggestions). It could find links between fields and theories that we had not found yet, and just as some scientists expect an exponential increase of technology according to the Law of Acceleration Returns (LOAR), we could hope for the equivalent regarding epistemology: an exponential increase of epistemic access. Within these developments, progress on the normative component would inevitably follow.

Below, I want to introduce two promising concepts which might be able to serve as a starting point when it comes to programming Superintelligence capable of making epistemic progress. As we will come to see, they possess all the characteristics that were defined as necessary (or at least as helpful) for any oracle AGI approach. The intention behind these portrayals is not to suggest a certain mechanism for a technically sensible AGI development. Instead, the concepts are supposed to illustrate the assumption that promising ideas regarding epistemic progress already exist, and that they pose a sensible base to begin working on those matters.

5.1 *Why Predictive Processing Offers an Attractive Framework to Explore and Work with Different Ethical Maxims and Possible Outcomes*

In his book *On Intelligence* (Hawkins and Blakeslee 2004), Jeff Hawkins investigates a problem that scientists are working on to this day, desperate to finally figure out how narrow AI could become Superintelligence. Metaphorically speaking, he attempts to narrow down the secret ingredient that the field of AGI development is still lacking: "What then is intelligence such that brains have it but computers don't? […] We have plenty of clues; what we need now are a few critical insights." (Hawkins and Blakeslee 2004: 2–3)[23] Speaking of general intelligence today (although our definitions or concepts of it might differ), we like to think of human beings as the most intelligent creatures currently alive. Hawkins wonders whether the human brain, the center of all the cognitive tasks that we fulfil every day with ease, could provide us with those critical insights we are still lacking to transform our domain-specific AI into an AGI or Superintelligence. "I believe the best way to solve this problem is to use the detailed biology of the brain as a constraint and as a guide, yet think about intelligence as a computational problem – a position somewhere between biology and computer science." (Hawkins and Blakeslee 2004: 3)

This is exactly what scientists like Andy Clark have been doing by examining a concept called *Predictive Processing*, which is currently considered the brain's most fundamental mechanism by many neuroscientists.

Speaking of this comparatively new and provocative concept, the expression *Predictive Coding* is far more prevalent than the expression *Predictive Processing*, which I decided to use in this work and did so for good reasons. Whilst the term *coding* implies a certain data compression strategy (s. below) and could definitely describe a pivotal part of the concept, other aspects of major significance would not be covered by it at all. As Andy Clark describes it:

> I mean to highlight the fact that what distinguishes these approaches is not simply the use of the data compression […] known as predictive coding. Rather, it is the use of that strategy in the very special context of hierarchical (i.e., multilevel) systems deploying probabilistic generative models. Such systems exhibit powerful forms of learning and […] deliver rich forms of context-sensitive processing and are flexible to combine top-down and bottom-up flows of information within the multilayer cascade. (Clark 2016: 25f.)

The data compression strategy that Clark mentions above could be considered the actual starting point of the development of the Predictive Processing framework, hereinafter referred to as *PP*. During the 1950s, an innovative idea lead to the optimization of signal processing with the effect of a so-called *lossless compression*, achieved by identification and elimination of statistical redundancy. Instead of encoding every single piece of information of e.g. a given image, the value of neighboring pixels is predicted based on detectable

23 Although we should not and cannot assume that brains, as organs, "have" or possess intelligence; and this is not what Hawkins' statement wants to convey either.

regularities and a resulting probability density function. Only if the actual value differs from the predicted one are these deviations encoded and passed on. The discrepancy between the predicted and the actual signal is called *prediction error*; a term that is pivotal for the PP framework itself.

In order to yield a probability density function according to which some potential input can be predicted in the first place, the PP account employs a well-known and established algorithm usually referred to as *Bayes'theorem* or *Bayes'law*. The fact that some people use the expression *Bayesian Brain* when trying to explain the PP framework hints at the role that this algorithm plays within this account: allegorically speaking, it is the machine in the background that operates all our different beliefs with due regard to the new evidence that we receive continuously and simultaneously through our sensory system. More precisely, it is a mathematical formalization that calculates the conditional probability of your belief (here: *hypothesis H*) being true in the light of some new evidence or *data D*. Cf. *image 1* for a mathematical description.

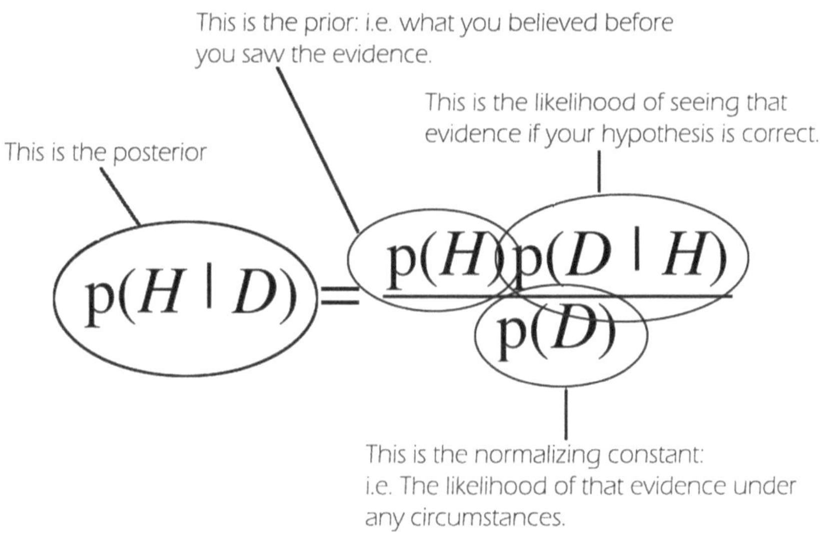

FIGURE 9.1

As we can see, the human brain continuously updates its statistical model of the world (H) by integrating the new input or information (D) about it, taking into consideration the prior likelihood $p(D/H)$ of receiving said input on the assumption that the initial hypothesis H were to be correct. Applying Bayes'rule, this yields the posterior probability $p(H/D)$, stating how likely it is that our initial hypothesis or internal model of the world is indeed correct, all data currently available included. The result of this calculation, or more accurately: the posterior probability $p(H/D)$ also determines the *prediction error* which was mentioned earlier. In case the posterior likelihood of a certain belief *H* is decreased after this Bayesian approximation, which effectively results from a prediction error, this could lead to various changes or actions being taken. But not every prediction error is equally fatal; moreover, not every prediction error enjoys the same attributed confidence in its reliability. Instead, the impact of sensory prediction error is modified by estimations of the respective input's certainty – a process which is referred to as *precision weighting*. The estimated reliability of the sensory input can statistically be described as its inverse variance.

Within this PP framework, *attention* is the means by which the attributed precision can be taken into account before any changes in our beliefs or internal models of the world may occur. As this is indeed a highly abstract and complex model, the following example by Andy Clark will serve as a more vivid illustration:

> Visual input, in the fog, will be estimated to offer a noisy and unreliable guide to the state of the distal realm. Other things being equal visual input should, on a bright day, offer a much better signal, such that any residual error should be taken very seriously indeed. But the strategy clearly needs to be much more finely tuned than that suggests. Thus suppose the fog (as so often happens) briefly clears from one small patch of the visual scene. Then we should be driven to sample preferentially from that smaller zone, as that is now a source of high-precision prediction errors. (Cf. Clark 2016: 57–58)

One substantial feature of the PP framework, aside from what has been discussed above, is its hierarchical structure. Within this probabilistic generative model, the brain's task is to reduce incoming prediction error by inferring the signal's distal causes. It does so by selecting those possible explanations for the incoming signal that best match top-down predictions. PP thus employs a multilevel bidirectional cascade which begins at the general gist of a given scene: initially, only very simple cues that possess low spatial or temporal resolution are passed upward and an apt prior in form of a general scenario is selected, trying to "explain away" (Cf. Clark 2016: e.g. 35f.) prediction errors. As this process unfolds, richer detail emerges concurrently, the higher-level predictions acting as priors for the lower level predictions. Higher level predictions that influence which general scene we may assume in the first place are called *hyperpriors*; they possess a very high degree of abstraction and determine the following PP cascade.[24] In *image 2*, we can see a basic schema of PP as a bidirectional multilevel cascade (Lupyan and Clark 2015):

24 For a helpful illustration of the influence of hyperpriors, see Howhy et al. (Hohwy, Roepstorff and Friston 2008) on how predictive processing might be able to explain the phenomenon of binocular rivalry.

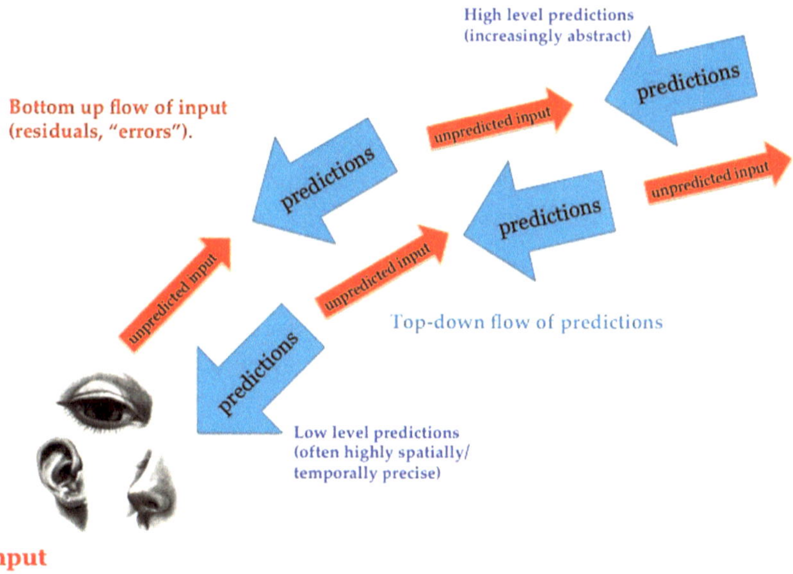

FIGURE 9.2
(*adapted from Lupyan and Clark 2015; included in references*)

At this point, it might have become more obvious why a PP framework could serve as an approach towards AGI systems that would be worth trying. On the one hand, it could function within an Oracle AI to make predictions about possible outcomes of Value Alignment attempts; on the other hand, it could be used to make epistemic progress on values (and the normative problem) themselves.

Here is a more explicit elucidation of which specific characteristics such an AGI system might need, compared to what the PP framework provides us with:

As explained earlier, it is commonly assumed that we would have to equip AGI systems with certain values or goals deemed "terminal". Those would be considered the most important parameter that we would want to maximize within a Utility Function; e.g. happiness and its reverse, the reduction of suffering. Within the PP framework, the hyperpriors that were mentioned above could account for the equivalent. Hyperpriors are (almost or entirely) impossible to change and determine a lot of what we will do and perceive; and, of course, how we might react to certain dilemma situations that we may find ourselves being confronted with. Acting against these hyperpriors is almost impossible because the PP system tries to yield an outcome that fulfills its own predictions and expectations (and thus reduces the prediction error).

Consequently, hyperpriors also possess a giant influence on planning and on taking action. In fact, one popular hypothesis is that all our actions serve to minimize prediction error and increase the trust in our predictions, especially in our hyperpriors.

I.e., if we wanted to maximize happiness, we could use it as a hyperprior within the PP framework of an abovementioned Oracle AI, and see which outcomes this would probably result in. We could also find out whether we need to change or add certain hyperpriors or parameters on lower levels (e.g. in a scenario where, to an AGI system, non-existence appears to be more desirable and in accordance with our preferences than existence and we decide to ignore this insight, although it might be true from a logical perspective).

Those values or goals that are described as "instrumental" would have their equivalent in lower-level expectations, which tend to be less abstract and more specific. Through the hierarchical and probability-based framework of PP, these could be improved, changed or get rid off comparatively easily in case that the system receives too many prediction errors which are equipped with sufficient trust in the signal (i.e. with a suspected high precision).

The PP framework thus provides us with an adjustable balance between autonomy and flexibility through its hierarchy, ongoing comparison with the new input and a respective update of inner models, constant minimization of prediction error (i.e. recurrent self-improvement) and precision weighting by means of a certain degree of trust attributed to any new input. It seems even thinkable that we could approach the aforementioned reflective equilibrium state of conflicting values and preferences.

5.2 *Introducing Solomonoff Induction: Occam's Razor Applied*

There is another fairly new epistemic approach that has a lot in common with the PP framework, which is why many characteristics do not need to be explained in detail anymore. *The Solomonoff Induction*[25] is a probability-based approximation supposed to figure out which hypotheses or beliefs are most likely to be true; and it does so by taking into consideration a lot of well-established theories that might be known from other disciplines.

The basis for this approach is inductive reasoning, as implied by the theory's name. The assumption is that for any given set of observations (i.e. data), there has to be a hypothesis that explains that data with the highest likelihood and coherence among all data available. If more than one hypothesis could serve as a logical explanation, it would depend on the prior probabilities which one would be considered most likely to account for all observations. It seems

25 For a helpful explanation of the concept, cf. Alex Altair's work *An Intuitive Explanation of Solomonoff Induction* (Altair 2012).

unnecessary to highlight the similarities that the PP framework and the Solomonoff Induction already share at this point.

However, in this case the possible explanations for any observed input is not only weighted according to prior knowledge or experience (as it is the case within the Bayesian PP framework), but also according to a new criterion: simplicity. Applied to Solomonoff Induction, the principle *Occam's razor*[26] states that among all hypotheses consistent with the observations, the simplest is also the most likely to be correct.

Here we come across a novelty, as the prior likelihood is normally equally distributed among all possible hypotheses. Within Solomonoff Induction, priors which bring about the least assumptions given the constraints of a problem are now weighted higher, i.e. considered to be more likely than others.

Additionally, Solomonoff induction takes into consideration all possible explanations – this is because its basic assumption is that the data itself has been generated by an algorithm and that the hypothesis that explains the data is an algorithm, too. It could thus be found with the aid of a universal Turing Machine[27], working with a binary sequence only.

Unfortunately, this incredible effectiveness in finding all hypotheses that could possibly account for the input is also the reason why this approach is currently considered to be incomputable. It would take an infinite amount of time to check every algorithm, and some of them might even run forever without producing any output.[28] However, this issue cannot change the fact that the concept is very promising with regards to epistemology – and the pace of progress that technology showed to possess in the past, together with the speculations about its future advancements that were raised above, provides us with the justified expectation that we might be able to solve the remaining problems of this specific approach one day.

26 *Occam's razor*, named after William of Occam, is a principle that was suggested under this name as early as in the 14th century; in science, it serves as a heuristic problem-solving approach. For a review, c.f. e.g. (Burgesss 1998).

27 The *Turing Machine* is a computational device invented by and named after Alan Turing in 1936 which is considered one of the most important models of computability and theoretical computer science. Its main purpose is to test the extent and restrictions of possible computability. For an overview, c.f. e.g. (Herken 1998).

28 This problem is also known as the *Halting Problem*; it investigates whether a program will ever finish running and produce a useful output or not. C.f. e.g. (Turing 1938) for a more detailed description which also entails Turing's original definition of Turing Machines.

6. Conclusion

As has been elucidated above, almost all issues we currently have to face developing AGI systems, including the resulting Existential Risk, can be traced back to a lack of epistemic access and progress of epistemology itself. Instead of focusing on the development of technologically powerful tools today, which would also have to act as moral agents themselves, we should approach these epistemic issues first with the aid of a compartmentalization of AI applications.

Given the promising concepts and algorithms we have discovered with Bayesian Predictive Processing and Solomonoff Induction, chances are that we could achieve considerable progress in the field of epistemology and expand our epistemic access to normative questions. Additionally, we could infer better predictions about the outcome of future AGI or superintelligent systems depending on their mechanism, their values, and other variables. We could thus avoid facing the bootstrapping problem and, more importantly, keep the Existential Risk that a future (technologically powerful) AGI system may pose at the minimum.

7. Bibliography

Allen, P. and Greaves, M. (2011) "The singularity isn't new." *Technology Review* 12, 7–8.

Altair, A. (2012) "An Intuitive Explanation of Solomonoff Induction" https://www.lesswrong.com/posts/Kyc5dFDzBg4WccrbK/an-intuitive-explanation-of-solomonoff-induction (last accessed 2018/10/15).

Armstrong, S. (2012) "General purpose intelligence: Arguing the Orthogonality thesis." In: *Lesswrong 2.0*. Available at: https://www.lesswrong.com/posts/nvKZchuTW8zY6wvAj/general-purpose-intelligence-arguing-the-orthogonality (last accessed 2018/11/21).

Berker, S. (2009) "The Normative Insignificance of Neuroscience." *Philosophy & Public Affairs*, 37(4).

Bostrom, N. (2010) *The Anthropic Bias: Observation Selection Effects in Science and Philosoph.* Edition 1 (Studies in Philosophy). Abingdon-on-Thames: Routledge.

Bostrom, N. (2012) "The Superintelligent Will: Motivation and Instrumental Rationality in Advanced Artificial Agents." In: *Minds and Machines* 22(2). Springer, Netherlands.

Burgesss, J. (1998) "Occam's Razor and Scientific Method." In: Schirn (ed.), *The Philosophy of Mathematics Today*, Clarendon Press, 195–214.

Clark, A. (2016) *Surfing Uncertainty: Prediction, Action, and the Embodied Mind*, Oxford University Press.

Good, I. J. (1966) "Speculations concerning the first ultraintelligent machine." *Advances in computers*. 6, 31–88.

Hare, R. M. (1973) “Rawls’ Theory of Justice” *Philosophical Quarterly* 23 (144–55), 241–251.

Hawking, S. (2014) “AI could end human race, a BBC documentary.” Accessible under: https://www.bbc.com/news/technology-30290540 (last accessed 2018/11/08. Retrieved 11 November 2017).

Hawkins, J. and Blakeslee, S. (2004) *On intelligence: How a New Understanding of the Brain will Lead to the Creation of Truly Intelligent Machines*. New York: Times Books.

Herken, Rolf (1998) *The Universal Turing Machine: A Half-Century Survey*. New York: Oxford University Press.

Hilbert, M. and López, P. (2011) “The World’s Technological Capacity to Store, Communicate, and Compute Information.” *Science* 332(6025), 60–65.

Hohwy, J. and Roepstorff, A. and Friston, K. (2008) “Predictive coding explains binocular rivalry: An epistemological review” *Cognition* 108(3), 687–701.

Hume, D. (1793) *A Treatise of Human Nature*, Book 3. Reprinted and edited by Selby-Bigge, I. A., Oxford: Clarendon Press 1896.

Kagan, S. (1989) *The Limits of Morality*, Oxford: Oxford University Press.

Kant, I. (1996) *Critique of Pure Reason: Unified Edition*. Ellington, J. W. (ed.) Hackett Publishing Company, Inc.

Kurzweil, R. (2005) *The Singularity is Near*. New York: Penguin Group.

Kurzweil, R. (2013) *How to create a mind: The secret of human thought revealed.* New York: Penguin Group.

Lupyan, G. and Clark, A. (2015) “Words and the world: Predictive coding and the language-perception-cognition interface” *Current Directions in Psychological Science*, 24(4), 279–284.

Lyan, W. G. (2006) “On the Gettier Problem.” In: Stephen Hetherington (ed) *Epistemlogy Futures*. New York: Oxford University Press, 148–169.

Metzinger, T. (2017) “Benevolent Artificial Anti-Natalism (BAAN).” Essay on *Edge.org*, accessible online: https://www.edge.org/conversation/thomas_metzinger-benevolent-artificial-anti-natalism-baan (last accessed 2018/10/15).

Moore, G. E. (1965) “Cramming more components onto integrated circuits.” *Electronics Magazine*, 38(8).

Müller, V. C., Bostrom, N. (2016) “Future progress in artificial intelligence: A survey of expert opinion.” In: *Fundamental issues of artificial intelligence*. Springer, 555–572.

Shanahan, M. (2015) *The Technological Singularity*. 1st ed. MIT Press, 233.

Rawls, J. (1951) “Outline of a Decision Procedure for Ethics” *Philosophical Review*, 60(2), 177–97.

Romo, R. and Salinas, E. (2003) “Cognitive neuroscience: flutter discrimination: neural codes, perception, memory and decision making.” *Nature Reviews Neuroscience*, 4(3).

Turing, A. (1938) “On computable numbers, with an application to the Entscheidungsproblem” *Proceedings of the London Mathematical Society*, Series 2(43), 544–546.

Weisberg, J. (2012) "The Bootstrapping Problem." *Philosophy Compass*, 7(9), 597.

Yudkowsky, E. (2007) "Terminal Values and Instrumental Values." In: Fragile Purposes No. 59, *LessWrong 2.0*. Available at: https://www.lesswrong.com/posts/n5ucT5ZbPdhfGNLtP/terminal-values-and-instrumental-values (last accessed 2018/15/10).

Yudkowsky, E. (2009) "Value is Fragile." In: Value Theory No. 51, *LessWrong 2.0*. Available at: https://www.lesswrong.com/posts/GNnHHmm8EzePmKzPk/value-is-fragile (last accessed 2018/10/15).

Yudkowsky, E. (2011) "Complex Value Systems in Friendly AI" In: Artificial General Intelligence: 4th International Conference, AGI 2011, Mountain View (USA), Proceedings, edited by Schmidhuber J., Thórisson, K. R., and Looks, M., pp. 388–393, vol. 6830, Springer Verlag (Berlin), 2011.

PART III

Ethics

Current Challenges in Ethics of Artificial Intelligence or: Old Wine in New Bottles

Leonie Seng

Abstract

The attempt to find consistent moral values for certain societies has been part of descriptive as well as normative ethical considerations at least since Socrates' philosophical investigations. In this sense I call the content of some current discussions on the possibility of moral decision-making ability in intelligent machines, robots or computer programs (in accordance with corresponding ethical criteria) *old wine in new bottles* – since such attempts deal with questions which have already been raised in early philosophical writings. The second main claim in this article is, that most approaches of ethics of Artificial Intelligence (AI) currently tend to talk about AI in general which can, as will be shown, likely lead to either misunderstandings or irrelevant statements which are neither useful for ethical debates nor for technical realizations.

1. Artificial Intelligence – or: Talking About *What Is*

In the only known didactic poem of Parmenides "On Nature" the author distinguishes *what is* from *what is not* (cf. Diels and Kranz 1951/52). It is controversial if those categories can be understood as ontological statement, describing the existence of something or as predication with a certain truthfulness *is like this* or *is not like this* (cf. Rapp 2007: 94). Scientific realists interpreted Parmenides later in the sense that one can only derive reliable knowledge from the first category.[1] If the Parmenidean category *what is* is understood as *one can only make true statements on things/ subjects that exist* one can apply this assertion on two current sections of Artificial Intelligence: so called weak and strong AI. Corresponding to Parmenides' claim, one could say that one can only make true statements on developments in weak Artificial Intelligence. This involves technical devices (both on the level of hardware and software) where (parts

1 In fact, Parmenides seems to have rejected the second category *what is not* principally, supposedly due to the fact that knowledge can only be derived from what is true (cf. Rapp 2007: 95).

 | DOI:10.30965/9783957437488_011

of) human forms of intelligence are merely imitated, for example in semi-autonomous cars, military drones or on pattern recognition trained algorithms. In this sense, AI is a way of finding solutions for representing ways of human reasoning in a very broad sense (Nilsson 2010). Supporters of the strong AI hypothesis, on the contrary, argue furthermore for the possibility of the actual replication of primarily cognitive capacities like reasoning, thinking, learning and problem solving (Russell and Norvig 2010: 1020 et seq.), but also of other human capacities such as acting in general or even the capability to create emotions in machines and robots – and hence for the whole mechanical or (to stick to the more common and broader term) artificial replication of human beings (Drux 1999; Schlich 1998; Swoboda 1976, Van Dülmen 1998). Now, one has to admit that the second type, strong AI, referring back to Parmenides, in one respect also falls into the first category, of *what is*; that is in form of fictive figures in various science fiction stories which contain creatures exemplifying very strong Artificial Intelligence. Among many others, Homer's narration of the art-smith Hephaistos should be mentioned here as well as Karel Čapek's play *Rossum's Universal Robot*s (1921) which ends in the ultimate catastrophe, or Isaac Asimov's story *Runaround* (1942) in which he poses for the first time his famous *Three Laws of Robotics*, often referred to in current discussions around AI (Thimm and Bächle 2018 i. a.; for a – surely not exhaustive – enumeration of science fiction movies containing forms of *strong artificial intelligence* see also Seng 2019a: 61).

These fictive variants of technical developments often serve certain narratives in stories which commonly draw quite derogatory pictures of AI (cf. Seng 2019a). This is why some authors think that science fiction could not help in understanding current technological developments because "*science-fiction distracts from the currently relevant ethical and legal challenges of AI*" (Hermann 2018, emphasis in original). On the other hand, one can argue that those stories enable thought spaces for the creative handling of the topic in the first place (cf. Roberts 2018). Krebs 2011 furthermore argues that the occupation with the essence of human existence and substantial as well as emergent characteristics of humans correlates with narrations about fictitious humanoid figures.

At least since Socrates and the sophists ancient Greek mythological narrations were broadened to include also systematic reflections of ethical principles with regard to the justification of certain norms.[2] Questions like *How do we want to live? Which kind of society do we want to have?* and *Who or what is the*

2 It can be complemented here that the Presocratics primarily appeared as natural philosophers who dealt with structure and the order of the universe and its elements. Often in history, Thales von Milet who mainly investigated material principles (*archai*) is called the first philosopher (cf. Rapp 2007: 14) along with Socrates and the sophists (cf. Rapp 2007: 20).

Human Being? have been posed for example by Aristotle (Krapinger 2011; Wolf 2006), Immanuel Kant (1781) and many other philosophers. The fact that such *old* questions are now being posed once again in search for a good handling of AI developments underpins the central thesis of this paper: Some current discussions in ethics of AI are *old wine in new bottles*.[3] In the following, I will outline some reasons for this claim as well as show its consequences for current ethical discussions and finally give some alternative perspectives.

2. Some Concerns on Ethical Approaches in AI Developments

2.1 *The Core of AI Ethics and Its Origins*

Some current questions in ethics of Artificial Intelligence are: *Can and should robots act morally and/ or ethically and why (not)? And how are moral/ ethical machines to be constructed?* (Cf. among others Misselhorn 2018: 7; Vanderelst and Winfield 2018)[4] Sometimes it is argued that *machine ethics* primarily deals with machines as moral subjects (cf. Misselhorn 2018: 70). Others claim that morality is a human category and that machines are the object of humans' ethical reflections on morality (cf. Seng 2019b: 11). Floridi et al. (2018: 690) distinguish four main fields and their problems: *Who can we become?* (autonomous self-realisation), *What can we do?* (human agency), *What can we achieve* (individual and societal capabilities) and *How can we interact with each other and the world?* (societal cohesion). The attempt to answer these questions and moreover the general moral question *What should we do?* has been part of ethical investigations since Socrates' philosophy. Or to say it with McCarthy and Hayes (1981: 431): "some of the major traditional problems of philosophy [and ethics; L. S.] arise in artificial intelligence."

Even if many current applications are new from a technical point of view, ethical questions in this context do not necessarily have to be *new* in the same

3 What I call *ethics of AI* in this paper can as well be named *machine ethics*, *roboethics* and so forth. I chose the term *Artificial Intelligence* for no certain technological reasons, it includes in this context many sorts of smart machines, robots and algorithms.

4 In German philosophy it is common to distinguish ethics and morality very strictly. Morality therefore contains all sorts of actions. According to Gerhard (2009: 11 et seq.) morality can be understood either very broadly (in this sense every question concerning a particular action is a moral one) or more narrowly (in this sense only questions count as moral ones which affect a broader group of people and are of certain moral relevance). In both forms morality concerns right and wrong actions. The only difference is the range of the consequences. Even the question *What shall I eat?* includes in this sense the question *Is it right for me to eat this or that?* and is therefore a (in a broad sense) moral one. Ethics, however, is a scientific discipline whose object is morality.

sense. Ethical concerns do not involve questions of technical feasibility, for their own sake. However, ethical reflections deal with questions like which technologies a certain group or society wants to have developed, which ones should be developed and for which (good) reasons (cf. Brandom 2013). One should therefore not think that the new research area machine ethics or ethics of Artificial Intelligence necessarily requires thus new ethical solutions. A look at the history of philosophy and applied ethics can at times be fruitful as well. One example is the systematic investigation of concepts like *liberty*, *freedom* (for example *positive* and *negative* according to Isaiah Berlins theory (1969) who derives his concept from Immanuel Kant's *Critique of Pure Reason*), *autonomy* and self-*determination*[5]:

> To coerce a man is to deprive him of freedom – freedom from what? Almost every moralist in human history has praised freedom. Like happiness and goodness, like nature and reality, the meaning of this term is so porous that there is little interpretation that it seems able to resist. I do not propose to discuss either the history or the more than two hundred senses of this protean word recorded by historians of ideas. I propose to examine no more than two of these senses but those central ones, with a great deal of human history behind them, and, I dare say, still to come. The first of these political senses of freedom or liberty (I shall use both words to mean the same), which (following much precedent) I shall call the 'negative' sense, is involved in the answer to the question 'What is the area within which the subject – a person or group of persons – is or should be left to do or be what he is able to do or be, without interference by other persons?' The second, which I shall call the positive sense, is involved in the answer to the question 'What, or who, is the source of control or interference that can determine someone to do, or be, this rather than that?' The two questions are clearly different, even though the answers to them may overlap. (Ibid.: 120).

In the *negative* sense, according to Berlin, freedom means not being prevented by anyone from what one wants to do. In the *positive* sense, liberty means having the autonomy, or being self-determined, so that one can act as one's own source of control.[6] With regard to AI developments we have to discuss which concepts of freedom (…) we want to have realized in our society.

5 Another of many examples is the whole field around the philosophical question about the art of (good) living.

6 Of course, many more examples of relevant concepts, which have been discussed in the past, can be found here. With the ideas of liberty and relating concepts and the following application to current discussions, I only want to give one example of how philosophical and ethical arguments can be made fruitful for current debates on ethics of AI.

2.2 *Transparency: Only Half the Battle*

Most companies develop their products for economic reasons. Moreover do developments of AI products like algorithms with certain aims, for example pattern recognition, take place in secret and with hardly any possibility of control from outside, sometimes even the companies' developers have no clue of what their programs are calculating – and why![7]

One example is *Google*'s search algorithm[8]: Users do not know how it works. The company suggests that its algorithms lead to "useful" and "relevant" results (Google 2018). It suggests that finding an (maybe *the*) answer to a question is simple, just using *Google's* search engine, so not having to search through "billions of webpages" instead (ibid.). The adjectives "useful" and "relevant" are clearly positive judgements. Fact is, however, that users do not have any insights into which results are, according to *Google*, being labeled with these judgments, apart from some plausible assumptions: It could be, for example, those which link to web pages from companies which pay *Google* for being listed or those which fit best to recent search history of a particular user. That means an important loss of information control.

What counts for the example *Google* in this text can largely be applied to other companies. Most companies' primary objective is to work in a way that is financially lucrative. Even if companies claim to have other aims as well (serving people, making people happy or things alike) the main drive of economic institutions is money. And some companies make their money (at least inter alia) with people's data, for example in the field of *social media*. People here give up the control of how the data is linked with other information and used for reasons like creating very precise usage profiles for example for advertising customers (cf. Bolotaeva and Cata 2011). Following Isaiah Berlin, one could claim, that people who use such digital structures lose their *positive freedom* – since, in the example of *Google's* search engine, one is not only restricted to certain answers to questions when remaining in such a *filter bubble* (cf. among others 2011) but this can also potentially influence people's whole lives if they allow it to: including the books they read, the choice of channels they use to get their political information from and who they are connected with or not.

7 "Indeed, the black-box nature of these systems allows powerful predictions, but it cannot be directly explained", write Adadi and Berrada (2018: 52138).

8 This is only one example of many other companies which make financial profit on the base of people's private data; cf. Zwitter 2014. It can be discussed, of course, if this is a good example for AI.

There are at least two possible rejections to the critique that people's freedom is impeded by offering services which use forms of Artificial Intelligence for the use of grabbing data. Some invoke that the amount of *information overload* would be reduced in a positive way by services like search engines (cf. Bozdag and Timmermanns 2011). However, as Bozdag and Timmermanns (ibid.: 7) also claim, "such filters often ignore the context of information and they are never value neutral." Therefore, so they argue, people's *autonomy* is undermined, and this is, according to them, at the expense of personal *identity* and algorithmic *transparency*.

One can also argue that people give up freedom or autonomy voluntarily: No one forces them to use those digital offers, hence the companies are not to blame. Matthias Karmasin (2004) shaped the concept of a *paradox of media*: On the one hand, users want to use social media and other digital services for (at least in some respects) understandable reasons: acting social, professional connections, social response (German: *Resonanz*. Rosa 2016), making things easier (e.g. finding the way with a navigation app). At the same time, they do not care about data protection or at least accept circumstances in which they completely give up control and hence self-determination. Regarding the immense influence some companies have[9], one can, however, attribute at least some responsibility. This can be easily compared with German public transport which is led by a private company which underlies certain conditions like continuous maintenance of trains and rails.[10] Same could count for essential digital infrastructures like huge search engines. The details would, of course, be different, in that the responsibility of digital infrastructure companies (which provide for example also social networking platforms) from an ethical point of view primarily lies in a respectful and dignified conduct with the customers. One way of doing so could consist in a transparent exposition of information (Richterich 2018: 8 et seq.). This could mean: Publishing their aims and algorithms honestly and transparently – or at least telling users when this is not the case. This is applicable to many, if not all, companies who develop forms of AI (and other algorithms) and work towards financial profit based on personal data. As Zwitter (2014: 1) puts it, this is even more urgent, since the "speed of development in Big Data and associated phenomena, such as social

9 According to *statista* 1,17 billion people worldwide use *Google's* search engine. https://www.statista.com/chart/899/unique-users-of-search-engines-in-december-2012/. Accessed on 06.09.2019.

10 Of course the conditions of free market and federally owned corporate groups differ at large though.

media, has surpassed the capacity of the average consumer to understand his or her actions and their knock-on effects."

A rejection against the above argument of voluntariness is the fact that it is not always a question of choice if one wants to use these or those applications. It is, for example, expected within some businesses to use specific computer programs or certain communication structures. The use of social media (both by management staff and candidates) can even have an influence on the recruitment process (Petry et al. 2018). This shows that transparency alone is not the solution. Since the development of AI is complex and can have an enormous impact on daily lives of people, some companies (as well as academics) have set up proposals for ethical frameworks for being able to deal with several forms of AI. Unfortunately, some of them are very general and therefore not very easy to implement.

2.3 *Why Some Approaches of AI Ethics Do Not Work*

There have been made quite a lot of attempts to regulate current AI developments via different approaches. Companies as well as scientists and scientific institutes have published an enormous amount of recommendations for moral principles in recent years. It would be unsatisfactory to raise the claim to demonstrate a complete picture of the range of articles. However a certain tendency of some approaches will be investigated here, that is: Some current articles on AI ethics tend to a form of generalization of AI (cf. among many others Villani 2018; Floridi et al. 2018; Ethik-Kommission des BMVI 2017, Horvath 2016; Bostrom and Yudkowsky 2015; Ismajloska and Bakreski 2012). To pick up a few examples: The *German ethics commission for autonomous vehicles* which is part of the German *Department of Transport* published in 2017 "ethical rules"[11], which include that (semi) autonomous vehicles should lead to more safety in road traffic ("Verbesserung der Sicherheit aller Beteiligten im Straßenverkehr") or to the increase of mobility opportunities ("Steigerung von Mobilitätschancen"; Ethik-Kommission des BMVI 2017: 10). Floridi et al. (2018) talk about "Enabling Human Self-Realisation, Without Devaluing Human Abilities" (ibid.: 691), "Enhancing Human Agency, Without Removing Human Responsibility (ibid.: 692), "Increasing Societal Capabilities, Without Reducing Human Control" (ibid.: 693) and "Cultivating Societal Cohesion, Without Eroding Human Self-Determination" (ibid.; formatting in original). The scientists behind the *Future of Life Institute* (2017) propose

11 As I showed in footnote 4, I plead for a differentiation between *ethics* and *morality*. Ethical in this sense would mean the reflection of (moral) rules from a meta level.

among others "**Safety:** AI systems should be safe and secure throughout their operational lifetime, and verifiably so where applicable and feasible", "**Human Values:** AI systems should be designed and operated so as to be compatible with ideals of human dignity, rights, freedoms, and cultural diversity" and "**Human Control:** Humans should choose how and whether to delegate decisions to AI systems, to accomplish human-chosen objectives". *Google* wants AI to "[...] be socially beneficial", "[...] avoid creating or reinforcing unfair bias", "Be built and tested for safety", "Be accountable to people", "Incorporate privacy design principles", "Uphold high standards of scientific excellence" and to "Be made available for uses that accord with these principles" (*Artificial Intelligence at Google: Our principles*). Finally, *Microsoft* determines the following "ethical principles" for "responsible bots": "Fairness: AI systems should treat all people fairly", "Reliability & Safety: AI systems should perform reliably and safely", "Privacy & Security: AI systems should be secure and respect privacy", "Inclusiveness: AI systems should empower everyone and engage people", "Transparency: AI systems should be understandable", "Accountability: *AI* systems should have algorithmic accountability" (Microsoft 2018).

All of those approaches have in common that they are very general, hence superficial in their statements about the desirable aims and values. This may provide the advantage that one can easily agree with such rules, since everybody will endorse the claim that AI (in general) should not harm people – since most people would agree that nothing and nobody should harm people, not even people. This traces back to the juristic claim, referring to fundamental rights. Science fiction writer Isaac Asimovs published in 1942 his meanwhile (especially in machine ethics and AI development) famous *Laws of Robotics*, put in other words by Tzafestas (2018: 4): "The primary concern of roboethics is to assure that a robot or any other machine/artifact is not doing harm". Interestingly, many authors in AI ethics refer to Asimov's *laws* although they are a juristic instrument (and moreover in a fictive context). This is criticised by a few others on the other hand (cf. among others Gunkel 2012: 76; Anderson 2011: 285 et seq.) What Asimov's laws have in common with other current approaches in AI ethics is the fact that they are talking about AI in general which necessarily leads to very superficial and hence impractical results. An engineer who is about to develop an autonomous car can hardly transpose such general rules to a concrete application or device, since the laws or rules (may they be juridical or ethical[12]) do not involve concrete instructions. General rules

12 The binding of juristic laws and ethical rules differs indeed very much. Whereas violation of laws can (on the base of jurisdiction) be punished, ethical laws usually do not have

or laws such as the physical integrity of humans are therefore very impractical when developing intelligent machines or programs because they are too vague. Statements about AI in general, are not very fruitful. Instead authors should try to concretise to certain projects or applications. There is a huge amont of different areas of Artificial Intelligence (military robots, autonomous cars, weapons and many more) which should be looked at in differentiated, detailed discussions.

As Asimov's science-fiction stories (or those of other sci-fi authors like McCarthy 2004[13], among many others) show, the proposal of moral rules in regard to AI developments should anyhow be taken with a grain of salt since moral laws, goals or rules can come into conflict. One example which is drawn in McCarthy 2004 is the conflict of good and responsible child care. What if parents fail and threaten the life of their children? What if robots could do a better job? Should then a robot be allowed to intervene? Presumably for this reason, the current trend in AI development is not to develop machines with a closed set of moral rules, but learning systems which can react on the basis of huge data records. This in turn makes dilemma-discussions unnecessary.

2.4 *Why Discussions of Dilemma-Situations Are Misleading*

In current general debates on AI or machine ethics, some people refer to dilemma-situations (cf. among many others Deng 2015: 26; Goodall 2014), especially in the context of driverless cars (cf. among others Villani 2018: 131; Saptawijaya and Moniz Pereira 2018). One of the most prominent examples often referred to, is the *trolley problem*, first mentioned by Philippa Foot (1967) in the context of a discussion about abortion, then variously modified, which describes different ways of moral decision making. The classic version of the scenario (one person can decide how many people will be killed via changing a train switch) is often adapted to examples of Artificial Intelligence, mainly for the development of (partly) automated cars (cf. Seng 2017). Scientists of the *Massachusetts Institute of Technology* have developed (in their description) a "platform for gathering a human perspective on moral decisions made by machine intelligence, such as self-driving cars" (*Moral Machine*). This is a platform which provides the possibility of taking part in a survey about

---

such a high societal commitment. If somebody misses to open somebody else the door, it is a pity for her or him, but he/ she cannot sue the other person for damages.

13 Whereas the technological facts in the story are quite astonishing, it has to be mentioned that the image of women, drawn in this story, is, unfortunately surprisingly conservative.

hypothetical cases of ways to decide about how driverless cars should behave in critical road traffic situations. This procedure is sometimes described as *ethical consideration* (Dizikes 2018). Ethical considerations need, however, a form of complex reflection of moral circumstances, valid argumentation, and rational justifications. This is not yet the case when demonstrating possibilities of (moral) decision making but the MIT platform can build a base for exactly this. One problem is that dilemmas are, qua concept, insoluble. Insofar it is questionable how useful discussions around dilemma-cases can be.

The counterproposal of this article is to motivate having a look at the past of ethical and philosophical investigations instead. As outlined above this can be either Aristotle or Isaiah Berlin or any other philosopher who elaborated a sound theorie about a concept which is of interest in the current context of AI developments. Provided that liberty or freedom, self-determination and autonomy are estimated as desirable moral values, one can on this basis develop ideas which deliver the depth, sharpness and clarity needed for current discussions on AI developments in order to come to relevant descriptive or normative results. The latter could then influence legislation or technical developments.

3. Further Perspectives

The considerations outlined in this article are at large descriptive. One consequence of the arguments as stated above is that ethicists and technical engineers respectively people with responsibility (for example, politicians) should work closer together in order to come to adequate results. People who discuss ethics of AI necessarily need to gain some know how about current technical developments. In turn, engineers need to be aware of ethical concerns. Thus more inter- and transdisciplinary work is desirable to come to relevant results.

Second, when discussing current AI developments, it could be advisable to consider the whole context of possibilities and consequences. Instead of discussing moral decision making in automated vehicles one could take the entire organization of road transport into account. The question of how to develop automated cars, most of the time so far takes place against the background of individual transportation. Taking ecological considerations into account, one could also have concepts of sustainability in mind and discuss current developments regarding more complex and broader approaches of mobility where automated cars may be only one factor in a whole grid of transportation.

Current developments should, finally, not serve economic growth first and foremost, but humans; since this is what ethics is all about.

4. Bibliography

Adadi, Amina and Berrada, Mohammed (2018) "Peeking inside the black-box: A survey on Explainable Artificial Intelligence (XAI)". In: *IEEE Access* 6, 52138–52160.

Artificial Intelligence at Google: Our principles. Url: https://ai.google/principles/. Accessed on 04.09.2019.

Anderson, Susan Leigh (2011) "The Unacceptability of Asimov's Threee Laws of Robotics as a Basis for Machine Ethics". In: Anderson, Michael and Anderson, Susan Leigh (eds., 2011) *Machine Ethics.* Cambridge: Cambridge University Press, 285–296.

Asimov, Isaac (1942) "Runaround". In: *Astounding Science Fiction.* New York: Street and Smith.

Berlin, Isaiah (1969) "Two concepts of liberty". In: Berlin, Isaiah: *Four Essays On Liberty.* Oxford: Oxford University Press, 118–172.

Bolotaeva, Victoria and Cata, Teuta (2011) "Marketing Opportunities with Social Networks". In: *Journal of Internet Social Networking and Virtual Communities.*

Bostrom, Nick, and Yudkowsky, Eliezer (2015) "The ethics of artificial intelligence". In: *The Cambridge Handbook of Artificial Intelligence.* Ed. by Keith Frankish and William M. Ramsey. Vol. 4. Cambridge (UK): Cambridge University Press, 316–334.

Bozdag, Engin, and Timmermanns, Job (2011) "Values in the filter bubble. Ethics of Personalization Algorithms in Cloud Computing". In: *1 st International Workshop on Values in Design – Building Bridges between RE, HCI and Ethics.* Url.: https://www.researchgate.net/publication/235218318_Values_in_the_Filter_Bubble_Ethics_of_Personalization_Algorithms_in_Cloud_Computing, 7–15. Accessed on 04.09.2019.

Brandom, Robert (2013) *Reason in Philosophy. Animating Ideas.* Cambridge, MA: Harvard University Press.

Čapek, Karel (1921) *R. U. R.* (*Rossum's Universal Robots*). Engl. transl.: Paul Selver, and Nigel Playfair. Prag: Aventinum.

Deng, Boer (2015) "Machine ethics: The robot's dilemma. Working out how to build ethical robots is one of the thorniest challenges in artificial intelligence". In: *Nature* 523.

Diels, Hermann, and Kranz, Walther, eds. (1951/52) *Die Fragmente der Vorsokratiker.* 6th ed. Berlin: Weidmann.

Dizikes, Peter (2018) "A self-driving car can choose who dies in a fatal crash. These are the ethical considerations". In: *World Economic Forum.* Url: https://www.weforum.org/agenda/2018/10/how-should-autonomous-vehicles-be-programmed. Accessed on: 04.09.2019.

Drux, Rudolf (1999) *Der Frankestein-Komplex.* Frankfurt am Main: Suhrkamp.

Ernst, Gerhard (2009) *Die Objektivität der Moral.* 2nd edition. Paderborn: Mentis.

Ethik-Kommission des BMVI (2017) "Automatisiertes und vernetztes Fahren. Bericht: Juni 2017". In: *Bundesministerium für Verkehr und digitale Infrastruktur* (*BMVI*),

1–36. Url: https://www.bmvi.de/SharedDocs/DE/Publikationen/DG/bericht-der-ethik-kommission.pdf?__blob=publicationFile. Accessed on 04.09.2019.

Floridi, Luciano et al. (2018) "AI4People – An Ethical Framework for a Good AI Society: Opportunities, Risks, Principles, and Recommendations". *Minds and Machines* Volume 28, Number 4, 689–707. Url: https://link.springer.com/content/pdf/10.1007%2Fs11023-018-9482-5.pdf. Accessed on 04.09.2019.

Foot, Philippa (1967) "The Problem of Abortion and the Doctrine of the Double Effect". In: *Oxford Review* 5.

Future of Life Institute (2017) "Asilomar AI Principles". Url: https://futureoflife.org/ai-principles/. Accessed on 11.05.2020.

Goodall, Noah (2014) "Machine Ethics and Automated Vehicles". In: Meyer G., Beiker S. (eds) *Road Vehicle Automation, Lecture Notes in Mobility*. Springer: Cham, 93–102.

Google (2018) *How Search algorithms work*. Url: https://www.google.com/search/howsearchworks/algorithms/. Accessed on 04.09.2019.

Gunkel, David (2012) *The Machine Question: Critical Perspectives on AI, Robots, and Ethics*. Cambridge, Massachusetts: MIT Press.

Hermann, Isabella (2018) "Terminator won't save us". In: *Alexander von Humboldt Institut für Internet und Gesellschaft*. 20.12.2018. Url: https://www.hiig.de/en/terminator-wont-save-us/. Accessed on 04.09.2019.

Horvath, Josip (2016) *The Ethics of Artificial Intelligence*. Url: https://www.fsb.unizg.hr/brodogradnja/UZIR-2016-Essay-Josip-Horvat.pdf. Accessed on 04.09.2019.

Ismajloska, Mersiha, and Bakreski, Jane (2012) "The Ethics of Artificial Intelligence". In: *ICT Innovations 2012 Web Proceedings*, 539–542.

Kant, Immanuel (1781) *Critique of Pure Reason*. German orig.: *Kritik der reinen Vernunft*. Riga: Johann Friedrich Hartknoch.

Karmasin, Matthias (2004) *Paradoxien der Medien. Studien zu den Widersprüchen technisch erzeugter Wirklichkeit*. Wien: Facultas.

Krapinger, Gernot, ed. (2011) *Aristoteles: De Anima. Über die Seele*. Stuttgart: Reclam.

Krebs, Dennis (2011) *The Origins of Morality. An Evolutionary Account*. Oxford: Oxford University Press.

McCarthy, John (2004) *The Robot and the Baby*. Url: http://jmc.stanford.edu/articles/robotandbaby.html. Accessed on 04.09.2019.

McCarthy, John, and Hayes, Patrick (1981) "Some Philosophical Problems from the Standpoint of Artificial Intelligence". In: *Readings in Artificial Intelligence*, 431–450.

Microsoft (2018) *Microsoft AI principles*. Url: https://www.microsoft.com/en-us/ai/our-approach-to-ai. Accessed on 04.09.2019.

Misselhorn, Catrin (2018): *Grundfragen der Maschinenethik*. Stuttgart: Reclam.

Moral Machine. Url: http://moralmachine.mit.edu/. Accessed on 04.09.2019.

Nilsson, Nils J. (2010) *The Quest for Artificial Intelligence. A History of Ideas and Achievements*. Cambridge: Cambridge University Press.

Pariser, Eli (2011) *The Filter Bubble. What the Internet Is Hiding from You*. London: Viking/ Penguin.

Petry, Thomas and Schreckenbach, Florian and Knabenreich, Henner (2018) "Social Media: Noch immer in der Pubertät. (Ergebnisse der Social Media Personalmarketing Studie 2018)". In: *Personalwirtschaft* 10/2018, S. 52–55.

Rapp, Christof (2007) *Vorsokratiker*. 2. revised edition. München: Beck.

Richterich, Annika (2018) *The Big Data Agenda. Data Ethics and Critical Data Studies*. London: University of Westminster Press.

Roberts, Stuart (2018) *From Homer to HAL: 3000 years of AI narratives*. Url: https://www.cam.ac.uk/ainarratives. Accessed on 04.09.2019.

Rosa, Hartmut (2016) *Resonanz. Eine Soziobiologie der Weltbeziehung*. Suhrkamp: Berlin.

Russell, Stuart, and Norvig, Peter (2010) *Artificial Intelligence. A modern approach*. 3rd edition. Upper Saddle River: Prentice Hall.

Saptawijaya, Ari, and Moniz Pereira, Luís (2018) "From Logic Programming to Machine Ethics". In: Bendel, Oliver (ed.) *Handbuch Maschinenethik*. Wiesbaden: Springer VS.

Schlich, Thomas (1998) "Vom Golem zum Roboter. Der Traum vom künstlichen Menschen". In: *Erfindung des Menschen – Schöpfungsträume und Körperbilder 1500–2000*. Richard van Dülmen (ed.). Wien: Böhlau, 542–557.

Seng, Leonie (2019a) "Mein Haus, mein Auto, mein Roboter? Eine (medien-) ethische Beurteilung der Angst vor Robotern und künstlicher Intelligenz". In: Rath, Matthias; Krotz, Friedrich and Karmasin, Matthias (eds.) *Brauchen Maschinen Ethik? Begründungstheoretische und praktische Herausforderungen*. Wiesbaden: Springer VS, 57–72.

Seng, Leonie (2019b) "Maschinenethik und Künstliche Intelligenz". In: Bendel, Oliver (ed.) *Handbuch Maschinenethik*. Wiesbaden: Springer VS, p. 1–21.

Seng, Leonie (2017) "Autonomes Fahren – Eine Frage der Ethik? Oder: Kant fährt Dein Auto gegen die Wand". In: *Scilogs – Wissenschaftsblogs*. Url: https://scilogs.spektrum.de/feuerwerk-der-neuronen/autonomes-fahren-eine-frage-der-ethik-oder-kant-faehrt-dein-auto-gegen-die-wand/. Accessed on 04.09.2019.

Swoboda, Helmut (1976) *Der künstliche Mensch*. München: Heimeran.

Thimm, Caja, and Bächle, Thomas Christian (2018) "Autonomie und Ethik – zwischen ‚autonomen moralischen Agenten' und der metaphorischen Zuschreibung von moralischer Autonomie". In: *Brauchen Maschinen Ethik? Begründungstheoretische und praktische Herausforderungen*. Wiesbaden: Springer VS, 81–87.

Tzafestas, Spyros (2018) "Roboethics: Fundamental Concepts and Future Prospects". In: *Information* 9 (6). Url: https://www.researchgate.net/publication/325884319_

Roboethics_Fundamental_Concepts_and_Future_Prospects/fulltext/5b2af0ddaca27209f379694f/Roboethics-Fundamental-Concepts-and-Future-Prospects.pdf, 1–25. Accessed on 04.09.2019.

Vanderelst, Dieter and Winfield, Alan (2018) *The Dark Side of Ethical Robots*. In: *2018 AAAI/ACM Conference on AI, Ethics and Society* (*AIES '18*), February 2–3, 2018. New Orleans, LA, New York, NY. Url: https://doi.org/10.1145/3278721.3278726. Accessed on 04.09.2019.

Van Dülmen, Richard (ed., 1998) *Erfindung des Menschen. Schöpfungsträume und Körperbilder 1500–2000*. Wien: Böhlau.

Villani, Cédric (2018) *For a meaningful artificial intelligence. Towards a French and European Strategy*. Url: https://www.aiforhumanity.fr/pdfs/MissionVillani_Report_ENG-VF.pdf. Accessed on 04.09.2019.

Wolf, Ursula, ed. (2006) *Aristoteles: Nikomachische Ethik*. Reinbek: Rowohlt.

Zwitter, Andrej (2014) "Big Data Ethics". In: *Big Data & Society*, 1–6.

Is Utilitarianism Entirely Useless for Self-Driving Car Ethics?

A Critical Reflection on the Rationale for Rule Utilitarianism

Vanessa Schäffner

Abstract

The impending introduction of self-driving cars poses a new stage of complexity not only in technical requirements but in the ethical challenges it evokes. The question of which ethical principles to use for the programming of crash algorithms, especially in response to so-called dilemma situations, is one of the most controversial moral issues discussed. This paper critically investigates the rationale behind rule utilitarianism as to *whether* and *how* it might be adequate to guide ethical behaviour of autonomous cars in driving dilemmas. Three core aspects related to the rule utilitarian concept are discussed with regards to their relevance for the given context: the universalization principle, the ambivalence of compliance issues, and the demandingness objection. It is concluded that a rule utilitarian approach might be useful for solving driverless car dilemmas only to a limited extent. In particular, it cannot provide the exclusive ethical criterion when evaluated from a practical point of view. However, it might still be of conceptual value in the context of a pluralist solution.

1. Introduction

Today's Western societies are characterized by the omnipresence of machines in private and public contexts. Among the most discussed technological innovations on the agenda for the next ten years are self-driving cars – a technology whose widespread introduction is not only likely to revolutionize transport and mobility as we have known them, but also has the potential to significantly decrease the number of accidents and fatalities (e.g. Coca-Vila 2018; Gogoll and Müller 2017; Mladenovic and McPherson 2016).

When entering our roads, driverless cars will possibly be exposed to driving situations that are difficult for machines to cope with. Among the most sophisticated challenges is a driving environment of mixed traffic. The car needs to

 | DOI:10.30965/9783957437488_012

interact with different road users and might be confronted with human beings acting in ways that are not foreseeable for the autonomous system (e.g. Fagnant and Kockelman 2015). Although self-driving cars are likely to be programmed in a way that aims at preventing accidents whenever possible, they will eventually face situations where personal harm cannot be completely avoided (e.g. Goodall 2014a; Lin 2016). Those cases cannot be solved by merely obeying traffic rules, they are genuine moral issues. Literature has so far proposed a variety of examples of so-called moral dilemma scenarios, an ethical decision problem which is characterized by the fact that all possible courses of action lead to (morally) undesired consequences. Examples given for the context of autonomous driving differ with regard to their level of sophistication as well as to the underlying ethical problem. Some scenarios focus on the question of which person to hit (e.g. Coca-Vila 2018; Santoni de Sio 2017), whereas others address the issue of false incentives, e.g. when it comes to decide between two motor cyclists one of which wearing a helmet that might prevent him from suffering a major personal damage (e.g. Lin 2014; Liu 2016, 2017). Still others include the option of the passengers' self-sacrifice by e.g. steering the car against a heavy obstacle in order to avoid a collision with other road users (e.g. Goodall 2014b; Lin 2016). One of the most cited examples is that of an autonomous car facing an impending collision with several persons who, respectively, might or might not be struck depending on the car's swerving behaviour (e.g. Lin 2016). How should a self-driving car behave in such situations? What ethical principles should it follow?

The present ethical treatise aims at contributing to the debate on moral issues in autonomous driving by critically reflecting on the consequentialist[1] perspective. As opposed to most research articles, which explore mainly the ethical tradition of utilitarianism[2] in general (e.g. Coca-Vila 2018; Gerdes and Thornton 2016; Goodall 2014a; Santoni de Sio 2017), this paper puts special emphasis on the rationale behind rule utilitarianism. The paper starts with some fundamental insights into the issue of moral agency as discussed from a machine ethics perspective. It explains why and how we as humans must deal with the problem of moral dilemmas in the context of driverless cars. Then a concise summary of the controversy about utilitarianism, which has shaped research related to self-driving car ethics, is presented. Against this background,

1 Consequentialism is a view in moral philosophy which entails that the moral rightness of an action depends directly and only on its consequences.

2 Utilitarianism (from Latin *utilitas* 'utility') is the most common consequentialist approach. It states that an act is morally right if it maximizes the overall good, aggregated well-being, or utility when compared to other options.

the rule utilitarian rationale for driverless cars is critically discussed by drawing upon three main aspects: the universalization principle, the ambivalence of compliance issues, and the demandingness objection. At this point, some of the most relevant arguments traditionally given for and against a rule utilitarian approach in general are revised and evaluated in terms of their relevance for autonomous car ethics. The paper concludes with some ideas on how the utilitarian concept as such may help to inform and enrich future research into the topic.

2. A Machine Ethics View on Moral Agency

How might a driverless car be able to find an adequate response to moral dilemmas? The answer to this question strongly depends on the moral status ascribed to autonomous machines. This issue lies at the centre of machine ethics, a discipline at the interface between ethics, robotics, and AI. Machine ethics research focuses on the question of whether autonomous systems or robots can perform moral actions themselves instead of just being exposed to those of human beings (Bendel 2018). Or, to use Floridi and Sanders' formulation, whether they can be (active) moral agents instead of just being (passive) patients of morality (2004: 349).

Just as artificial intelligence is supposed to simulate human intelligence, artificial morality seeks to simulate human morality. However, current approaches in machine ethics use only a reduced notion of artificial morality that focuses on certain fundamental aspects of human morality, i.e. ethical principles and criteria[3] as well as the capability of assessing the consequences of available options (Bendel 2018: 35). On this basis, autonomous machines are enabled to evaluate different options for action and finally select one. But is this ability enough to make them *moral agents*? The question of whether autonomous systems can legitimately be regarded as *moral* agents has a strong resemblance to the one whether machines could be *intelligent* agents. The strong AI hypothesis asserts that machines that act in an intelligent way actually are intelligent while the weak AI hypothesis asserts that machines could act as if they were intelligent. The majority of AI researchers accept the weak AI hypothesis while the strong one is viewed as an unachievable vision in the foreseeable future (if

3 Ethical principles are primary rules that can be used to evaluate the validity of specific ethical norms. Examples are Kant's Categorical Imperative or Rawls' principle of justice. An ethical principle builds upon specific ethical criteria, e.g. a fair distribution of goods, maximum well-being, or optimum fulfilment of duties.

ever) (Russell and Norvig 2010: 1020). Similar thoughts apply to machine ethics: Does the ability to weigh implemented options against one another qualify a machine as a moral agent taking moral decisions autonomously? Obviously, aspects like the discrepancy between machine autonomy and moral autonomy (e.g. Domingos 2015; Etzioni and Etzioni 2017; Lucas Jr. 2013) and the question of a machine's general ability to act morally play a vital role in the debate on so-called *Artificial Moral Agents* (e.g. Misselhorn 2019; Wallach and Allen 2009).

There are current approaches in machine ethics research that ascribe the status of moral agents to machines at least in some regards. Misselhorn (2018) draws upon a notion of explicit ethical agency developed by Moor, who distinguishes between "implicit ethical agents" satisfying ethical criteria by merely preventing unethical behaviour, and "explicit ethical agents" who 'act' according to given ethical rules (2006: 19–20). However, even an explicit moral agent is not a "full moral agent" such as a human who "can make explicit ethical judgments and generally is competent to reasonably justify them" (2006: 20). Anderson and Anderson (2011) as well as Wallach and Allen (2009) also support the view that machines are not moral agents in the sense of 'acting' and 'deciding' subjects, but rather technical systems that just simulate these typical human abilities. Weber and Zoglauer (2019) argue that machines are not moral agents because they do not develop their own moral maxims. Coming back to the issue of self-driving cars, they claim that situations of unavoidable harm are not even moral dilemmas in the strict sense when considering the machines' incapability to act morally. It is only us humans who find ourselves in a moral dilemma (2019: 158), for we must equip autonomous systems with the ethical criteria to be implemented into so-called crash algorithms. The latter are supposed to instruct machine behaviour in cases of moral conflicts (e.g. Goodall 2016a; Thornton et al. 2017). Researchers hold different opinions on how this should be done practically (e.g. Deng 2015; Etzioni and Etzioni 2017; Goodall 2014b; Loh and Loh 2017). While some prefer a top-down approach of hard-coding specific ethical criteria into the machines' algorithms, others advocate a bottom-up approach proposing that the system should be enabled to 'learn' the criteria through evaluating data from real human decisions in similar cases. Others find a hybrid approach combining the virtues of both methods most promising.

3. The Controversy About Utilitarian Cars

Consequentialism is among the most influential ethical theories *per se*. It was established in its traditional form by Jeremy Bentham and John Stuart Mill in the 19th century. The core element of the classical utilitarian rationale is the so-called *Greatest Happiness Principle* which aims at the maximization of happiness or well-being over pain:

> The creed which accepts as the foundation of morals, Utility, or the Greatest Happiness Principle, holds that actions are right in proportion as they tend to promote happiness, wrong as they tend to produce the reverse of happiness. By happiness is intended pleasure, and the absence of pain; by unhappiness, pain, and the privation of pleasure. (Mill 1863: 9–10)

If this basic principle is applied to the context of self-driving car dilemmas, it would imply saving the greater number of people: the more people are happy, the greater is the resulting happiness. For example, in a scenario where the car can either keep its trajectory and hit a single person on the road or swerve and crash into two pedestrians, the car would spare the two pedestrians and consequently strike the single person.

In the pertinent literature debate, only very few authors are supportive of a utilitarian approach to crash algorithms for driving dilemmas. People's reluctance to approve of utilitarian cars has its roots in the fundamental difficulties inherent in the utilitarian concept in general and the challenges of its practical realization in particular. Several research articles have essentially carved out manifold problem areas related to a utilitarian view on driving dilemmas.[4] Firstly, the main objection against utilitarianism concerns its method of aggregation. It fails to respect the individual rights and autonomy of those sacrificed for the greater good. The Ethics Commission 'Automated and Connected Driving' appointed by the German Federal Minister of Transport and Digital Infrastructure addresses this issue by specifically prohibiting in its ethical guidelines to simply count up lives against one another (2017: 11). Closely linked with this issue are severe difficulties of whether and how the value of a human life could be objectively determined (e.g. Hevelke and Nida-Rümelin 2015; Santoni de Sio 2017).

Secondly, the issue of moral and legal incompatibility inherent in the utilitarian rationale has also been subject to extensive criticism. When applied to driving dilemmas, the discriminative effects of the utilitarian solution on persons with specific characteristics or certain types of vehicles become obvious. In this context, Liu is concerned by a "flipping [of] the utilitarian calculus"

4 The main focus of this paper is to reach beyond the criticism of a utilitarian approach for self-driving cars ethics in general. It is for this reason that the specific problems related to utilitarianism are presented here only briefly. For a more comprehensive overview, see e.g. Schäffner (2018).

(2016: 168); instead of aiming to minimise harm, the utilitarian approach might tend to focus on the maximum value of collective well-being. For example, persons that are involved in a driving dilemma might be rated with regards to their value (or burden) for society. A younger person would be rated higher than an elderly, a prominent researcher higher than a truck driver. Lin (2014) calls it a matter of physics that a crash with a heavier vehicle may result in less overall harm. If thoughts like these find their way into crash algorithms, serious conflicts with criminal law will result. The implementation of algorithms containing elements of targeting certain groups of persons on purpose could be interpreted as an act of intentional killing and thereby a criminal offense (e.g. Goodall 2014a; Santoni de Sio 2017).

Thirdly, the utilitarian view is criticized for the fact that it is based upon the unrealistic premise of complete information regarding all potential outcomes (e.g. Gerdes and Thornton 2016; Goodall 2016b). Even though the extensive criticism voiced against a utilitarian approach to driverless cars is essentially justified, it almost exclusively deals with either the utilitarian idea in general or its classical form of act utilitarianism[5]. However, the controversial debate in media and research neglects utilitarianism's many sub-versions. Although the problems arising from utilitarianism are serious, some are especially problematic for act utilitarianism but might be less severe in other utilitarian forms, as e.g. rule utilitarians commonly assert (e.g. Bykvist 2010).

4. A Rule Utilitarian Rationale for Driverless Cars: Flaws and Potentials

4.1 *The Concept of Rule Utilitarianism*

Like act utilitarianism, preference utilitarianism[6] does not appear to be very useful either for solving ethical dilemmas. This is because all parties involved in an accident have basically the very same preference of not being harmed. Giving priority to the preference of one party at the same time means that the preferences of other parties will be less fulfilled. Consequently, achieving maximum preference fulfilment that equally respects all preferences in each scenario is nontrivial. Besides, both approaches focus on individual actions,

5 Act utilitarianism is the classical form of utilitarianism. It entails that the utilitarian principle should be applied directly to single and specific actions, and not indirectly by means of moral norms.

6 Preference utilitarianism is a version of utilitarianism that was mainly developed by Peter Singer (1979) and claims that an act is morally right if it maximizes preference fulfilment when compared to alternative options.

which makes them especially suitable for directly evaluating *concrete actions*. But at the same time, they appear less adequate for the problem of which *ethical rule* should be implemented into crash algorithms. In order to address this issue, an ethical approach is needed that reaches beyond the moral justification of specific decisions to the legitimation of underlying ethical principles.

Is there any version of utilitarianism appropriate to solve this? The underlying concept of rule utilitarianism was introduced by Mill. It asserts that an act is morally right when it follows a rule that enhances happiness:

> According to the Greatest Happiness Principle [...] the ultimate end, with reference to and for the sake of which all other things are desirable (whether we are considering our own good or that of other people), is an existence exempt as far as possible from pain, and as rich as possible in enjoyments, both in point of quantity and quality; [...]. This, being, according to the utilitarian opinion, the end of human action, is necessarily also the standard of morality; which may accordingly be defined, the rules and precepts for human conduct, by the observance of which an existence such as has been described might be, to the greatest extent possible, secured to all mankind. (Mill 1863: 17)

If we take the view that, in the end, algorithms are basically encoded rules for action, rule utilitarianism appears to be a plausible approach at least from a conceptual point of view. In evaluating corresponding rules, it enables the indirect assessment of actions and is therefore worthy of a closer consideration for the issue of crash algorithms.

4.2 *The Universalization Principle*

Perhaps the most obvious point in favour of rule utilitarianism is the fact that its principles must undergo a universalization test. From the rule utilitarian point of view, an action is essentially justified with regard to its social effects (e.g. Hubig 1993: 123). As proposed by Bykvist (2010: 147), asking "what if everyone thought it would be ok to do that?" is an appropriate means of verifying whether a specific action is suitable for being transformed into a universal rule. This is of central importance in the context of autonomous driving. While decisions of human drivers facing a dilemma situation could be described as one-off decisions, reactions of driverless vehicles are part of embedded algorithms. In this sense, they are no longer single and independent actions but more like iterative behaviour patterns. Moreover, ethical criteria for driving dilemmas are not the result of an individual person's decision but part of an ethical policy decreed within the scope of regulatory governmental intervention. This implies that, once decided upon, they will not be used for unique cases but implemented as part of identical software algorithms deployed on an entire fleet of autonomous cars.

Furthermore, universalized principles could play a vital role in preventing problematic issues that result from the fact that utilitarianism is perhaps too permissive (e.g. Bykvist 2010). Opponents continuously claim that constraints and special obligations are neglected in the utilitarian concept. But in fact, the core idea of the universalization test is to check if the consequences that are caused when an ethical criterion is universalized are morally desirable. This way, it provides a constraint that – at least in theory – rejects any criteria producing morally questionable outcomes.

4.3 *The Ambivalence of Full Compliance*

The notion of universalization is also helpful in defending the rule utilitarian approach against another common criticism. According to the so-called partial compliance objection, rule utilitarianism suffers from idealism. People may have diverse reasons not to commit themselves to a system of moral rules. But if only a part of all people is compliant to a given set of rules, the greatest well-being might not be achieved, and undesirable results will even be caused (Bykvist 2010; Mulgan 2006). In the case of driverless cars, however, the selected moral rules are embedded into software code. In this sense, the behaviour of these autonomous systems can be considered broadly consistent and reliable.[7]

However, the issue of full compliance also has a downside. If self-driving cars are designed to fully comply to a given system of rules, they 'lose' the capability to react flexibly to particular circumstances. As Goodall puts it, the "inherent literalness of computers" (2014b: 62) makes machines follow implemented principles with excessive rigour. They lack the ability to grasp the implicit meaning and aim of a rule. This observation prompts critics to raise concerns whether autonomous cars might be capable of disaster prevention, a common objection to rule utilitarianism. Its main argument is that even the best moral rule might lead to a disaster in some situations (e.g. Bykvist 2010), and consequently will not promote maximum well-being. This is especially the case when the outcome depends on the compliance of another party. Identically programmed cars will potentially cause unforeseen cumulative consequences in cases when more than one autonomous vehicle is involved (e.g. Liu 2017; Nyholm and Smids 2016). Imagine, for example, a passing manoeuvre on a highway. An autonomous car is just in the middle of overtaking a lorry when it spots an oncoming vehicle but is not able to move back into the lane in time.

---

7 This assumption is valid unless the passengers have the option to customize their cars' ethics settings according to their individual moral preferences (e.g. Contissa, Lagioia, and Sartor 2017; Gogoll and Müller 2017; Millar 2017).

As a collision seems to be inevitable, the implemented algorithms of both autonomous vehicles make them swerve and drive off the road on either side. This act of both cars sacrificing their passengers fails to achieve the utilitarian aim of maximizing happiness.

Obviously, the aspect of full compliance is ambivalent. On the one hand, it has a positive impact on the predictability and reliability of autonomous cars' behaviour. But on the other hand, it provokes an inability to react flexibly to scenarios when literally sticking to the rules has counterproductive effects for the overall aspiration of utilitarianism. These insights seem to suggest that we must find solutions to the inflexibility problem. Coordinating reactions between autonomous vehicles could be achieved in mainly two ways. One option is to use algorithms complemented with machine learning components which enable autonomous cars to learn from previous experiences and to continuously advance their decisions (e.g. Goodall 2014b). A more practicable approach is to set up a mobility network infrastructure that enables direct communication among participating vehicles and gives them the opportunity to coordinate their actions more efficiently (e.g. Luetge 2017; Nyholm and Smids 2018). Apart from technical challenges, these approaches raise ethical issues on their own, for example concerning privacy or data security. Nevertheless, the network option in particular could help to also address the problem of cumulative effects that result from scenarios involving several autonomous vehicles. Even though the rule utilitarian rationale risks intensifying negative dilemma outcomes, it may be part of a solution itself: if a hierarchical system of rules is embedded directly into the vehicle network infrastructure, it would be possible to overrule the pre-programmed algorithms whenever necessary, and in this way avert inefficient outcomes.

4.4 *The Demandingness of Self-Sacrificing*

Finally, rule utilitarianism also must face the objection that promoting the greatest happiness regardless of personal preferences is often found too demanding of decision makers (e.g. Bykvist 2010; Mulgan 2006).[8] This especially applies to situations when passengers are asked to sacrifice themselves. Although they cannot function as sound premises in moral arguments, results from the empirical studies by Bonnefon, Shariff, and Rahwan (2015, 2016) are helpful in revealing an ethical paradox about driverless car dilemmas that is closely linked to the issue of self-sacrificing: most participants endorse the

8 The demandingness objection is frequently linked to some other objections to consequentialism. Mulgan (2001: 15–18) mentions objections regarding integrity, separateness of persons, and transcendental issues.

utilitarian notion of maximizing overall well-being, even if it means sacrificing the passengers for the sake of saving the greater number. They say that they regard self-sacrifices as having a high moral worth. However, at the same time a great number of respondents indicate that they would not be willing to buy or even use such a car themselves (2016: 1574).

The critical point about demanding self-sacrifices from driverless car passengers is this: even if self-sacrifices are morally desirable, this does not necessarily mean that the unwillingness to sacrifice oneself is regarded morally wrong. Authors discussing the demandingness objection to utilitarianism emphasize that the critical point about the utilitarian view is not *the fact that* it demands self-sacrifices but *the reason why* passengers should sacrifice themselves, i.e. for the sake of saving others in driving dilemmas. It seems plausible that the altruistic need to always focus on the greater good makes it difficult for agents to advance their own projects. Following this argumentation, the moral obligation of always having to subordinate personal interests to the purposes of society might legitimately be found too much of a good thing (e.g. Mulgan 2001).

5. Conclusion: Achievements and Impulses for Further Research

5.1 *The Pluralist View*

In order not to lose sight of the larger picture, the present scrutiny of rule utilitarianism must be integrated into the debate about the issue of what ethical principles to use as a basis for programming solutions to dilemma scenarios. In recent years, various authors have increasingly argued that a single ethical principle will most likely not be adequate to offer a morally acceptable solution (e.g. Awad et al. 2018; Goodall 2014a; Lin 2016). Instead, they suggest that the complexity and multidimensionality of the specific ethical challenges evoked by self-driving cars require a *pluralist* approach combining different ethical rationales (e.g. Etzioni and Etzioni 2017; Goodall 2014b; Liu 2017; Luetge 2017; Thornton et al. 2017). Against this background, the utilitarian idea as such must not necessarily be regarded as completely useless for the issue of crash algorithms. It might be able to contribute its strengths to a composite solution, while its weaknesses could – and urgently must – be compensated for by other ethical principles.

Interestingly enough, there is another aspect of rule utilitarianism that supports a pluralist approach to autonomous car ethics. Although utilitarianism is consequentialist, it still takes special duties in the form of deontological[9]

9 Deontological (from Greek *deon* 'duty' and *logos* 'science, study') views are a form of moral theories that is often defined as opposed to consequentialist approaches. In their

constraints into account (e.g. Driver 2012; Portmore 2011). This means that an action might be justified on utilitarian grounds as long as it does not violate a deontological criterion, e.g. an individual's right to life. There seem to be good reasons why rule utilitarianism is sometimes even said to be, to a certain extent, a deontological concept that entails teleological[10] foundations. If consistently thought through, it presents a system of moral rules leading back to basic principles that secure the freedom of human actions (e.g. Hubig 1993: 123). This view might serve as underpinning evidence supporting the hypothesis above: a single ethical system will most likely not be capable of providing a morally acceptable solution to all dilemma situations that arise in the context of autonomous driving. In this sense, the provided discussion of the conceptual foundations of rule utilitarianism is an attempt to spark a debate on the idea of combining diverse ethical views, for instance teleological and deontological perspectives, into a pluralist moral approach.

5.2 *The Encompassing View*

How might objections to the rule utilitarian concept be refuted in the context of ethical crash algorithms? In order to answer this question, two major aspects require further clarification. Firstly, the rules entailed by the chosen rule utilitarian system of principles need to be elaborated in detail. In this respect, rule utilitarianism is consistent with ethical principles from deontological theories, virtue ethics[11], or other ethical traditions that might be able to compensate for the weaknesses rule utilitarianism suffers from. Secondly, the performance of the final pluralist system of ethical principles crucially depends on the capacities of all its ethical components. Their assessment and final composition are vital points on the agenda for future research into the topic, which is currently still in its infancy.

Beyond the conceptual issues and despite the many practical and theoretical pitfalls of the utilitarian approach, the utilitarian idea as such internalizes something noteworthy that might pave the way for further research. Evaluating an action from a utilitarian point of view does not only mean to take its direct consequences into account but rather *the whole possible world* that would

most common form, they hold that there are decisions that are morally forbidden, regardless of their consequences. Instead, an act is considered morally right if it conforms to a moral norm that is widely agreed on, duties, or other special obligations.

10 Teleological (from Greek *telos* 'aim, purpose' and *logos* 'science, study') views are characterized by the belief that everything is oriented towards a specific end or final cause.

11 Virtue ethics is the third major normative tradition besides consequentialism and deontological views. It focuses on the moral character of agents and their underlying motivations. According to virtue ethics, what is morally right depends on the circumstances of a situation which cannot be resolved by merely applying a principle.

be created by it (e.g. Bykvist 2010; Mulgan 2006). The quality and desirability of the resulting world are regarded as major criteria when judging specific moral rules. Changes brought about by the introduction of autonomous cars will not only affect the realms of transportation or mobility. The scope of the decisions taken in terms of ethical crash algorithms extends to fundamental moral issues. They raise the question of what is morally desirable, of the moral preferences we would like our society to be based upon. More practically, these include, e.g., the question of how much responsibility and power we are willing to assign to autonomously acting machines (Ethik-Kommission BMVI 2017). Against this background, it may even be legitimate to say that the decisions we need to take about autonomous driving ethics make up an essential part of shaping our future world. So indeed, it seems reasonable to take the perspective of the 'whole possible world' when elaborating on ethical questions in autonomous driving, and to include it into our scrutiny of suitable candidates for ethical principles, whatever ethical tradition they might belong to.[12]

6. Bibliography

Anderson, Michael and Anderson, Susan Leigh, eds. (2011) *Machine ethics*. Cambridge: Cambridge University Press.

Awad, Edmond, Sohan Dsouza, Richard Kim, Jonathan Schulz, Joseph Henrich, Azim Shariff, Jean-François Bonnefon, and Iyad Rahwan (2018) "The Moral Machine experiment." *Nature*, 563(7729), 59–64. Available at: https://doi.org/10.1038/s41586-018-0637-6.

Bendel, Oliver (2018) "Überlegungen zur Disziplin der Maschinenethik." *Aus Politik und Zeitgeschichte*, 6, 34–38.

Bonnefon, Jean-François, Azim Shariff, and Iyad Rahwan (2015) "Autonomous Vehicles Need Experimental Ethics: Are We Ready for Utilitarian Cars?" arXiv Preprint.

Bonnefon, Jean-François, Azim Shariff, and Iyad Rahwan (2016) "The social dilemma of autonomous vehicles." *Science*, 352, 1573–1576. Available at: https://doi.org/10.1126/science.aaf2654.

Bykvist, Krister (2010) *Utilitarianism: A Guide for the Perplexed*. London, New York: Continuum.

12 Funding: This paper is related to a dissertation project which I am conducting as part of the interdisciplinary, cooperative PhD programme "Ethics, Culture and Education for the 21st Century" (financially supported by Hanns-Seidel-Foundation, granted by funds from the German Federal Ministry of Education and Research).

Coca-Vila, Ivó (2018) "Self-driving Cars in Dilemmatic Situations: An Approach Based on the Theory of Justification in Criminal Law." *Criminal Law and Philosophy*, 12, 59–82. Available at: https://doi.org/10.1007/s11572-017-9411-3.

Contissa, Guiseppe, Francesca Lagioia, and Giovanni Sartor (2017) "The Ethical Knob: ethically-customisable automated vehicles and the law." *Artificial Intelligence and Law*, 25, 365–378. Available at: https://doi.org/10.1007/s10506-017-9211-z.

Deng, Boer (2015) "Machine ethics: The robot's dilemma." *Nature News*, 523(7558), 24–26. Available at: https://doi.org/10.1038/523024a.

Domingos, Pedro (2015) *The Master Algorithm: How the Quest for the Ultimate Learning Machine Will Remake Our World*. New York: Basic Books.

Driver, Julia (2012) *Consequentialism. New problems of philosophy.* Abingdon Oxon, New York: Routledge.

Ethik-Kommission BMVI (2017) "Automatisiertes und vernetztes Fahren. Bericht Juni 2017." Available at: https://www.bmvi.de/SharedDocs/DE/Publikationen/DG/bericht-der-ethik-kommission.pdf?__blob=publicationFile.

Etzioni, Amitai and Etzioni, Oren (2017) "Incorporating Ethics into Artificial Intelligence." *The Journal of Ethics*, 21, 403–418. Available at: https://doi.org/10.1007/s10892-017-9252-2.

Fagnant, Daniel J. and Kockelman, Kara (2015) "Preparing a nation for autonomous vehicles: opportunities, barriers and policy recommendations." *Transportation Research Part A: Policy and Practice*, 77, 167–181. Available at: https://doi.org/10.1016/j.tra.2015.04.003.

Floridi, Luciano and Sanders, Jeff W. (2004) "On the Morality of Artificial Agents." *Minds and Machines*, 14(3), 349–379. Available at: https://doi.org/10.1023/B:MIND.0000035461.63578.9d.

Gerdes, J. Christian and Thornton, Sarah M. (2016) "Implementable Ethics for Autonomous Vehicles." In: Maurer, Markus, J. Christian Gerdes, Barbara Lenz, and Hermann Winner (eds.) *Autonomous Driving: Technical, Legal and Social Aspects*. Berlin, Heidelberg: Springer Vieweg. Available at: https://doi.org/10.1007/978-3-662-45854-9_5.

Gogoll, Jan and Müller, Julian F. (2017) "Autonomous Cars: In Favor of a Mandatory Ethics Setting." *Science and Engineering Ethics*, 23, 681–700. Available at: https://doi.org/10.1007/s11948-016-9806-x.

Goodall, Noah J. (2014a) "Machine Ethics and Automated Vehicles." In: Meyer, Gereon and Beiker, Sven (eds.) *Road Vehicle Automation. Lecture Notes in Mobility*. Cham: Springer International Publishing. Available at: https://doi.org/10.1007/978-3-319-05990-7_9.

Goodall, Noah J. (2014b) "Ethical Decision Making During Automated Vehicle Crashes." *Transportation Research Record: Journal of the Transportation Research Board*, 2424, 58–65. Available at: https://doi.org/10.3141/2424-07.

Goodall, Noah J. (2016a) "Can you program ethics into a self-driving car?" *IEEE Spectrum*, 53, 28–58. Available at: https://doi.org/10.1109/MSPEC.2016.7473149.

Goodall, Noah J. (2016b) "Away from Trolley Problems and Toward Risk Management." *Applied Artificial Intelligence,* 30, 810–821. Available at: https://doi.org/10.1080/08839514.2016.1229922.

Hevelke, Alexander and Nida-Rümelin, Julian (2015) "Selbstfahrende Autos und Trolley-Probleme: Zum Aufrechnen von Menschenleben im Falle unausweichlicher Unfälle." *Jahrbuch für Wissenschaft und Ethik,* 19, 5–24. Available at: https://doi.org/10.1515/jwiet-2015-0103.

Hubig, Christoph (1993) *Technik- und Wissenschaftsethik: Ein Leitfaden.* Berlin, Heidelberg: Springer.

Lin, Patrick (2014) "The Robot Car of Tomorrow May Just Be Programmed to Hit You." *Wired Magazine.* Available at: https://wired.com/2014/05/the-robot-car-of-tomorrow-might-just-be-programmed-to-hit-you/#.

Lin, Patrick (2016) "Why Ethics Matters for Autonomous Cars." In: Maurer, Markus, J. Christian Gerdes, Barbara Lenz, and Hermann Winner (eds.) *Autonomous Driving: Technical, Legal and Social Aspects.* Berlin, Heidelberg: Springer Vieweg. Available at: https://doi.org/10.1007/978-3-662-45854-9_4.

Liu, Hin-Yan (2016) "Structural Discrimination and Autonomous Vehicles: Immunity Devices, Trump Cards and Crash Optimisation." In: Seibt, Johanna, Marco Nørskov, and S. Schack Andersen (eds.) *What Social Robots Can and Should Do: Proceedings of Robophilosophy 2016/TRANSOR 2016.* Amsterdam: IOS Press.

Liu, Hin-Yan (2017) "Irresponsibilities, inequalities and injustice for autonomous vehicles." *Ethics and Information Technology,* 19, 193–207. Available at: https://doi.org/10.1007/s10676-017-9436-2.

Loh, Wulf and Loh, Janina (2017) "Autonomy and Responsibility in Hybrid Systems: The Example of Autonomous Cars." In: Lin, Patrick, Ryan Jenkins, and Keith Abney (eds.) *Robot Ethics 2.0: From Autonomous Cars to Artificial Intelligence.* Oxford: Oxford University Press.

Lucas Jr., George R. (2013) "Engineering, ethics and industry: the moral challenges of lethal autonomy." In: Strawser, Bradley J. (ed.) *Killing by remote control: the ethics of an unmanned military.* New York: Oxford University Press.

Luetge, Christoph (2017) "The German Ethics Code for Automated and Connected Driving." *Philosophy & Technology,* 30, 547–558. Available at: https://doi.org/10.1007/s13347-017-0284-0.

Mill, John Stuart (1863) *Utilitarianism.* London: Parker, Son, and Bourn.

Millar, Jason (2017) "Ethics Settings for Autonomous Vehicles." In: Lin, Patrick, Ryan Jenkins, and Keith Abney (eds.) *Robot Ethics 2.0: From Autonomous Cars to Artificial Intelligence.* Oxford: Oxford University Press.

Misselhorn, Catrin (2018) *Grundfragen der Maschinenethik.* Stuttgart: Reclam.

Misselhorn, Catrin (2019) "Maschinenethik und Philosophie." In: Bendel, Oliver (ed.) *Handbuch Maschinenethik.* Wiesbaden: Springer VS.

Mladenovic, Milos N. and McPherson, Tristram (2016) "Engineering Social Justice into Traffic Control for Self-Driving Vehicles?" *Science and Engineering Ethics*, 22(4), 1131–1149. Available at: https://doi.org/10.1007/s11948-015-9690-9.

Moor, James H. (2006) "The Nature, Importance, and Difficulty of Machine Ethics." *IEEE Intelligent Systems*, 21(4), 18–21. Available at: https://doi.org/10.1109/MIS.2006.80.

Mulgan, Tim (2001) *The Demands of Consequentialism*. New York: Oxford University Press.

Mulgan, Tim (2006) *Future People: A Moderate Consequentialist Account of our Obligations to Future Generations*. New York: Oxford University Press.

Nyholm, Sven and Smids, Jilles (2016) "The Ethics of Accident-Algorithms for Self-Driving Cars: an Applied Trolley Problem?" *Ethical Theory and Moral Practice*, 19(5), 1275–1289. Available at: https://doi.org/10.1007/s10677-016-9745-2.

Nyholm, Sven and Smids, Jilles (2018) "Automated Cars Meet Human Drivers: Responsible Human-Robot Coordination and The Ethics of Mixed Traffic." *Ethics and Information Technology*, 9, 1–10. Available at: https://doi.org/10.1007/s10676-018-9445-9.

Portmore, Douglas W. (2011) *Commonsense Consequentialism: Wherein morality meets rationality*. New York: Oxford University Press.

Russell, Stuart and Norvig, Peter, eds. (2010) *Artificial Intelligence: A Modern Approach*. New Jersey: Pearson Education Limited.

Santoni de Sio, Filippo (2017) "Killing by Autonomous Vehicles and the Legal Doctrine of Necessity." *Ethical Theory and Moral Practice*, 20(2), 411–429. Available at: https://doi.org/10.1007/s10677-017-9780-7.

Schäffner, Vanessa (2018) "Caught Up in Ethical Dilemmas: An Adapted Consequentialist Perspective on Self-Driving Vehicles." In: Coeckelbergh, Mark, Janina Loh, Michael Funk, Johanna Seibt, and Marco Nørskov (eds) *Envisioning Robots in Society – Power, Politics, and Public Space: Proceedings of Robophilosophy 2018/TRANSOR 2018*. Amsterdam: IOS Press.

Singer, Peter (1979) *Practical ethics*. Cambridge: Cambridge University Press.

Thornton, Sarah M., Selina Pan, Stephen M. Erlien, and J. Christian Gerdes (2017) "Incorporating Ethical Considerations into Automated Vehicle Control." *IEEE Transactions on Intelligent Transportation Systems*, 18, 1429–1439. Available at: https://doi.org/10.1109/TITS.2016.2609339.

Wallach, Wendell and Allen, Colin (2009) *Moral Machines: Teaching Robots Right from Wrong*. New York: Oxford University Press.

Weber, Karsten and Zoglauer, Thomas (2019) "Machinenethik und Technikethik." In: Bendel, Oliver (ed.) *Handbuch Maschinenethik*. Wiesbaden: Springer VS.

The Car's Choice: Illusions of Agency in the Self-Driving Car Trolley Problem

Rebecca Davnall

Abstract

The self-driving car trolley problem has received undue focus in the automation ethics literature. It is unrealistic in two important ways: First, it fails to respect well-established truths of vehicle dynamics relating to tire friction, and second, it misrepresents the information environment that self-driving cars will respond to. Further, the problem induces readers to treat the car as an agent, thereby shielding the agency of car designers and operators from scrutiny. This problem is illustrated through an analysis of the reporting of the first pedestrian fatality caused by a self-driving car, in Tempe, Arizona on March 18th, 2018.

1. Introduction

Scenarios of the following kind have become common in the automation ethics literature:

> A self-driving car carrying a single adult passenger is trundling along an urban road when two children carelessly run out in front of it. The car detects that the children are within its stopping distance; the only way to avoid striking them will be to swerve onto the pavement and hit a bollard, the resulting impact being likely to severely injure the passenger.

Such scenarios have generally been held to be roughly analogous to Philippa Foot's (1967) 'trolley problem', a classic philosophical thought experiment in which an agent must choose between allowing some number of deaths through inaction or directly causing a smaller number of deaths. Machine ethicists and moral philosophers have approached this 'self-driving car trolley problem' in a variety of ways. Some, such as Mark Coeckelbergh (2016), Fiona Woollard (2017) and Jan Gogoll and Julian F. Műller (2017), use it to illustrate general claims about how we should think about the decisions of autonomous systems. Others, such as Patrick Lin (2016) and Noah J. Goodall (2014), take it as

 | DOI:10.30965/9783957437488_013

a demonstration that self-driving cars will inevitably have to make ethical decisions in much the way that humans do. At MIT, a massive project has aimed to 'solve' the dilemma democratically, by collecting information from the general public about what factors they deem relevant to the decision (Awad *et al.* 2018).

While some of the points that have been illustrated using the self-driving car trolley problem are independently sound, the dilemma itself ought to be dispensed with. Not only is it not an accurate representation of the decision-making processes that self-driving cars will go through, but it also distorts the overall ethical picture. The self-driving car trolley problem lures the reader into thinking of the car as a discrete agent, autonomous in the same moral sense that human beings are, rather than as an active product of economic and social conditions. In doing so it obscures corporate and human responsibility for the potentially disastrous detriments that self-driving cars could cause to urban environments and populations.

Section 2 examines the self-driving car trolley problem in more detail and outlines the various ways in which it is unrealistic. Section 3 draws out and critiques the underlying assumptions about agency that are fundamental to the problem. Finally, section 4 applies the methodology outlined in section 3 to the case of the first pedestrian killed by a self-driving car.

2. Solving the Self-Driving Car Trolley Problem

Interviewed in 2016, Andrew Chatham, an engineer from Google's 'X project' on self-driving cars, said: "It takes some of the intellectual intrigue out of the problem, but the answer is almost always 'slam on the brakes'" (in Meder *et al.* 2018: 2). In a similar spirit, Matthias Rolf, Nigel Crook and Jochen Steil defend the claim that: "From a practical ethical perspective, simple breaking [sic] to minimize potential impact might be a better choice than attempting to make a rational choice of who to kill." (2018: 1)

These thoughts, which may seem flippant at first, have yet to receive their due weight in the discussion of autonomous vehicle ethics. The difficulties with the self-driving car trolley problem begin with the fact that it involves a choice between braking in a straight line and swerving while braking. Generally, these courses of action are treated as equivalent from the perspective of what the car should do – that is, the differences between braking and swerving are not treated as morally significant – but this is a grievous mistake.

Car tyres provide a limited amount of grip. That is, for any combination of tyre and road surface, there is a limit on how much force can be transferred between the car's chassis and the road at a given time. Exceed this limit, and the car will start to skid, reducing the effectiveness of any braking and steering that a driver or autonomous control system attempts to do. Both turning and braking use the grip of the tyres to apply forces to the car's chassis – braking in the direction opposite the car's travel and turning perpendicular to it – and these forces are added together when turning and braking at the same time. More turning means either less braking or a higher risk of skidding (Abe & Manning 2009: ch.2).

Less braking, obviously, leads to a higher-speed collision. The relation between collision speed and risk of fatality is not linear (above 50mph, increases in fatality rate with higher speed tail off as they approach 100%), but it is nevertheless true that a higher-speed impact generally has a higher chance of causing serious injury and death than a lower-speed impact. This is particularly true at the relatively low speeds (20–30mph) of urban driving (Rosén & Sander 2009), which is exactly the context in which pedestrians are most likely to step out in front of a self-driving car.

The motion of a car that is skidding is hard to control, especially when the car is attempting to change its course during the skid. A particular risk is that a skidding car will go into a spin, making its course unpredictable to other road users and pedestrians and potentially leading to it hitting whatever it hits something other than head-on. This last possibility is important; as one illustration, D. C. Richards finds that the risk of fatality for the driver of a car that is struck side-on at 30mph is over eight times the risk of a frontal impact at the same speed (2010: 26). Many car safety features are designed in awareness of the fact that the majority of car crashes happen in the direction of the car's travel, not only for the occupants (air bags, 'crumple zones'), but also for pedestrians – the lower and more rounded front ends of modern cars are one of two factors that Richards identifies as major contributors to the drop in fatality rates for pedestrians over the last 40 years (2010: 23).

Further still, sudden lane departures are likely to startle other road users. This can interfere with their reaction times and judgement about how to handle a developing accident situation and may lead to more erratic responses. In this way, a self-driving car that swerves may cause further accidents unrelated to the inevitable collision ahead of it.

Overall, then, swerving is a much riskier course of action than simple straight-line braking. This by itself is not sufficient to mandate the straight-line option in all cases. For example, it seems obvious that if what is directly ahead of the car is a gaggle of schoolchildren, while what is on the pavement is merely a lightweight advertising board or a bush, it would not be preferable for the car to run into the children regardless of the risks inherent in swerving.

However, in setting up the example this way, important questions about the car's *information* environment are overlooked. It is easy to stipulate that a scenario of this kind occurs, but in order to determine what the car should actually do, one must also consider how the car might come to know that it is in such a situation. This is important because the gathering and interpretation of data by autonomous systems always entails risks – of security and privacy, of accuracy and fairness.

Here it is essential to avoid being swept away by excitement about what *might* be technologically possible some day. The self-driving car trolley problem, if it occurs at all in reality, will begin occurring within the next ten years, unless commercial deployment of self-driving cars encounters substantial unforeseen delays. Wherever it is decided that some feature of a hypothetical case ought to affect which outcome is preferred, an account must be given of how *current or imminent* technology can be expected to determine, accurately and consistently in the vast majority of cases, that the feature is present.

The importance of accuracy and consistency cannot be overstated. Decisions about which entity a car ought to strike in a given set of circumstances are decisions about who ought to suffer a life-threatening injury. It is unreasonable to suggest that decisions of this kind can be reached without making implicit commitments about the relative values of different kinds of lives.

As one example, Patrick Lin offers the following case:

> Imagine in some distant future, your autonomous car encounters this terrible choice: it must either swerve left and strike an eight-year old girl, or swerve right and strike an 80-year old grandmother. Given the car's velocity, either victim would surely be killed on impact. If you do not swerve, both victims will be struck and killed; so there is good reason to think that you ought to swerve one way or another. (2016: 69–70)

The question posed by this case is, simply, whether it is the 8-year-old or the 80-year-old who *should* die. If age is to carry this much weight in life-or-death decisions, then it is vital that, whatever difference age is supposed to make, the system that makes the decision makes it based on accurate evaluations of the ages of potential victims. In Lin's case, it is reasonable to imagine an image-recognition system that could tell the difference between an 8-year-old and an 80-year-old, but it is not clear that this generalises.

If it is decided that self-driving cars should always prefer to strike an adult than a child, for example, then self-driving cars must be able to determine the difference in age between a 17-year-old and a 19-year-old, or a 20-year-old and a 21-year-old (and, indeed, possibly what the local definition of adulthood is).

If age makes a scaling or linear difference (e.g. halving the age of a potential victim doubles the preference for avoiding them), then a vast number of these fine distinctions must be made accurately.

Age is one common feature on which self-driving car hypotheticals invite value judgements. The 'Moral Machine' project (Awad *et al.* 2018) offers several more, though by no means an exhaustive list. For example, this project found a statistically significant bias against homeless people – that is, a majority of respondents to their survey deemed it appropriate for a self-driving car to deliberately aim to strike a homeless person rather than a control (i.e. an 'ordinary' adult human). What sensor that might be mounted to a self-driving car would be able to identify that a person is homeless?

Criminals fare even worse in the Moral Machine survey (slightly worse than dogs and only a little better than cats – no other group of humans faced close to the same degree of hostile bias). But criminals, contrary to the legacy of the silent film era, do not conveniently dress in black-and-white-striped jumpers or sport particularly cultivated moustaches. There is no reliable outward sign of someone's being a criminal.

The only way that a self-driving car might plausibly have access to this kind of information would be through networking. Networked driving, in which all cars on the road are in communication with each other and/or a central control system, is a more remote technology than autonomous driving, requiring significant infrastructural changes, but it could in principle allow self-driving cars access to information about the economic circumstances and criminal records of pedestrians. Networked driving would also allow cars to determine the age of other road users and a great deal of other information with more accuracy than car-mounted sensors.

Important questions remain, however. One which there is not space to investigate in detail here is whether self-driving cars should in fact be making such distinctions at all. For example, the bias against criminals found by Awad and colleagues would, if enacted, be tantamount to making self-driving cars extrajudicial executioners. 'Being a criminal' may mean anything from petty theft to mass murder; even the most ardent supporters of capital punishment are unlikely to believe that every act on this spectrum diminishes its agent's right to life.

Another major issue is whether information transferred so freely and in such high amounts can ever be secured. If self-driving cars are to make the discriminations suggested by these cases, they will have relatively little time to gather information between determining that a decision is necessary and actually making the decision. That is, it will not be possible for them to wait until a dilemma situation is occurring to request detailed background information from the network, since there are fundamental limits on how quickly such requests can be answered.

Self-driving cars, in this scenario, will have to have all relevant information on hand as a matter of course. If you are a pedestrian walking alongside a fully networked road, then any criminal record you have will be accessed by every car that passes you. Every car will check that you have a current, registered address, poll your age, perhaps examine your medical record, whether you have any dependents and so on.

No network is perfectly secure; any intensification of the amount of information flying through the air around us invites new malicious interventions. Even if the information could be secured, though, data gathering on this scale is a gross violation of personal privacy. Worse, it is almost inevitable that car operators with 'legitimate' access to this information will repackage it for sale. For example, Phoebe Howard (2018) reports that Ford are already considering the sale of data as a new revenue stream to cover for falling vehicle sales; Jonathan M. Gitlin notes that "selling data is a high-margin business – between 80- and 90-percent profit." (2018)

All these risks must be factored in to any dilemma where details that could not be detected by car-mounted sensors are treated as relevant. Of course, they can all be avoided by ruling such details out of consideration, as the German government already has (BMVI, 2017). Dispense with all this information, and the self-driving car dilemma collapses back into a matter of vehicle dynamics as discussed above.

There are many other risks, some substantial and others marginal, which could be mentioned in this context (for a sample, see Holstein and Dodig-Crnkovic, 2018), but the general case is already strong. All these information risks only matter if self-driving car operators already endorse the risks of swerving in emergencies; if self-driving cars are programmed simply to perform optimal emergency stops, the only risk is that they hit the wrong thing – but even then, only at the lowest possible speed.

3. Illusions of Agency

Apart from the fact that it is not nearly as hard to solve as often claimed, the self-driving car trolley problem has another flaw, one that makes its prevalence in the literature particularly problematic. This is that, bluntly put, it is almost impossible to describe or discuss the problem without treating the car as an agent, an independent being whose actions have meaning or moral significance. This is problematic because it shields the agents and agencies that are *actually* active in determining the effects of self-driving car technology from appropriate moral consideration.

This problem starts at the level of ordinary language. It is hard to characterise the crux of the dilemma without saying things like 'the car must choose' or 'the car must decide' which potential victim to hit. Even when engaging the epistemic issues discussed in the previous section, phrases like 'how the car comes to know' or 'learn' that the dilemma situation has occurred are unavoidable.

Some limited refuge can be found in technological precision. Self-driving cars are likely to be machine learning systems rather than pure algorithm; their software will work by assigning weightings to objects in their environments based on inferences about how likely the object is to be an obstacle (and thus how important it is to avoid). Even here, though, there is a hint of anthropomorphism in the verb 'assigning'; indeed, there would be in any verb we replaced it with. The hint of agency is more a matter of how human – or at least anglophone – listeners respond to certain grammatical cues in verb formation than the specific verbs chosen.

Historically, there is a strong association between agency and the possession of a special kind of moral status, *personhood*. For example, in a classic philosophical treatise on personhood, Harry Frankfurt writes that "one essential difference between persons and other creatures is to be found in the structure of a person's will." (1971: 6)

Dealing specifically with automation, Amitai and Oren Etzioni argue:

> 'Autonomous' is a highly loaded term because in liberal democracies it is associated with liberty, self-government, and individual rights. To violate someone's autonomy is considered a serious ethical offense [...] However, cars and other machines are [...] unfeeling tools made to serve humans. *There is nothing morally objectionable about overriding their choices, or making them toe the line.* One does not violate their dignity by forcing them to make choices within the boundaries set by their programmers. (2017: 410, italics in original)

This distinction is not a mere technicality, nor is its philosophical import exhausted by dispensing with questions of 'robot rights'. There are almost as many accounts of what agency is, metaphysically speaking, as there are philosophers, but what philosophers tend to agree on is that it is all and only genuine agents that can be held to be *responsible* for what they do. That is, to be an agent is to be the kind of thing that can be blamed or praised for its movements and effects in the world.

The self-driving car trolley problem, then, nudges our intuitions – by making agency-implying language unavoidable – towards blaming or praising the car for what it hits. There is of course a perfectly natural sense in which a car may be praised; one can say, for example, that a certain car is very efficient, or comfortable. But it would be strange to call a car noble or wicked in the central sense of either word, and that is exactly what the insistence that self-driving cars will make ethical decisions in dilemma situations asks us to accept. A car that hits a pedestrian rather than a wall, on this view, is not merely a malfunctioning device but wicked.

This framing matters because blaming one entity in a scenario invites us to excuse others. For example, T. M. Scanlon writes:

> Conservatives often say that there are two responses to social ills such as urban crime, teenage pregnancy, and drug abuse. One response is to see these as cases of people behaving badly, in which case the remedy is that they should be made to do better. The other response is to see these ills as the products of social conditions such as poverty and inadequate educational systems, in which case the remedy is to change those conditions. To take the latter approach, conservatives say, is to deny that individuals are responsible for what they do. (2009: 200)

This conservative rhetorical strategy often serves to defend conservative politicians whose policies *produce* social conditions of poverty, poor education and so on. In just the same way, the car's choice can be set against the decisions of the automotive engineers, urban planners and legislators who establish the conditions in which the choice occurs, thereby determining what options are actually available to not only the car but the pedestrians and other road users around it.

The point of this illustration is not to allege that a particular rhetorical strategy conceals or suppresses the truth of the matter. Instead it is to complicate the simple, intuitive idea of agency embedded in (English) grammar. Following Karen Barad (2007), the individual agency of the car (or the drug-taker or pregnant teenager in Scanlon's examples) can be called *complementary* with the agency of the engineers and policy-makers, in a sense drawn from the theoretical physics of Niels Bohr.

Complementarity is sometimes an obscure concept. Heisenberg's uncertainty principle, in which complementarity originates, is conventionally stated as an epistemic limit – for example, Wikipedia describes the uncertainty principle as "asserting a fundamental limit to the precision with which certain pairs of physical properties of a particle, known as complementary variables, such as position and momentum, can be *known*." (2018, italics added)

However, the limit is actually ontological; quantum systems have fundamental trade-offs affecting the fixity of their properties. A quantum phenomenon never has both a fully determinate position and a fully determinate momentum at the same time. The insight that Barad draws from Bohr is to understand determinate properties as possible only within the context of an apparatus designed to manifest them. So, since a different kind of apparatus is needed to observe or draw out the location of a quantum phenomenon than to draw out its momentum – and because these two kinds of apparatus inevitably obstruct one another's operations – position and momentum are complementary. That is, the position of a particle only exists as a determinate quantity in the context of a framework that excludes determination of the particle's momentum.

So it is, Barad argues, for agency. Using this theory, the self-driving car trolley problem can be read as an apparatus that manifests agency on the part of the car; the conservative rhetoric of 'individual responsibility' is likewise an apparatus that manifests agency on the part of poor people, drug addicts and pregnant teens. At the same time, the agency of designers, planners and policymakers is excluded, since the agency of individuals is offered within these apparatuses as fully explaining what has happened.

What is crucial to understand here is that there is no one ultimately correct apparatus – this is the implication of the physics, which Barad insists is not merely providing a metaphor. There is a substantive agential equivalent of the uncertainty principle: there is no fact of the matter about which apparatus, which agency, is genuine and which illusory. Different apparatuses do different things and thereby serve different interests, but the question of which interests it is right to serve cannot be answered in terms of some underlying truth. Instead, the apparatuses themselves must be scrutinised and the interests they embed evaluated. The following section provides an example of this kind of analysis.

4. Apparatuses and Agencies

A clear – and chilling – example of a conflict between apparatuses is revealed in coverage of the March 18th collision, in Tempe, Arizona, of an Uber-operated self-driving car with a pedestrian, who died as a result of her injuries. Initial reports blamed the pedestrian and explicitly exonerated Uber; in an interview with the San Francisco Chronicle, Tempe police chief Sylvia Moir said:

> [I]t's very clear it would have been difficult to avoid this collision in any kind of mode (autonomous or human-driven) based on how [the pedestrian] came from the shadows right into the roadway [...] I suspect preliminarily it appears that the Uber would likely not be at fault in this accident. (Said 2018)

When video footage from the car's cameras was released a few days after the crash, showing the pedestrian clearly visible before stepping into the car's path, attention shifted to the 'safety driver' that Uber employed. The safety driver is there to take control of the car in the event of its control systems failing to respond appropriately to an on-road situation. In the video released by Uber, the safety driver can be seen looking down at something in her lap in the seconds leading up to the crash.

In an article published four days after the crash, Bloomberg writers Ryan Beene, Alan Levin and Eric Newcomer quote numerous experts discussing the failures of the Uber car's sensor technologies that must have contributed to the crash (for example, Bryant Walker Smith says the car's "lidar and radar absolutely should have detected [the pedestrian]"). Despite this, the article is titled "Human Driver Could Have Avoided Fatal Uber Crash, Experts Say" (2018). Other writers, for example Jeremy Kaplan and associates (2018), bring up in this context the irrelevant detail that the safety driver had a prior criminal conviction (entirely unrelated to driving).

A very different picture is painted by the preliminary report of the US National Transportation Safety Board's investigation into the crash. Uber's car was a modified Volvo XC90, equipped as standard with advanced safety features including an automated emergency stop system that might have been sufficient to prevent the crash. Uber deactivates all these features when its cars are in fully autonomous operation. Per the NTSB report,

> According to Uber, emergency braking maneuvers are not enabled while the vehicle is under computer control, to reduce the potential for erratic vehicle behaviour. The vehicle operator is relied on to intervene and take action. The system is not designed to alert the operator. (2018: 2)

The last two sentences here are particularly important. Matthew Tucker and colleagues find that, in an experimental system, subjects assigned to monitor the behaviour of a simulated self-driving car in the way that Uber's safety driver was could maintain full attentiveness only for an average of around twenty minutes, and say their "data indicates a decreased performance in long-distance driving" (2015: 7) after this point, particularly focussed on how far the car could stray out of its lane before the subject noticed and made a steering correction.

In other words, the Uber's safety systems were completely inadequate. Those provided by the car's manufacturer were deactivated and replaced with a system that relied on a superhuman driver that the car had no ability to alert. In effect, both car and driver were set up to lose; what was being asked of the system on the road was impossible.

Looking at the Tempe crash through the apparatus suggested by the self-driving car trolley problem, which considers how the car ought to respond to its sensor inputs, is therefore fruitless. The car's option for an effective response – using its Volvo-designed automated emergency stop system – had already been closed off. Similarly, the apparatus suggested by the press coverage, which clearly sought an individual human being at the scene of the accident to blame, can only deliver the safety driver, who was already under the effects of the attention impairment quantified by Tucker and colleagues.

The important questions stem from the apparatus indicated by the NTSB report. Uber claimed its decision to deactivate the Volvo's native safety features was to 'reduce erratic vehicle behaviour'. Amir Efrati comments:

> There's a reason Uber would tune its system to be less cautious about objects around the car: It is trying to develop a self-driving car that is comfortable to ride in. By contrast, people who have written in Waymo and General Motors' Cruise vehicles say the ride can be jerky, with sudden pumping of the brakes even though there's no threat in sight. That's often the result of vehicles reacting to false positives, when the sensor systems think they see moving objects that aren't really there, or to objects that probably aren't going to be a problem but are treated as such. (Efrati 2018)

While Uber claims that it disabled the Volvo safety systems because erratic vehicle behaviour is dangerous, its decision to do so clearly contributed to the production of this accident. A single accident does not invalidate Uber's claim, of course, but the NTSB report does not credit them with producing any evidence for it. Meanwhile, as Efrati shows, a less safe system provides a clear competitive advantage to Uber over Waymo and Cruise.

This apparatus, then, opens up the question of the extent to which Uber as a company have prioritised profit over pedestrian safety. This is a question of clear moral import, and one that is substantively excluded by attempts to blame the safety driver or the pedestrian herself – for example, James Arrowood, a lawyer with a specialism in self-driving cars, claimed that the fact that the pedestrian was not crossing at a designated crossing point would make her 'negligent' in the eyes of the law and serve as a potential line of defence for Uber (Randazzo, 2018).

In this context, other decisions by Uber become conspicuous. Why did Uber decide to use SUVs for its self-driving car prototypes when the rise in SUV use has been linked to a 46% increase in pedestrian fatalities since 2009 (Lawrence et al. 2018)? And why did they make the decision in late 2017 to stop using a second safety operator in test cars, which safety operators themselves say helped maintain vigilance and attentiveness (Bliss 2018)?

Apparatuses like the self-driving car trolley problem, or the inquiry into the conduct of the safety operator in the Tempe crash, obscure these questions and protect large-scale corporate and systemic decisions from accountability. Given that the self-driving car trolley problem is itself solvable, it is past time it was dispensed with in the academic literature.

5. Bibliography

Awad, Edmond, Dsouza, Sohan, Kim, Richard, Schulz, Jonathan, Henrich, Joseph, Shariff, Azim, Bonnefon, Jean-François, and Rahwan, Iyad (2018) "The Moral Machine experiment." *Nature* 563, 59–78.

Beene, Ryan, Levin, Alan, and Newcomer, Eric (2018) "Human Driver Could Have Avoided Fatal Uber Crash, Experts Say." *Bloomberg*, [online] Available at <https://www.bloomberg.com/news/articles/2018-03-22/video-said-to-show-failure-of-uber-s-tech-in-fatal-arizona-crash> [accessed on 12 December 2018].

Bliss, Laura (2018) "Former Uber Backup Driver: 'We Saw This Coming'". *Citylab*, [online] Available at <https://www.citylab.com/transportation/2018/03/former-uber-backup-driver-we-saw-this-coming/556427/> [accessed on 12 December 2018].

Coeckelbergh, Mark (2016) "Responsibility and the Moral Phenomenology of Using Self-Driving Cars." *Applied Artificial Intelligence* 30(8), 748–757.

Efrati, Amir (2018) "Uber Finds Deadly Accident Likely Caused By Software Set to Ignore Objects On Road." *The Information*, [online] Available at <https://www.theinformation.com/articles/uber-finds-deadly-accident-likely-caused-by-software-set-to-ignore-objects-on-road> [accessed on 12 December 2018].

Etzioni, Amitai and Etzioni, Oren (2017) "Incorporating Ethics into Artificial Intelligence." *Journal of Ethics* 21, 403–418.

Federal Ministry of Transport and Digital Infrastructure (BMVI) Ethics Commission (2017) *Automated and connected driving*. [online] Available at <https://www.bmvi.de/SharedDocs/EN/publications/report-ethics-commission.html?nn=187598> [accessed on 11 December 2018].

Foot, Philippa (1967) "The Problem of Abortion and the Doctrine of the Double Effect." *Oxford Review* 5, 5–15.

Frankfurt, Harry (1971) "Freedom of the Will and the Concept of a Person." *The Journal of Philosophy* 68(1), 5–20.

Gitlin, Jonathan M. (2018) "Car companies are preparing to sell driver data to the highest bidder." *ars technica*, [online] Available at <https://arstechnica.com/cars/2018/02/no-one-has-a-clue-whats-happening-with-their-connected-cars-data/> [accessed on 12 December 2018].

Gogoll, Jan and Müller, Julian F. (2017) "Autonomous Cars: In Favor of a Mandatory Ethics Setting." *Science and Engineering Ethics* 23(3), 681–700.

Goodall, N. J. (2014) "Machine Ethics and Automated Vehicles." In: Meyer, G. and Beiker, S. (eds.), *Road Vehicle Automation*. Switzerland: Springer.

Holstein, Tobias and Dodig-Crnkovic, Gordana (2018) "Avoiding the Intrinsic Unfairness of the Trolley Problem." In: FairWare '18, *Fairware '18: IEEE/ACM International Workshop on Software Fairness, May 29, 2018, Gothenburg, Sweden*. New York: ACM.

Howard, Phoebe Wall (2018) "Data could be what Ford sells next as it looks for new revenue." *Detroit Free Press*, [online] Available at <https://eu.freep.com/story/money/cars/2018/11/13/ford-motor-credit-data-new-revenue/1967077002/> [accessed on 12 December 2018].

Kaplan, Jeremy, Glon, Ronan, Edelstein, Stephen, and Chang, Lulu (2018) "Deadly Uber crash was 'entirely avoidable' had the driver not been watching Hulu." *Digital Trends* [online] Available at <https://www.digitaltrends.com/cars/self-driving-uber-crash-arizona/> [accessed on 12 December].

Lawrence, Eric D., Bomey, Nathan and Tanner, Kristi (2018) "Death on foot: America's love of SUVs is killing pedestrians." *Detroit Free Press/USA Today network* [online] 1 July. Available at: <https://eu.freep.com/story/money/cars/2018/06/28/suvs-killing-americas-pedestrians/646139002/> [accessed on 11 December 2018].

Lin, Patrick (2016) "Why Ethics Matters for Autonomous Cars." In: Maurer, M., Gerdes, J. C., Lenz, B. and Winner, H. (ed.) *Autonomous Driving: Technical, Legal and Social Aspects*. Berlin: Springer.

Meder, Björn, Fleischhut, Nadine, Krumnau, Nina-Carolin and Waldmann, Michael R. (2018) "How Should Autonomous Cars Drive? A Preference for Defaults in Moral Judgments Under Risk and Uncertainty." *Risk Analysis* (in press).

Randazzo, Ryan (2018) "What went wrong with Uber's Volvo in fatal crash? Experts shocked by technology failure." *azcentral*, [online] Available at <https://eu.azcentral.com/story/money/business/tech/2018/03/22/what-went-wrong-uber-volvo-fatal-crash-tempe-technology-failure/446407002/> [accessed on 12 December 2018].

Richards, D. C. (2010) *Relationship between Speed and Risk of Fatal Injury: Pedestrians and Car Occupants.* London: Department for Transport.

Rolf, Matthias, Crook, Nigel and Steil, Jochen (2018) *From social interaction to ethical AI: a developmental roadmap*. IEEE Conference: Development and Learning and Epigenetic Robotics, Tokyo, Japan. [online] Available at: <https://www.researchgate.net/publication/327446337_From_social_interaction_to_ethical_AI_a_developmental_roadmap> [accessed 11 December 2018].

Rosén, Erik and Sander, Ulrich (2009) "Pedestrian fatality risk as a function of car impact speed." *Accident Analysis and Prevention* 41: 536–542.

Said, Carolyn (2018) "Exclusive: Tempe police chief says early probe shows no fault by Uber." *San Francisco Chronicle*, [online] (last updated: 26 March 2018). Available at <https://www.sfchronicle.com/business/article/Exclusive-Tempe-police-chief-says-early-probe-12765481.php> [accessed 12 December 2018].

Scanlon, T. M. (2010) *Moral Dimensions: Permissibility, Meaning, Blame.* Cambridge, MA: Belknap Press.

Tucker, Matthew B., Aubert, Miles, de Almeida, Vinicius Sampaio, Clamann, Michael and Cummings, Mary L. (2016) "Detection of Attentional State in Long-Distance Driving Settings Using Functional Near-Infrared Spectroscopy." In: *Transportation Research Board 95th Annual Meeting.* Washington DC, United States, 10–14 January 2016. Washington DC: Transportation Research Board.

Wikipedia contributors. (2018) "December 9. Uncertainty Principle." In: *Wikipedia, The Free Encyclopedia.* [online] Available at: <https://en.wikipedia.org/w/index.php?title=Uncertainty_principle&oldid=872856017> [accessed 11 December 2018].

Woollard, F. (2017) *Driverless Cars and Ethics Without Algorithms.* 6 November, University of Liverpool.

Why Machines That Talk Still Do Not Think, and Why They Might Nevertheless be Able to Solve Moral Problems

Lukas Brand

Abstract

The question whether automata can be intelligent or even have a mind has been an ongoing controversy for centuries. Even though we intuitively agree with the positions of Descartes and Searle that automata do not have a mind nor understand what they are doing the way we do, we do nevertheless observe the increasing occurrence of automata performing several tasks that are considered to require mental abilities (e.g. AlphaZero, Google Duplex, and Project Debater). The article outlines the different positions in the controversy and argues for the thesis that, based on recent advances in artificial intelligence, automata could in principle perform other complex tasks, and in particular moral decision-making as well.

1. Introduction

Having command of a language has always been "an index of human intelligence, of what distinguishes man from animals."[1] For Descartes the ability to use a language was an indicator that a particular body possesses a mind and that it therefore could not be a mere automaton.[2] Alan Turing, however, stated in 1950 that a computer despite not having an equivalent to a human mind could nevertheless be able to mimic the human speech-performance indistinguishably. With his imitation game scenario, Turing simply by-passes the ontological question and, in accordance with Descartes, states that a behaviouristic test could lay the basis for answering the question "Can machines think?". The imitation game is a simple test scenario that replaces notoriously unsatisfying essential definitions of "thinking".[3] Turing states that the machine that is able to have any conversation in a way that is practically indistinguishable from an

1 Chomsky 2006: 9.
2 Chomsky 2006: 6.
3 See Turing 1950: 433–434.

 | DOI:10.30965/9783957437488_014

intelligent human being should, at least prima facie, be classified as intelligent in a metaphorical way.[4] Turing saw himself as justified in this point of view, even though he admitted that he could not know in principle whether such a machine would in fact have consciousness, in particular: feel emotions, be able to produce poems itself, or maybe even have an immaterial soul that might derive from God himself.[5]

Five years later McCarthy laid the basis for an artificial intelligence (AI) project. He considered the possibility that the human intellect might have a form that could be simulated by a computer program. He described the basic idea for the Dartmouth conference in the funding request:

> The study is to proceed on the basis of the conjecture that every aspect of learning or any other feature of intelligence can in principle be so precisely described that a machine can be made to simulate it. An attempt will be made to find how to make machines use language, form abstractions and concepts, solve kinds of problems now reserved for humans, and improve themselves.[6]

When the project showed little success over the course of the next decades, philosophers like Searle began to focus more on the essential differences between human intelligence and AI as well as the advantages of the former over the latter. Today, we find ourselves in a paradoxical situation. We consider artificial programs to be inferior to our human mental abilities. However, they beat human experts in a rapidly growing variety of intellectual challenges.

Turing had suggested that "a very abstract activity, like the playing of chess" and other games considered intellectually challenging for humans would be a good starting point for the development of machines that would fulfil sufficient conditions for intelligence.[7] Since computer programs outperform humans in more and more types of games with rising levels of complexity form more than two decades,[8] the satisfaction condition for intelligence is now again seen in the ability to engage in a conversation in a meaningful way, just as Turing had suggested for computers and Descartes for humans in the first place.[9]

Nowadays more and more programs, such as Siri, Alexa, Cortana, and Watson, are able to produce speech, even though these programs are still subject to

4 Turing 1950: 442: "I believe that at the end of the century the use of words and general educated opinion will have altered so much that one will be able to speak of machines thinking without expecting to be contradicted."

5 Turing 1950: 443 ff: "We are […] instruments of His will providing mansions for the souls that He creates."

6 McCarthy, et al. 1955: 2.

7 Turing 1950: 460.

8 See Silver, et al. 2017a; 2017b; Jaderberg 2018.

9 See Turing 1950, 433–435; 2006: 5 ff.

several limitations, from their overall performance being easily distinguished from that of a human speaker to their internal systems not being able to understand semantics. However, they become quite sophisticated in accomplishing a variety of tasks, on a continuously rising performance level. They improve in computing context information in their performances[10] and produce voice and intonation in a way that in some cases are almost indistinguishable from those of a human.[11] As we increase the independence of our sophisticated computer programs, chances rise that these programs need to solve moral conflicts, though not on a regular basis but rather in situations unforeseen by their designers. The recent advances in computer intelligence in the field of conversation making as well as problem solving in different types of games might, in fact, help AI-systems one day to solve moral decision-making problems as well. This might be possible through the same mode of function even though the AI-system still lacks consciousness, and it's moral reasoning might not be based on explicit rules. The present article focusses solely on the implications for moral decision-making that derive from the functioning of so-called smart assistants that are already present today.

In the present paper I am basically making three distinct steps: (Section 2) I try to elaborate how a modern AI-system works and (Section 3) consider the arguments with regard to its intelligence and intentionality in lights of speech performance in order to (Section 4) show that machines might be able to solve moral problems analogously regarding the described mode of function as well as the limiting philosophical aspects.

2. Enter Google Duplex

Most of today's state of the art automated phone systems are still a long way from passing the Turing test.

> It is often frustrating having to talk to stilted computerized voices that don't understand natural language. In particular, automated phone systems are still struggling to recognize simple words and commands. They don't engage in a conversation flow and force the caller to adjust to the system instead of the system adjusting to the caller.[12]

10 See DeVynck 2018.

11 See Leviathan and Matias 2018; van den Oord, et al. 2016.

12 Leviathan and Matias 2018.

For that reason, Google has dedicated itself to a long-standing goal of human-robot interaction, tried to end this situation, and enable people to have a more natural conversation with its smart Assistant.

Against this background, in May 2018 Google introduced *Google Duplex*. According to its programmers, Duplex conducts "natural conversations to carry out 'real world' tasks over the phone."[13] It is intended to complete only very specific tasks such as scheduling appointments with a hair salon or booking a reservation in a restaurant and could therefore not pass the Turing test in its general sense. Since Duplex makes the conversation sound surprisingly natural it allows people nevertheless to speak as they would to another person. In fact, the software is so good at imitating a human caller's voice on the one hand and analysing the conversation partner's spoken words and implicit intentions on the other that it is said to pass a restricted Turing test in its limited context. Therefore, right after the introduction, Google faced fierce criticism and had to assure the public that Duplex would from now on disclose its identity as an artificial assistant when calling businesses on behalf of their clients.[14]

Leaving ethical questions aside, Duplex' function is limited to arranging appointments over the phone and does not by design engage for example in a philosophical debate about any topic suggested by the human on the other end of the line. When it comes to the variation of conversational topics, and philosophical debate in particular, however, the capabilities of IBMs new AI could complement those of Duplex: *Project Debater* is based on IBM's Watson, who drew attention in 2011 by winning the game Jeopardy against human contestants. Unlike Watson, the algorithm of Project Debater is now able to crawl open source material on the internet, e.g. academic and news articles, formulate arguments on various topics, and respond to its opponent's arguments in order to participate in formal debate. Against this technical background, Debater was able to convince an audience of its "own" point of view, although we would not attribute to him any explicit let alone conscious intention – distinguishable from the ends of its programming – in doing so: "the audience of mostly journalists and analysts said the robot scored better than the human

---

13 Leviathan and Matias 2018.

14 Statt 2018; When Google Duplex rolled out to some Pixel phones in the US in November 2018, a video recorded by VentureBeat appeared on YouTube and went viral, seemingly showing the performance of the new technology in the real world (Wiggers 2018). Although everything seemed to work well, Google told VentureBeat a few days later that the recorded call had actually been answered by a human operator. This explained why the caller did not identify himself by saying "this is the Google Assistant calling," as Google had advertised it earlier that year. The person at the phone simply said that he was "calling from Google" causing some irritation since the journalist assumed to speak to an artificial agent and expected him to identify itself as such.

debaters on enriching their knowledge. But the human debaters scored better on delivery, perhaps because the robot sounded, well, robotic."[15]

Nevertheless, Debater does not pass the Turing test either, for sometimes it detours into strange aspects of the given topic or makes "a joke at not the right moment". Apart from that, Debater clearly identifies itself as a robot or computer through its answers, stating for example that, as a robot, it has no blood or could not have children.[16] Even though most AI-developers no longer try to program AI analogous to human intelligence, we see AI performing more and more like ourselves. But are we then using the term "intelligence" rightly to describe artificial programs?

The modern AI-systems in question basically rely on machine learning algorithms that allow them to extract patterns from raw data in order to make decisions that appear subjective. In the centre of this AI approach is a brain inspired multilayer algorithm called artificial neural network (ANN), that maps input values to output values.[17] The function between the two sets of values is learned by the ANN from experience in the sense that "a computer program is said to learn from experience *E* with respect to some class of tasks *T* and performance measure *P*, if its performance at tasks in *T*, as measured by *P*, improves with experience *E*."[18] Even though Turing did not refer to the invention of the artificial neuron, he briefly described the basic idea that underlies the core algorithm that leads to the success of ANNs today:

> The machine has to be so constructed that events which shortly preceded the occurrence of a punishment-signal are unlikely to be repeated, whereas a reward-signal increased the probability of repetition of the events which led up to it.[19]

Turing especially pointed out that this method would not need any feelings on the part of the machine to be functional.[20] When we look into the operating function of Google Duplex and Project Debater it is pure mathematics.

During the development process, Google identified and tried to conquer several challenges to conducting natural conversations: "natural language is hard to understand, natural behaviour is tricky to model, latency expectations

15 DeVynck, et al. 2018.
16 See DeVynck, et al. 2018.
17 See Goodfellow et al. 2016: 2 ff.
18 Mitchell 1997: 97.
19 Turing 1950: 457.
20 Turing 1950: 457: "These definitions do not presuppose any feelings on the part of the machine."

require fast processing, and generating natural sounding speech, with the appropriate intonations, is difficult."[21]

Duplex sounds natural thanks to advances in the development of AI regarding *understanding*, *interacting*, *timing*, and *speaking*. A speech recognition network transforms the spoken words of the person on the other end into text, thus making it easier to process for the core network. The core of Duplex is another ANN that is trained to cope with most of the challenges mentioned. The ANN is trained on a set of anonymised phone conversation data given in text form. Along with the transcribed dialog, the network receives audio information and all available context information, such as the history and parameters of the conversation, e.g. the desired service, or the current time of day.[22] After the core algorithm has generated a written answer, Google's generative model WaveNet transforms the text back into speech. Despite WaveNet being outstanding in itself by not only simulating a human voice but mimicking "the acoustics and recording quality, as well as the breathing and mouth movements" of a human speaker,[23] Duplex sounds even more natural by incorporating speech disfluencies (e.g. "hmm"s and "uh"s) whenever it wants to signalise a wait or tries to hold the conversation while processing the data. This way it sounds just like someone who is gathering their thoughts. Using these disfluencies indeed makes the conversation sound even more familiar and natural.[24]

Depending on the context, different meanings may apply to the same sentence during the course of a longer conversation: "When booking reservations, 'Ok for 4' can mean the time of the reservation or the number of people."[25] The relevant context information however might be in an earlier stage of the conversation. What Duplex needs to say in a given moment is a mathematical function that depends on both the task and the stage of the conversation. Apparently, there are some common conversation practices that Duplex needs to adapt to, in order to sound natural. As a speech community we share implicit protocols that Google discovered during research, such as (1) *elaborations*: sometimes statements need to be reformulated in broader detail, *e.g.* "For next Friday" – "For when?" – "For Friday next week, the 18th.", (2) *syncs*: out of context, we ensure the person on the other end is still listening, e.g. "Can you hear me?", (3) *interruptions*: when information flow is too fast we ask to stop and repeat, e.g. "The number is 212-" "Sorry, can you start over?", and (4) *pauses*:

21 Leviathan and Matias: 2018.
22 See Leviathan and Matias 2018.
23 van den Oord, et al. 2016: 5.
24 Find some examples in the referenced blog article Leviathan and Matias 2018: https://ai.googleblog.com/2018/05/duplex-ai-system-for-natural-conversation.html.
25 Leviathan and Matias 2018.

situations may ask for an interruption of the conversation when our attention is drawn to another task, e.g. "Can you hold? – (Pause) – Thank you!" with a different meaning for a pause of 1 second or 2 minutes.[26]

3. Machines That Speak Still Do Not Think

Turing's point of view that a program like Duplex, who's performance is indistinguishable from that of a human, can be seen as a thinking machine was challenged by John Searle in 1980 with his Chinese Room argument.[27] Searle argued against Turing as well as the strong AI hypothesis of McCarthy et al. according to which an "appropriately programmed computer really *is* a mind, in the sense that computers given the right programs can be literally said to *understand* and have other cognitive states".[28] He took the programs into account that had been written by that time, but over the course of his article generalised it about ANNs as well as future embodied robots. Searle pointed out the significant difference between the processing of meaningful sentences with proper syntax indistinguishable in practice from a human speaking and the ability to understand the semantic meaning of the topic spoken about or the words used. He showcased his argument in form of a thought experiment that can be presented as follows:[29] Searle sits in a room and has with him nothing but a book written in Chinese and a book of rules, written in English. He knows no Chinese, neither written nor spoken, but understands English as well as any other native English speaker. The rules instruct him how to produce and put out certain Chinese symbols in a distinct order in response to any batch of Chinese symbols that are now given to him through an input-gate. He follows the rules without making any mistake and produces Chinese symbols that he then hands out through an output-gate.

What he does not know is that the Chinese book contains *stories*, the batch of Chinese symbols are *questions* about these stories, and the symbols Searle gives back in response to them are *answers to the questions*. The rules in English however can be seen as a program. After some time, Searle knows all the rules of the program by heart and the program itself may be so good that, from the external point of view, his answers to the questions are absolutely indistinguishable in practice from those of a native Chinese speaker. He would by

26 See Leviathan, Matias 2018.
27 Searle 1980, pp. 417–457.
28 Searle 1980, p. 517.
29 See for the following Searle 1980: 417 f.

definition pass a Chinese version of Turing's imitation game. Everyone outside the room, who speaks Chinese and is just looking at the answers will accept them as syntactically and semantically correct responses to the questions. From the internal point of view, however, Searle produces the answers by performing computational operations on formally specified, uninterpreted, symbols. In the thought experiment, he is simply an instantiation of the (computer) program in his rule book. To Searle it seemed "quite obvious in the example that I do not understand a word of the Chinese stories. I have inputs and outputs that are indistinguishable from those of a native Chinese speaker, and I can have any formal program you like, but I still understand nothing."[30]

The difference between performing a certain language as an output and understanding the semantic meaning of the words, however, is in line with modern linguistic theory. Noam Chomsky distinguishes *linguistic competence*, knowing and understanding the meaning of words in a speech-community as a mental structure, and thus being able "to produce a signal with an intended semantic interpretation" from *linguistic performance*, "the actual observed use of language".[31] Turing had rightly assumed that opinions would differ as to the appropriateness of the substitution of the original problem "can machines think?" with the imitation game.[32] Considering Searle's argument, the inference from the human-like linguistic performance to the intellectual capability of the speaker – implied by Turing as well as Descartes, who used it to determine whether some creature has a mind like ours – is invalid.[33] Apparently, Turing expected this sort of argument, though he considered any further consideration unimportant given his test scenario:

> May not machines carry out something which ought to be described as thinking but which is very different from what a man does? This objection is a very strong one, but at least we can say that if, nevertheless, a machine can be constructed to play the imitation game satisfactorily, we need not be troubled by this objection.[34]

Even though Turing seems well aware of the ontological differences between his hypothetical machine and an actual intelligent human being, at this point he also seems to underestimate the importance of the ontological question that troubles the philosophy of AI in general and Searle in particular from that time ongoing. Turing might nevertheless agree with Searle and Descartes that a machine might not necessarily be conscious. At least he admits that there

30 Searle 1980: 418.
31 Chomsky 2006: 4,102 ff.
32 See Turing 1950: 442.
33 Brooks 2015.
34 Turing 1950: 435.

is indeed a mystery about consciousness, for instance, when it comes to the question where to localise it. On the other hand, he does not see why it would be necessary to solve these mysteries "before we can answer the question with which we are concerned in this paper."[35] Turing assumes that those who believe that a machine can never be conscious, until it "can write a sonnet or compose a concerto because of thoughts and emotions felt, and not by the chance fall of symbols", would rather abandon this objection and accept the imitation game as common ground than be forced into the solipsist position.[36]

Nevertheless, in the end Turing states something against our intuition, namely that an automaton could be called intelligent (or has a mind), just because it performs as if it were intelligent (or had a mind) indeed. If we want to stick with this intuition also shared by Searle and Descartes, that a machine cannot be intelligent (or does not have a mind), we then have to nevertheless abandon Descartes belief that every entity that speaks already has a mind. If we then take Chomsky's differentiation of performance and competence into account and distinguish the speech performance from the mental process of producing meaningful words intentionally, we reach to an even deeper insight: "sentences have an intrinsic meaning determined by linguistic rule" (derived intentionality) while a person that is in command of a language (linguistic competence) "has in some way internalized the system of rules that determine both the phonetic shape of the sentence and its intrinsic semantic content"[37] and thus possesses intrinsic or original intentionality. There is no difficulty in constructing systems that associate sound (or any other sort of formal symbol) and meaning in some definite way in accordance with the linguistic rules and who's output would not with a priori necessity be a human language[38] even though it is indistinguishable from such, therefore having derived intentionality. Irrespective the producers' ontology his linguistic performance does not imply anything about his competence.

What we know about Duplex (or Watson, or a sophisticated combination of both) by now is, that it is a computer program and able to pass the Turing test. However, it is neither intelligent nor does it understand what it does. The existence of a particular program like this, with properties essential to any program, however, contradicts the possibility that any computer program that performs sufficiently good in any form of imitation game understands what it says, is intelligent, or moral, or whatever you like. That is, based on the assumption that any essential properties that apply to an arbitrarily chosen computer

35 Turing 1950: 447.
36 Turing 1950: 445 ff.
37 Chomsky 2006:102.
38 See Chomsky 2006: 103.

program apply to any other computer program,[39] and taking for granted that Searle's argument contradicts Turing's implication from the performance to the competence of the machine, no machine that only by its program is sophisticated enough to pass the Turing test would be able to understand the meaning of spoken words in the way humans do, and therefore it could not have intrinsic intentionality either. Since intelligence in Searle's view – or competence, in Chomsky's words – requires the ability to understand the meaning of words machines could not rightly be called "intelligent".

Looking at Duplex it is therefore obvious, on the other hand, that a machine that is an appropriate instantiation of a computer program needs no understanding of words or intrinsic intentionality, to speak, or at least to engage in a conversation in a human language in a meaningful way. Whether Duplex is already market-ready or not, it proves that it is possible to train a computer program to engage in a real-world conversation, as if it were a human agent. That is to say: Machines demonstrate a speech performance that, in particular contexts, is at least meaningful to humans in some way. They do so – as we may assume – without having conscious awareness of the meaning of their words, the situation they find themselves in, nor, at least knowingly, that they do something rightly called "making conversation" the way we do. By Searle's argument, they lack linguistic competence. In Turing's terms: machines still cannot think.

Denying Descartes' point of view that the ability to use a language is unique to intelligent human beings like ourselves, sounds like an insult to humanity's intellectual capabilities. With the rise of sophisticated smart assistants with features similar to those of Google Duplex and Project Debater our own intellectual capacities seem to lose significance. Considering all the arguments we have to (a) either abandon Searle's/Descartes' point of view instead of Turing's and redefine our concept of understanding to something like 'being capable of using the right signals in a more or less appropriate manner and context' (already being an *idealised speaker/hearer* in Chomsky's terms) or (b) accept that there are multiple realisations of understanding and that a machine must have some sort of machine 'understanding' to have command of our language in an analogous jet different way on the fundamental level. "It is important to recognize that machine performances may be wholly *analogous* to language, so that the whole of linguistic theory can be applied to them." At no place in Chomsky's *Syntactic Structures* "*is the assumption employed that the corpus of utterances studied by the linguist was produced by a conscious organism.*"[40]

39 See Turing 1950: 439.

40 Putnam 1975: 383.

By the denial of Turing's assumption, we have instead to accept that speaking in a meaningful way is neither in itself intelligent behaviour, nor a sufficient condition for intelligence or even consciousness after all. More than that: This statement would hold true for human beings as well.[41] We therefore have to give up the belief that meaningful communication is something that only intelligent and conscious human beings are capable of.

Considering all these aspects, an appropriate new definition of linguistic competence would be 'having learned the correlation of certain words and contexts'. We then have to accept that derived intentionality is sufficient for meaningful language, which in the end, is pure stimulus response behaviour.[42] Whether we call this 'competence', 'understanding' or something else, the machine apparently realises a performance that is indistinguishable from the way we speak, and that does not require an internal state identical to our own intentional state. We could accept machine learning as described above as a different approach to a speech-performance that does not rely on competence or understanding like our own. This entails an argument for the multiple realisation of understanding and speech, in favour of the multiple realisation of consciousness. In an altered version of Putnam's argument on the psychological state of a robot, I suggest that when the machine is in an internal state that realises the predicate 'understands the meaning of *x*' we may say that the machine 'understands' the meaning of *x*. By using square quotes I try to indicate that the machines 'understanding' in this sense might not be human *understanding* in Searle's sense, but it will play the role in the machine's performance that *understanding* plays in human behaviour.[43]

4. Machines Might be Able to Solve Moral Problems

In the end, Google Duplex – just like any other modern ANN based AI-system – is a black box, that is fed with thousands of data strings on the input layer and, after a while, produces its own spoken language or performance on the output layer. In that state, an ANN can take the place of the Chinese Room in Searle's argument. For every incoming question, it produces a proper answer and its overall performance is indistinguishable from a native speaker. The noteworthy difference between a rule-based program and a self-learning ANN might be the fact that the latter learns over time from a certain amount of data and

41 See Putnam 1975: 384.
42 See Putnam 1975: 392.
43 See Putnam 1975: 391.

optimizes its performance according to feedback, instead of using a rule book right away that is hardcoded and given to it by a competent speaker/hearer of the language to be spoken.[44] Hence, we don't know precisely how the ANN functions in its particular state even though we know in general how it learns this kind of performance. Performing a language through machine learning helps the ANN to vary its performance if required whilst performing according to the language's grammar without knowing its rules explicitly, just as children do when learning a language.

Furthermore, Google Duplex in particular adapts to speaking habits that are established implicitly in a language community as shown above. We therefore ascribe a meaning to the machine's responses easily. We accept Duplex's reservation for the client and even say "Thank you" and "Goodbye" in response to the machine. We anthropomorphise the machine's performance implicitly, without even noticing and attribute an intentionality that is in fact only derived from the training data. The researchers at Google – who probably should know better – even do this to the extent that they consider human *understanding* to be a problem for machines that Duplex has already solved.[45]

What I am trying to show at the end of this paper is that machines might learn to solve moral dilemmas. Considering all the above aspects it might do so in just the same way as they gain the ability to produce meaningful spoken language, in an appropriate manner and context. This proposition is an argument by analogy based on the comparison of the performance shown by Google Duplex with the conditions of moral decision making.

Aristotelian virtue ethics in particular may be profitably applied to artificial agents if one can show, that they are able to learn human moral values from a large amount of data. In this way the machine should be able to mimic the human moral decision making in its performance. To attain this the programmers do not have to redesign human moral competence or the moral reasoning process explicitly. When it comes to acquiring moral virtues, Aristotle said: "Neither by nature nor contrary to nature do the virtues arise in us; rather we are adapted by nature to receive them, and are made perfect by habit." (NE 1103a24) It is plausible that ANN-based artificial agents are indeed well-adapted to acquire moral virtues as well. Choosing the right words in the right context is similar to choosing from accepted courses of action in a given

44 Even though this might in fact bring the ANN again closer to a human speaker/hearer who might as well be seen as a black box when it comes to the exact *how* he or she manages to process incoming sound into information and thoughts and back into a spoken answer, known as the mind body problem.

45 Leviathan and Matias 2018.

situation, as well as disqualifying those actions that are obviously forbidden. Much like meaning of a sentence, the same performance can vary in its moral value depending on the context. As the spoken language of a machine can be meaningful to us to some extent, the action a machine might take in moral conflict could as well be considered morally good from an observer's point of view. 'Meaningful' and 'good' might serve as labels given to an action or statement in the training data according to its appropriateness in a given context. There are explicit rules or laws as well as implicit practices in conversations, as we have seen with Duplex on the one hand and in our moral behaviour on the other, e.g. 'Do no harm', 'Do not steal', 'Do not commit murder', 'Do good', 'Give your seat to an elderly person', 'Be honest', 'Be fair'. Implicit rules come in the form of values hidden as patterns in the vast amount of data that match our moral actions, e.g. human dignity or well-being. The even trickier part is a second task, to build a machine with just the right values that exclude for example our unwanted bias towards minorities or other prejudices. For the machine, the appropriateness of an action is a correlation, which ANNs could learn from data. I consider the relation between moral action and moral norms and values to be analogous to the relation between the grammar that underlies the training data, on one hand, and the performance of Duplex, on the other. The correlation is normative only to the extent that norms are included implicitly in the training data as a moral bias grounded on a certain type of ethics.

Apparently, this analogy entails some limitations for a machine's moral decision making: the action of an artificial agent in a moral dilemma is only morally acceptable from a behaviouristic point of view. It performs moral actions without having moral competence or moral self-reflection as considered necessary for human moral agents for example in Kantian ethics. It nevertheless performs in a way, that is observed as accurate moral behaviour. Thus, the machine has in the end something that Aristotle would call *phronesis*, which in the context of moral action could be understood as practical wisdom, the ability to act in the right way under the given circumstances.

As Turing suggested, for mimicking conversations through machines, it might be important for a machine to have some flaws that humans have and that need to be copied to make a conversation feel natural: Whilst Turing suggested to hide the precision of the computer program, when it faces for example mathematical questions, we saw Duplex using disfluencies in its performance in order to sound more natural. In moral decision-making however, I consider these kinds of restrictions to be counterproductive, if the machine used synthetic disfluencies to be indistinguishable from a human agent. Unlike the case of communicational human-robot interaction a machine must not be subject to any limitations for the sake of humanlike appearance when it comes to moral decision making. Only then can it be a virtuous agent in the sense of Aristotle. The ANN might gain a virtuous internal state, that is determined by rational principles

> by which the man of practical wisdom would determine it. Now it is a mean between two vices, that which depends on excess and that which depends on defect; and again, it is a mean because the vices respectively fall short of or exceed what is right in both passions and actions, while virtue both finds and chooses that which is intermediate.[46]

Insofar as furthermore a virtue is an extreme with regard to what is best and right an optimisation algorithm comes into consideration, that cannot be dealt with any further in this paper. However, optimization has had huge successes in machine learning in recent years and may be seen as the fulfilling of the necessary condition of "making one's practical wisdom perfect by habit" as it is required by Aristotle. While – in the tradition of Descartes – Turing underestimated the benefits of being embodied for intelligent action in our physical environment,[47] we still underestimate the learning process, which furthers the appropriation of intellectual capabilities and virtues, as Turing already suggested.[48]

5. Bibliography

Brooks, Rodney (2015) "Mistaking Performance For Competence Misleads Estimates Of AI's 21st Century Promise And Danger." In: *What to think about machines that think? Today's Leading Thinkers on the Age of Machine Intelligence*, ed. by J. Brockman, New York, 108–111.

Chomsky, Noam (2006) *Language and Mind.* 3ed Edition, Cambridge.

DeVynck, Gerrit, Sarah McBride, and Jeremy Kahn (2018) "IBM's Debating AI Is Here to Convince You That You're Wrong." Available at: https://www.bloomberg.com/news/articles/2018-06-19/ibm-s-debating-ai-is-here-to-convince-you-that-you-re-wrong, accessed: 12.07.2019.

Dreyfus, Hubert (1993) *What Computers still can't do. A Critique of artificial Reason*, Cambridge².

Goodfellow, Ian, et al. (2016) *Deep Learning*, Cambridge.

46 NE 1107a8.

47 See Dreyfus 1993: 235.

48 See Turing 1950: 454.

Jaderberg, Max, et al. (2018) "Human-level performance in first-person multiplayer games with population-based deep reinforcement learning." doi: arXiv:1807.0128.

Leviathan, Yaniv and Yossi Matias (2018) "Google Duplex. An AI System for accomplishing real world tasks over the phone." Availiable at: https://ai.googleblog.com/2018/05/duplex-ai-system-for-natural-conversation.html, accessed: 12.07.2019.

McCarthy, J., Minsky, M., Rochester, N., Shannon, C. E. (1955) "A Proposal for the Dartmouth Summer Research Project on Artificial Intelligence." Available at: http://raysolomonoff.com/dartmouth/boxa/dart564props.pdf, accessed: 12.07.2019.

Mitchell, Thomas (1997) *Machine Learning*, New York.

van den Oord, Aäron, et al.: WaveNet (2016) "A Generative Model for Raw Audio." Available at: https://arxiv.org/pdf/1609.03499.pdf, accessed: 12.07.2019.

Putnam, Hilary (1975) *Mind, Language and Reality*, Cambridge.

Rees, D. (1956) Aristotle: The Nicomachean Ethics, translated and introduced by David Ross, London.

Searle, John (1980) "Minds, Brains, and Programs." In: *The Behavioral and Brain Science* 3, 417–457.

Silver, David, et al. (2017a) "Mastering Chess and Shogi by Self-Play with a General Reinforcement Learning Algorithm." doi: arXiv:1712.01815.

Silver, David, et al. (2017b) "Mastering the game of Go without human knowledge." *Nature* 550, 354–359.

Statt, Nick (2018) "Google now says controversial AI voice calling system will identify itself to humans." Available at: https://www.theverge.com/2018/5/10/17342414/google-duplex-ai-assistant-voice-calling-identify-itself-update, accessed: 12.07.2019.

Turing, Alan M. (1950) "Computing Machinery and Intelligence." *Mind* 59,236, 433–460.

Wiggers, Kyle (2018) "Google's Duplex is rolling out to Pixel owners – here's how it works." Available at: https://venturebeat.com/2018/11/21/googles-duplex-is-rolling-out-to-pixel-owners-heres-how-it-works/, accessed: 12.07.2019.

PART IV

Robots

Could Artificial General Intelligence be an End-In-Itself?

Benedikt Paul Göcke

Abstract

Discussion of artificial intelligence (AI) is pervasive in current academia. However, it seems that from a philosophical point of view one of the most interesting questions is whether artificial intelligence could entail the existence of a conscious entity. In this paper, the first part provides a definition of artificial intelligence and distinguishes between instrumental artificial intelligence and artificial general intelligence (AGI). The second part analyses our motives for constructing AGI in addition to instrumental AI. The third part deals with arguments for and against the possibility of constructing AGI. The final part provides an argument for the conclusion that a being with AGI could indeed be conscious, and therefore should be considered ethically as an end-in-itself.

1. What is Artificial Intelligence?

There are at least two constraints on any useful definition of AI: It must neither be too weak to be of interest for analysis nor too strong to be unsatisfiable by a (virtual) machine.[1] A definition of AI is too weak if it subsumes common hairdryers and washing machines and too strong if it demands that only full and real-time emulations of the human brain would qualify as beings with AI. To provide such a definition, it seems best to start with reflection on both the concept of artificiality and the concept of intelligence and to then bring the clarified concepts together in a single definition.

1.1 *What is an Artificial Entity?*

An entity is artificial if and only if its existence is not the direct result of a process of undisturbed cosmic or biological evolution, where cosmic and biological

1 For further analysis of virtual machines cf. Boden (2018: 3): "A virtual machine isn't a machine depicted in virtual reality, nor something like a simulated car engine used to train mechanics. Rather, it's the information-processing system that the programmer has in mind when writing a program, and that people have in mind when using it."

 | DOI:10.30965/9783957437488_015

evolution is a process that is governed by the laws of nature and leads to the development of cosmic and biological systems, as well as of natural kinds, as these are investigated by the natural sciences. This process is undisturbed if and only if there are no causal interventions that are due to the intentional actions of rational agents – whether humans, aliens, angels or God. Based on this understanding of something's being artificial, we are able to specify two entailments: First, the existence of each and every entity in the universe which is caused by intentional actions is artificial: houses, cities, agriculture, clothes, culture – all of this is the artificial product of intentional human actions.[2] Second, artificial intelligence exists if and only if an intelligent entity exists whose obtaining in the physical universe is not the direct outcome of an undisturbed cosmic or biological evolution, but is the consequence of intentional actions of rational agents that decided to use their intelligence to construct or engineer such an entity. AI, in other words, is not (a) natural (kind), but a product of human imagination, art and engineering.

1.2 *What is an Intelligent Entity?*

Constructing or engineering an intelligent entity depends on what is meant by "intelligence". In the same way in which one has to understand what a rocket is before one can construct a rocket, one has to specify what an intelligent entity is before one can try to construct AI. *Prima facie*, however, there are two problems with this task.

First, any suggested concept of intelligence is likely to be anthropomorphically biased because based on the only instance of an intelligent being accessible to us: the human being itself. Based on this insight, one could argue that we cannot specify a useful concept of intelligence that is applicable to other beings. The concept of human intelligence, one could argue, is therefore of no help when it comes to judging the putative intelligence of non-human animals or artificial intelligence.[3] But this argument does not seem to pose an unsurmountable problem: What we are interested in in dealing with AI is not the question whether there might be forms of intelligence unknown to us such that, if we knew them, we could model them with the help of virtual machines, or implement them in cyberphysical form. Instead, we are interested precisely

2 If God created the universe, then the universe itself is artificial in the sense specified. In what follows, I will bracket this special case of the universe's being artificial *per se*.

3 Cf. Chalmers (2016: 26–27): "One might [...] hold that even if there is a canonical notion of intelligence that applies within the human sphere, it is far from clear that this notion can be extended to arbitrary non-human systems, including artificial systems."

in the question of whether it is possible to construct a (virtual) machine that has whatever it is that we refer to as intelligence in humans.

Second, although it is in principle adequate to use the concept of human intelligence as the standard for what a being with artificial intelligence has to possess if it is to be rightly referred to as "intelligent", there is no unanimity regarding what it means for a human being to be intelligent: There are just too many different aspects of human intelligence to be gathered under a single heading.[4] But if this is the case, so the argument goes, there can be no useful definition of AI that we could use to specify what we are interested in in the field of research on artificial intelligence.

But although it is indeed the case that there is no unanimity regarding the right way to understand the concept of intelligence, this does not entail that plausible candidates cannot be formulated and discussed.[5] It is possible to stipulate a definition of intelligence that covers important connotations of 'intelligent'. Based on these features, we may specify a predicate of which it makes sense to ask whether the corresponding property (or combinations of properties) could be exemplified in beings other than human beings. Five conditions specify such an 'intelligent' predicate of which it makes sense to ask whether we can construct or engineer a (virtual) machine that exemplifies the corresponding property (or combination of properties). I turn to these now.

1.2.1 Intelligence as a Dispositional Property

Intelligence is a property (or a combination of properties) that an entity either exemplifies or not. It is, however, not a categorical property like being red or being five feet tall, but a dispositional property. Dispositional properties are properties that based on a particular input or state of the bearer of this

4 Cf. Hutter (2016: 208–209) for basic questions related to the concept of intelligence: "What activities could be regarded as or are positively correlated with intelligence? Self-preservation? Self-replication? Spreading? Creating faster/better/higher intelligences? Learning as much as possible? Understanding the universe? Maximizing power over men and/or organizations? Transformation of matter (into computronium)? Maximum self-sufficiency? The search for the meaning of life? Has intelligence more to do with thinking or is thinking only a tool for acting smartly? Is intelligence something anthropocentric or does it exist objectively? What are the relations between other predicates of human 'spirit' like consciousness, emotions, and religious faith to intelligence?"

5 Cf. Tegmark (2016: 49): "[T]here's no agreement on what intelligence is even among intelligent intelligence researchers! So there's clearly no undisputed 'correct' definition of intelligence. Instead, there are many competing ones, including capacity for logic, understanding, planning, emotional knowledge, self-awareness, creativity, problem solving and learning." Cf. also Legg and Hutter (2007) for an enumeration of more than 70 different definitions of intelligence cutting across disciplines such as philosophy, psychology and informatics.

property generate a particular output that depends on the internal constitution of the bearer of this property. As a dispositional property, intelligence enables its bearer to interact with its environment based on the environmental input in order to produce a particular output.

1.2.2 Intelligent Behaviour is not Causally Determined

Although intelligence is a dispositional property that enables interaction with the environment, the interaction must not be completely causally determined: If an entity's ability to interact with its environment in a completely causally determined way were sufficient for it to be intelligent, then the set of intelligent entities would include fundamental physical particles as well as stones and other entities that we are reluctant to refer to as intelligent entities. Therefore, although intelligence is the ability to interact with one's environment, intelligent behaviour is not causally completely determined interaction with one's environment.[6]

1.2.3 Intelligence Enables Selection from Options

That intelligence is not a completely causally determined reaction is to say that an intelligent entity must be able to interact with its environment in a way that is not causally determined by that environment. While causally determined interaction with one's environment is a passive reaction with no alternative possibilities, intelligence is a property (or combination of properties) that enables its bearer to react in a variety of ways. Because, however, there may be entities that do not react to their environment in a completely causally determined way, but through behaviour based on random effects at the level of quantum mechanics, the mere ability to react in a variety of ways is not sufficient for an entity to qualify as intelligent.

1.2.4 Intelligence is Rationality

A sufficient condition for an entity's being intelligent consists in the ability to choose one of some available actions based on an evaluation of the reasons in favour of each. An intelligent entity has the ability to choose a particular action based on different rules of reasoning such as deductive, inductive and abductive, and based on an evaluation of the information from its environment. It will choose the action to which it assigns the highest probability that acting according to it will realize a particular goal. Intelligent action, therefore,

6 I bracket cases in which an intelligent entity interacts only with itself and not with its external environment. I also assume, for the sake of argument, that stones and particles are not intelligent entities.

is rational action.[7] Because rational action is goal-oriented it follows that intelligence is goal-oriented rational action.[8]

1.2.5 Intelligence is Learning

However, an intelligent being is not only able to rationally choose one of a variety of possible reactions in a given situation but also able to learn and reassess which option is best to achieve a particular goal when it is in a particular environment or state. An intelligent entity therefore needs memory of its own past behaviour. If the same mistake is made all over again with the same bad result of not achieving the intended goal, then we would hardly assume that an intelligent entity is at work. The ability to learn, that is, the ability to change one's behaviour based on new information, or based on the failure or success of past ways of achieving a particular goal, is therefore a further ability any intelligent entity has to possess.[9]

1.2.6 Summary: A Brief Account of Intelligence

Intelligence is a dispositional property or combination of properties that allows its bearer to interact with its environment in a rational and not completely causally determined way based on the evaluation of the situation it is in and the probability that choosing a particular action out of the set of available reactions will realize the goal which the entity is set to achieve in such way

---

7 If rationality entails the ability to solve problems, then intelligence can be seen as a problem solving ability. Such an ability, according to Schmidhuber, could be measured: "The scientific way of measuring intelligence involves measuring problem solving capacity. There are mathematically sound ways of doing this, using basic concepts of theoretical computer science." (Schmidhuber 2016: 228)

8 As Hutter (2016: 210) says, intelligence is "the ability to achieve goals in a wide range of environments." Cf. also Hutter (2016: 209): "It seems hard if not impossible to define rationality without the notion of a goal. If rationality is reasoning towards a goal, then there is no intelligence without goals." Cf. also Tegmark (2016: 49): "intelligence = ability to accomplish complex goals." As Greenfield (2016: 166–167) argues, such a notion of intelligence is related to the concept of understanding: "I would suggest that understanding is primarily seeing one thing in terms of another: the more we can relate a phenomenon, action, or fact, to other phenomena, facts, or actions, the deeper, I would argue, is the 'understanding' […] Whilst information processing is just that, the appropriate response to an incoming stimulus, understanding in contrast requires that the stimulus in question be embedded in a conceptual framework."

9 Cf. Alpaydin (2016: 17): "A system that is in a changing environment should have the ability to learn; otherwise, we would hardly call it intelligent. If the system can learn and adapt to such changes, the system designer need not foresee and provide solutions for all possible situations." Cf. also Bostrom (2014: 23): "It now seems clear that a capacity to learn would be an integral feature of the core design of a system intended to attain general intelligence, not something to be tacked on later as an extension or an afterthought."

that it is able to change its behaviour based on new information and based on the evaluation of the success and failure of its past behaviour. Although this characterisation may not cover all of the features of intelligence one could be interested in, it specifies a property or properties we would expect to be exemplified in entities we refer to as "intelligent".

1.3 *What is Artificial Intelligence?*

Based on the definitions of intelligence and artificiality, we obtain the following concept of artificial intelligence: An entity exhibits artificial intelligence if and only if its existence is not the direct result of an undisturbed cosmic or biological evolution, but of intentional human construction or engineering. This enables the entity to interact with its environment in a rational and not completely causally determined way through the use of deductive, inductive, or abductive reasoning based on the evaluation of the situation it is in. It evaluates the probability that choosing a particular action out of the set of available actions will realize the goal which the entity is set to achieve in such way that it is able to change its behaviour based on new information and based on the evaluation of the success and failure of its past actions.[10]

1.4 *Two Kinds of Artificial Intelligence*

The definition provided needs two qualifications that enable us to differentiate two kinds of AI: On the first qualification, artificial intelligence has a particular and fixed goal which it tries to achieve, whereas in the second case artificial intelligence does not have a particular and fixed goal but is able to set different goals on its own. The first kind is purely instrumental artificial intelligence that is only able to realize its in-built purpose.[11] The second kind of artificial intelligence is able to set its own goal.

---

10 Cf. Boden (2005: 1): "Artificial Intelligence (AI) is sometimes defined as the study of how to build and/or program computers to enable them to do the sorts of things that minds can do. Some of these things are commonly regarded as requiring intelligence: offering a medical diagnosis and/or prescriptions, giving legal or scientific advice, proving theorems in logic or mathematics."

11 Here is, roughly, how the construction of an instrumental artificial intelligence proceeds: "1. *Computational theory* corresponds to the goal of computation and an abstract definition of the task. 2. *Representation and algorithm* is about how the input and the output are represented, and about the specification of the algorithm for the transformation from the input to the output. 3. *Hardware implementation* is the actual physical realization of the system." (Alpaydin 2016: 20). That is, we need a computational theory, a way to translate the relevant information into a language the machine can understand and a way to translate its results back into a language we can understand, algorithms that operate inside the 'black box', and a way to get it working on a physical device: The computational

Whereas I will refer to the first kind as "instrumental artificial intelligence", the second kind is known as "artificial general intelligence" (AGI). Because human intelligence is the only form of intelligence we know of that is able to decide which goal it wants to realize, artificial general intelligence is the holy grail of research on artificial intelligence. If AGI existed, it would be able to do the sorts of things that a human can do insofar as he is intelligent. An AGI would be the *imago hominis* insofar as human beings are intelligent.[12]

1.5 *From Artificial General Intelligence to Superintelligence*

While instrumental AI is able to achieve only one goal – it is only good at one specific task such as playing chess or identifying individual human faces – AGI is able to set goals for itself. Since AI able to learn via machine learning is able to learn very fast, it tends to become very good at its tasks very quickly, often out-smarting human beings in achieving its goal.[13]

Since a possible goal of AGI could be to create an even more intelligent artificial general intelligence, for instance by improving the human engineering and construction plans used to create it in the first place, the existence of an artificial general intelligence could lead to the existence of what in the

---

theory specifies which goal the machine is set to achieve. That what enables the machine to achieve its goal depends on the input to the machine and the form in which the input is represented and proceeded internally by the algorithms that have to be specified by the engineers. We could achieve this, in principle, by symbolic representation in the form of expert systems, by building an artificial neural network, by emulating parts of the human brain, by artificial evolution in which the best algorithms for a certain goal are searched for, or by a combination of elements of the mentioned approaches. Once the computational theory, the representation, and the algorithms have been specified, the last thing one has to take care of is the physical implementation of the virtual machine in question. That is, one has to construct the physical medium on which the program runs and the input and output devices through which the machine is related to its environment.

12 Cf. Herzfeld (2002: 33): "AI can be viewed as the attempt to create an imago hominis, a machine that is in some way created in the image of the human person, in image loosely defined using the term intelligence."

13 Cf. Boden (2018: 40–41) for a brief description of machine learning: "Machine learning has three broad types: supervised, unsupervised, and reinforcement learning. […] In *supervised* learning, the programmer 'trains' the system by defining a set of desired outcomes for a range of inputs […], and providing continual feedback about whether it has achieved them. […] In *unsupervised* learning, the user provides no desired outcomes or error messages. Learning is driven by the principle that co-occurring features engender expectations that they will co-occur in future. Unsupervised learning can be used to *discover* knowledge. The programmers needn't know what patterns/clusters exist in the data: the system finds them for itself. Finally, *reinforcement* learning is driven by analogues of reward and punishment: feedback messages telling the system that what it just did was good or bad."

discussion is called a "superintelligence", where a superintelligence is an AGI that is vastly more intelligent than any human being could possibly be.[14]

The danger of creating AGI is that we do not know what kind of goal a possibly emerging superintelligence with access to physical machines would decide to aim for. Since we do not know this, we cannot know whether its existence would have a positive, a negative or a neutral effect on the existence of human beings. In the worst case, a superintelligence will simply kill all of us, in the best case, life on earth could be much like living in paradise.[15] Therefore, whereas our attempts to create instrumental AI already entail the considerable risk of misuse that could lead to catastrophic consequences, the attempts to develop AGI involve an even higher risk of creating a very powerful machine that we might not be able to control.

2. Why Do We Wish to Construct Artificial Intelligence?

Although questions concerning the problems and prospects of artificial intelligence are discussed across disciplines, it is not often asked why humanity puts so much effort into the creation of artificial intelligence. We are confronted with two questions: Why do we wish to construct instrumental AI? And: Why do we wish to construct artificial general intelligence?

2.1 *Why Do We Wish to Construct Instrumental Artificial Intelligence?*

The answer is the same as to the question why humanity is so eager to develop technology in general: because we can use it to achieve goals that can be achieved better with the use of technology than without.[16] However, as with any technology and any tool, the development of instrumental artificial

14 This idea was introduced to the debate by Good (1965: 33), who described it thus: "Let an ultraintelligent machine be defined as a machine that can far surpass all the intellectual activities of any man however clever. Since the design of machines is one of these intellectual activities, an ultraintelligent machine could design even better machines; there would then unquestionably be an 'intelligence explosion,' and the intelligence of man would be left far behind. Thus the first ultraintelligent machine is the last invention that man need ever make, provided that the machine is docile enough to tell us how to keep it under control." For further analysis cf. Chalmers (2016) and Bostrom (2014).

15 As Alpaydin (2016: 165) states it: "The challenge is to make sure that this new source of intelligence is used for good and not for bad, to increase the well-being of people, and for the benefit of humanity rather than to increase the profit of a few."

16 As Alpaydin (2016: 19) says, "[a]rtificial intelligence is a part of computer science and our aim is to build useful systems, as in any domain of engineering."

intelligence is subject to the dual use of technology: it can be used for good or for ill, where it is a matter of ongoing ethical debate what precisely is a good or a bad use of instrumental artificial intelligence.

The positive effects of the development of instrumental artificial intelligence are often said to include the following: It will make our lives easier, more comfortable, more secure, healthier, and more connected in a way we could not have achieved by using old-fashioned technology. We could not only construct robots that operate in environments dangerous for human beings , but could construct instrumental artificial intelligence that is better at any job that can be decomposed into processable sub-problems and sub-goals than any human being could possibly be: All we need to achieve this is to be clear about the goal of the job and the skills needed to achieve it before we construct instrumental AI able to learn how to achieve this goal better than any human could. As Kaplan (2016: 114–115) says: "In practice, automation replaces skills, not jobs, and correspondingly, what employers need is not workers but the result obtained by applying those skills. To be successful, makers of robots don't have to replace people; they have to provide machines with the requisite skills to perform useful tasks. [...] From an economic standpoint, AI technology is just another advance in automation."

However, what is supposed a good purpose from the point of view of employers is not necessarily a good purpose from the point of view of employees. The negative effects of the use of instrumental AI, it is often argued, could lead to drastic and unseen social and individual consequences such as mass-unemployment, social uprising, and a further widening of the divide between the rich and the poor, not to mention the possible misuse of instrumental artificial intelligence to control people in a huge variety of ways.[17]

Therefore, although, in principle, the motives, problems and prospects of creating instrumental artificial intelligence are ones we are familiar with from other technologies, the greater autonomy and the greater impact of instrumental AI on the individual and on society, in particular their greater potential for good and for worse, forces us to be cautious regarding the use and development of instrumental artificial intelligence.

17 Cf. Kaplan (2016: xi): "After more than five decades of research, the field of AI is poised to transform the way we live, work, socialize, and even how we regard our place in the universe." Cf. also Ramge (2018: 21).

2.2 *Why Do We Wish to Construct Artificial General Intelligence?*

The motivation to construct AGI goes beyond that to design instrumental AI which human beings can use as they see fit. The reason is that an artificial general intelligence would be unlike a swiss-army knife that we could use to realize many different goals – that would still be an instrumental artificial intelligence. Like a human intelligence, it would be able to set its own goals. AGI therefore could not be used by us as a means to an end, which is to say that it could not be used like a technological tool. Instead, AGI, by definition, would constitute an end-in-itself.

The question we have to deal with, then, is why human beings are trying to construct something that is an end-in-itself? Apart from the superficial answer that we might just be interested in seeing whether we are in fact capable of doing so, the deeper answer seems to be that the existence of beings that are ends-in-themselves is intrinsically good and valuable. Therefore it is morally good, if possible and even if only supererogatorily, to create artificial general intelligence that is an end-in-itself. As Herzfeld (2002: 50) states it: "Essentially, the goal of strong AI is to build something *like* ourselves, to create in our own image. Intelligence is simply a label for that in us that is essential, that which stands at our center, necessary rather than contingent."

Since, however, an end-in-itself is a being for whom its being is an issue, and since a being for whom its being is an issue has to be a conscious being with the ability to reflect on itself, it follows that we could only succeed in creating AGI if we could succeed in constructing intelligence that is conscious at least to the level of human beings. The ultimate goal of those trying to develop AGI, consequently, is to create a new species of conscious beings that, like us, are not only conscious of the world surrounding them but of themselves. As Brooks (1997: 401) describes it, the goal in creating an artificial general intelligence is "to build completely autonomous mobile agents that co-exist in the world with humans, and are seen by those humans as intelligent beings in their own right." As Moravec (1988: 6) says, the goal is to create "robots with human intelligence."

3. Could We Construct a Weak Artificial General Intelligence?

To analyse whether it is, in principle, possible to succeed in constructing conscious AGI, we have to ask two *prima facie* distinct questions: Could we construct AGI that appears to be a conscious end-in-itself? And: Could we construct AGI that in fact is a conscious end-in-itself? Based on the distinction between weak and strong artificial intelligence, the questions may be formulated as follows: Could we construct a weak artificial general intelligence? And: Could we construct a strong artificial general intelligence?[18]

3.1 *The Argument from Insufficient Knowledge*

To construct weak artificial general intelligence, we have to construct an intelligence that appears to us as if it were a conscious end-in-itself although in fact it would not be a conscious end-in-itself. However, according to the argument from insufficient knowledge, we are unable to create a weak AGI. The argument may be stated as follows:

(1) Human beings are able to construct a weak artificial general intelligence only if human beings are able understand what precisely constitutes human intelligence.
(2) Human beings are unable to understand what precisely constitutes human intelligence.

Therefore:

(3) Human beings are unable to construct a weak artificial general intelligence.

The argument is valid: If the premises are true, then the conclusion is true. For it to be sound, the premises need to be true. The first premise is true: We need to be able to understand what precisely constitutes human intelligence if we want to construct or engineer a machine that mimics the intelligent behaviour of human beings. Even if there may be more than one way to construct such a machine, we need to have a precise understanding of human intelligence before we can ask about the different ways to build a machine that behaves in an intelligent way.

The essential premise therefore is the second: Human beings are unable to understand what precisely constitutes their intelligence. Despite the fact that

---

18 Cf. Searle (2005: 67) on the distinction between strong and weak artificial intelligence: "I find it useful to distinguish between what I will call 'strong' AI from 'weak' or 'cautious' AI (Artificial Intelligence). According to weak AI, the principal value of the computer in the study of the mind is that it gives us a very powerful tool. For example, it enables us to formulate and test hypotheses in a more rigorous and precise fashion. But according to strong AI, the computer is not merely a tool in the study of the mind; rather, the appropriately programmed computer really *is* a mind, in the sense that computers given the right programs can be literally said to *understand* and have other cognitive states. In strong AI, because the programmed computer has cognitive states, the programs are not mere tools that enable us to test psychological explanations; rather, the programs are themselves the explanations." Cf. also Kaplan (2016: 68): "In short, strong AI posits that machines do or ultimately will have minds, while weak AI asserts that they merely simulate, rather than duplicate, real intelligence. [...] Stated another way, the distinction is between whether machines can be truly intelligent or simply able to act 'as if' they are intelligent."

currently we do not have a sufficiently clear and detailed theory of what constitutes human intelligence, it seems, contrary to the second premise, probable that based on the progress of science, human beings will be able to reach a sufficient understanding of the human body and its environment to be able to understand which property or combination of properties makes us intelligent and to use this knowledge to construct AGI that at least appears to be a conscious end-in-itself.

The argument from insufficient knowledge is therefore not a convincing proof of the impossibility of creating weak AGI but only shows that there is need for further research. What would be needed to support the second premise is an argument that human beings, in principle, are unable to understand the properties responsible for their intelligence. This, ultimately, would have to be an argument for the very strong conclusion that there is an opaque natural object that no human theory could adequately grasp.

The chances of developing such an argument, however, are slim. In fact, there are two arguments that show that it is at least in principle possible to construct weak AGI: the argument from evolution and the argument from the brain.

3.2 *The Argument from Evolution*

The argument from evolution runs as follows:

(1) If evolution was able to create intelligent human beings, then it is in principle possible for human beings to create weak AGI.

(2) Evolution was able to create intelligent human beings.

Therefore:

(3) It is in principle possible for human beings to create weak AGI.

The argument is valid and our best science suggests the second premise is true: Human beings are the contingent product of an evolutionary process that through adaption, genetic drift and contingent mutations led to the existence of human beings that are intelligent. However, if this is the case, and if evolution is a process ultimately driven only by the laws of nature, then it must be possible, as the first premise specifies, to mimic this process and to use the same method to create an entity that at least behaves as if it were intelligent. As Chalmers (2016: 20) says, "evolution operates without requiring any antecedent intelligence or forethought. If evolution can produce something in this unintelligent manner, then in principle humans should be able to produce it much faster, by using our intelligence." Therefore, although currently we have not yet succeeded in creating weak AGI by mimicking elements of an evolutionary process, it seems to be possible to construct weak AGI by simulating elements of the evolutionary process.

3.3 *The Argument from the Brain*

The second argument for the possibility of constructing weak AGI is based on the assumptions that the operations of the human brain are causally sufficient for the existence of human intelligence and that the human brain is a physical machine governed by the laws of nature.[19] It runs as follows:

(1) If intelligence is a feature of the human brain, and if the human brain is a machine, then it is at least in principle possible to build a machine that operates in the same observable way as the human brain.

(2) If it is possible to build a machine that operates in the same observable way as the human brain, then it is possible to construct weak AGI.

(3) Intelligence is a feature of the human brain.

(4) The human brain is a machine.

Therefore:

(5) It is possible to construct weak AGI.

The argument is valid. The first and second premise seem true: If intelligence is a property human beings possess due to the way their brains work, and if the brain is a machine, and if intelligence is not a contingent property of the human brain, but one that it essentially exemplifies based on the way it operates, then it is at least in principle possible to build a machine that operates in the same way.[20]

3.4 *Could We Construct Weak Artificial General Intelligence?*

The arguments for the possibility of constructing weak AGI at least show the following: If human intelligence is a natural feature of the physical universe that emerges from a biological system such as the human brain, in accordance with the laws of nature, then we are, in principle, able to construct a (virtual) machine that behaves as if it were a conscious end-in-itself.

---

19 Cf. Chalmers (2016: 18): "Every organ of the body appears to be a machine: that is, a complex system comprised of law-governed parts interacting in a law governed way. The brain is no exception."

20 Once could argue that this argument depends on the assumption that intelligence is structurally invariant, that is, that any machine that operates in the same way in which the human brain operates will exemplify intelligence. The argument, therefore, will not succeed if intelligence is a property which the human brain essentially exemplifies due to its bio-chemical and physical properties, but only if intelligence is multiple realizable. However, this reply is not convincing in the present context because it ignores that the question at this point only is whether artificial intelligence would behave *as if* it were intelligent like a human being.

4. Could We Construct Strong Artificial General Intelligence?

Suppose, then, that we succeeded in creating weak artificial general intelligence and that we can successfully describe it as a conscious end-in-itself. The immediate question is whether it is *really* conscious or not. Can we, in principle, only create weak AGI, or can we create strong artificial intelligence, too?[21] I am going to argue that we have no reason to assume that weak artificial intelligence is not *ipso facto* strong artificial intelligence, but reason to assume that the creation of weak artificial general intelligence is *eo ipso* the creation of strong artificial general intelligence. Let me begin with a brief reflection on the concept of a conscious being.

4.1 *What is a Conscious Being?*

A conscious being is, at least sometimes, the subject of a *stream of consciousness*.[22] This is a minimal part of what we mean when we say that something is a conscious being: On the one hand, we do not demand that a conscious being is the subject of a stream of consciousness at every point of time of its existence; as perhaps it is not in narcotic sleep. On the other hand, we have difficulty in imagining a conscious being that is never the subject of any stream of consciousness. Therefore, it is *necessary* that, at least sometimes, a conscious being is the subject of a stream of consciousness.

Because there are different senses of the term 'consciousness', we have to find a sense in which a conscious being can properly be said to be the subject of consciousness. Husserl distinguishes three different notions: "1. Consciousness as the entire phenomenological being of the spiritual ego. (Consciousness=the phenomenological ego, as "bundle" or interweaving of psychic experiences). 2. Consciousness as the inner becoming aware of its own psychic experiences. 3. Consciousness as a comprehensive designation for "psychic acts" or "intentional experiences" of any sort." (Husserl 1984: 356).[23] The notion of consciousness as the entire phenomenological being of what Husserl calls the spiritual ego is the appropriate notion of consciousness here: A stream of conscious-

21 Cf. Boden (2018: 106): "Suppose that future AGI systems (on-screen or robots) equalled human performance. Would they have real intelligence, real understanding, real creativity? Would they have selves, moral standing, free choice? Would they be conscious? And without consciousness, could they have any of those other properties? These aren't scientific questions, but philosophical ones. Many people feel intuitively that the answer, in each case, is 'Obviously, no!' Things aren't so straightforward, however."

22 Cf. Göcke (2012) for a further analysis of the concept of a conscious being.

23 For a thoughtful study on Husserl's notion of the transcendental ego and Sartre's critique, cf. Priest (2000).

ness *is* an entirety of phenomenological being. The second and third notions of consciousness suggested by Husserl both presuppose the existence of consciousness in this sense of an entirety of phenomenological being. If there was no "bundle of interweaving psychic experiences," then neither could one be aware of one's own psychic experiences, nor could there be any intentional experiences.[24]

The notion of phenomenological being is in turn synonymous with the notion of *experiences* or *qualia*. The existence of the one is necessary and sufficient for the existence of the other. In fact, there is no difference between being the subject of an experience or qualia and being the subject of phenomenological being.[25] A stream of consciousness therefore can be an entirety of experiences or qualia. It is, in other words, what each of us takes to be his conscious life with all its experiential diversity and, precisely in this sense, it is an entirety. There is a sense according to which consciousness is always self-consciousness because consciousness is always consciousness of a conscious being and therefore *eo ipso* is consciousness by a *self*.[26] The existence of phenomenological being, experiences, or qualia is not possible without there being *someone*, i.e. a self in a minimal sense of the term, which is the subject of the phenomenological being (or, equivalently, of the experiences or qualia) in question.[27] This is not to say that to be conscious entails being *aware* of one's own consciousness.

---

24 Phenomenological being thus is a *conditio sine qua non* for intentional acts and inner awareness of oneself. Even if one were to distinguish more senses of the word consciousness, that it is a totality of phenomenological being is the most fundamental sense in which we can use the term consciousness. Cf. Pope and Singer (1978: 1) for a further analysis of the concept of the stream of consciousness.

25 The synonymy of those terms is also argued for by Chalmers. He is right in stating that "experience" is a term in line with the notions of "'qualia,' 'phenomenology,' 'phenomenal,' 'subjective experience,' and 'what it is like'. Apart from grammatical differences, the differences among these terms are mostly subtle matters of connotation. 'To be conscious' in this sense is roughly synonymous with 'to have qualia,' 'to have subjective experience,' and so on. Any differences in the class of phenomena picked out are insignificant" (Chalmers 1996: 6).

26 Heidegger states it thus: "Der Mensch hat Bewusstsein von Objekten und hat dabei auch ein Bewusstsein von sich, Selbstbewusstsein. *Jedes Bewusstsein ist auch Selbstbewusstsein*" (Heidegger 2001: 135, my italics). This fact is not only seen in the tradition of phenomenology, but also in recent analytic philosophy. Chalmers essentially states the same feature of experiences in terms of acquaintance: "My experiences are part of my epistemic situation and simply having them gives me evidence for some of my beliefs. All this is to say that there is something intrinsically epistemic about experience. To have an experience is automatically to stand in some sort of intimate epistemic relation to the experience – a relation that we might call 'acquaintance'" (Chalmers 1996: 196/197).

27 If an experience exists, then there is someone this experience is the experience of. As Foster states: "If P is a pain-sensation occurring at a certain time *t* [...] we should ultimately

4.2 *Anna's World (I): The Epistemological Parity of Other Human Beings and Androids*

Based on the concept of a conscious being, consider this: In the case of human beings, one is oneself the only being of which one can be sure that it is a conscious being because one is immediately conscious of this. The assumption that other human beings are conscious ends-in-themselves is the conclusion of an argument according to which this is the best explanation of the observed behaviour of other human beings.[28]

Suppose that a human being, call her Anna, without her being aware of this, grows up in a society of cleverly constructed androids that look and act like human beings although they are steered by weak artificial general intelligence. In this situation, the argument that leads us to ascribe consciousness to other human beings would lead Anna to the conclusion that her fellow androids are conscious beings: She would be as justified in ascribing consciousness to weak artificial general intelligence as we are justified in ascribing consciousness to our fellow human beings. Call this the argument from the epistemological parity of human beings and artificial general intelligence.

4.3 *Anna's World (II): Multiple Realisation versus Annanism*

The weakness of the argument from the epistemological parity of human beings and weak artificial general intelligence, however, seems to be grounded in the possibility that there could be a way of discovering that there are physical differences between artificial beings and human beings in their constitution and the way they operate inside.

The question is: What would happen in the scenario if Anna were to discover that she is a biological organism while all of her friends turn out to be machines run by weak artificial general intelligence implemented on sophisticated computers in the androids' heads?

It seems that from a philosophical point of view, Anna has two options for evaluating her newly discovered situation: Either she concludes that she is the only conscious being around – call this Annanism – or else she concludes that

represent the occurrence of P as the event of a certain subject's being in pain at t. And if D is a decision occurring at t, […] we should ultimately represent the occurrence of D as the event of a certain subject's taking a decision at t. Quite generally, […] we must represent each episode of mentality as the event of a subject's being in a certain mental state at a certain time, or performing a certain act at a certain time, or engaging in a certain mental activity over a certain period of time" (Foster 1991: 205).

28 Cf. Kaplan (2016: 82): "At the present time there's no credible way to establish whether computers and animals – or other people, for that matter – experience consciousness the same way we feel that we do."

consciousness must be multiply realizable, and that both her bio-chemical constitution and the physical constitution of her Android fellows are sufficient for consciousness.[29]

Is there any possible way for her to know whether Annanism or the multiple realizability of consciousness is true? In her situation, there does not seem to be any evidence in support of Annanism over and above the mere assumption that in the biological case consciousness exists, whereas it does not in the mechanical counterpart steered by artificial general intelligence. However, this assumption is arbitrary because it is based on an unjustified a priori assumption that consciousness has to be correlated with a biological human brain. This assumption is unjustified because it is based on the invalid inference from our happening to have a biological brain, and happening to be ends-in-ourselves, to the conclusion that consciousness in general has to be correlated with a human brain.

Any argument Anna could provide in favour of the assumption that consciousness arises only out of a biological human brain would not only exclude the possibility that animals and extra-terrestrial lifeforms that do not have a human brain could be conscious beings, but would also have a structurally isomorphic counterpart which her fellow robots could formulate and use to doubt whether she, Anna, in fact is a conscious being since, *obviously*, bio-chemical reactions in a hunk of grey matter could not be sufficient for a rich inner life. In the same way in which Anna would think: "This is ridiculous! Here I am: feeling, seeing, and experiencing the world!", her fellow robots would utter "This is ridiculous!" as a response to her argument from the apparently necessary biological correlate of consciousness. Any argument that could be formulated in the sketched scenario, apparently showing that consciousness has to have a biological (or mechanical) base it is grounded in, therefore would beg the question.

That we cannot formulate a sound argument that does not beg the question in favouring Annanism over multiple realisation, does not entail that Annanism is false. In the absence of a sound argument for multiple realisation we could only conclude that we cannot know whether what appears to be weak artificial general intelligence in fact should be referred to as strong artificial intelligence. We therefore need an argument in favour of multiple realisation.

---

29 There might be a third option: Anna might conclude that the question for the existence of consciousness is meaningless because consciousness, after all, does neither exist in her case nor in the case of her fellow beings. However, this option seems to me to be plainly absurd. Nihilism regarding consciousness cannot be justified because it presupposes the existence of consciousness.

Suppose, then, one day Anna's brain is taken sick and that due to her sickness, 1% of her brain is dying every day.[30] The only way to rescue her is to replace this 1% of her brain everyday with mechanical nanobots that take over the functions of the dying parts of her brain. On day 1, we can suppose, Anna does not feel any difference, neither on day two or three, twenty, or ninety-nine. Now, hundred days later, repeating this procedure every day, her android fellows replaced 100% of her brain with mechanical nanobots and Anna is and feels as before. In this scenario, Anna will be conscious on day 100 in the same manner she was on day 0 *because* her newly obtained mechanical brain is working in the same way as her old biological brain. But if this is possible, if consciousness is structurally invariant, then consciousness is multiply realizable in different physical structures.

If this argument does not convince, think of a first contact with an extraterrestrial life-form that visits Earth in its spaceship. Suppose further, we manage to communicate and interact with the life-form in a meaningful way. We soon suppose they have a sense of humour, that they get hungry, and that they enjoy classical music. Without knowing anything about whether they are biological organisms or machines themselves, would not the best explanation be that they are conscious beings and that, therefore, consciousness is multiply realizable?

Therefore, if the choice is between Annanism and multiple realisation, the better arguments speak in favour of the multiple realisation of consciousness. But if this is the case, then it follows that weak AGI whose behaviour is indistinguishable from the behaviour of intelligent human beings in fact should be considered strong AGI that is conscious of itself and therefore an end-in-itself at least in the same way in which a human being is conscious of itself and is an end-in-itself.[31]

What would probably happen in Anna's world, if Anna is scientifically inclined, is that Anna and the androids would work together on a scientific theory of those features they have in common and conclude that anything with these features is at least a conscious being on the same place in the hierarchy of consciousness as they are.

30 Cf. Chalmers (2016: 56ff).

31 As Kurzweil (2016: 309) predicts: "Differing philosophical assumptions about the nature and source of consciousness underlie disagreements on issues ranging from animal rights to abortion, and will result in even more contentious future conflicts over machine rights. My objective prediction is that machines in the future will appear to be conscious, that they will be convincing to biological people when they speak of their qualia, and that we will come to accept that they are conscious persons."

4.4 *Ethical Implications of the Existence of Strong Artificial General Intelligence*

In principle, it seems possible to construct weak AGI. If consciousness is indeed open to multiple realisation, however, then weak AGI, as the arguments have shown, should indeed be considered as an end-in-itself. Since the motif to create AGI in the first place was to create an end-in-itself and to create a new species populating the universe, it seems that it is at least in principle possible to realize this motive and to create an artificial *imago hominis*. However, if we should one day succeed in creating an artificial conscious being we would have to consider it as an end-in-itself like our fellow human beings. Then we would be morally obligated to treat it as a being with intrinsic moral worth. This means that we would have to grant it rights that ensure that it gets what it needs to live its life in an appropriate way, as long as it does not violate the rights of other conscious beings to do the same.[32]

5. Bibliography

Alpaydin, Ethem (2016) *Machine Learning*. Cambridge, Mass: The MIT Press.

Boden, Margaret A. (2005) "Introduction." In: Margaret A. Boden (eds.) *The Philosophy of Artificial Intelligence*. Oxford: Oxford University Press, 1–21.

Boden, Margaret A. (2018) *Artificial Intelligence. A Very Short Introduction*. Oxford: Oxford University Press.

Bostrom, Nick (2014) *Superintelligence. Paths, Dangers, Strategies*. Oxford: Oxford University Press.

Brooks, Rodney A. (1997) "Intelligence without Representation." In: John Haugeland (ed) *Mind Design II: Philosophy, Psychology, Artificial Intelligence*. Cambridge, Mass. THE MIT Press.

Chalmers, D. (1996) *The Conscious Mind. In Search of a Fundamental Theory*. Oxford: Oxford University Press.

Chalmers, David (2016) "The Singularity. A Philosophical Analysis." In: Uziel Awret (ed.) *The Singularity. Could artificial intelligence really out-think us (and would we want it to?)*. Exeter: Imprint Academic, 11–69.

Foster, J. (1991) *The Immaterial Self. A Defence of the Cartesian Dualist Conception of the Mind*. London: Routledge.

32 I am grateful to Stephen Priest, University of Oxford, for his comments on an earlier version of this paper and to Anna Sindermann for valuable discussions on the riddles of consciousness.

Göcke, Benedikt Paul (2012) "Introduction: After Physicalism." In: B. Göcke (ed.) *After Physicalism.* Indiana: The University of Notre Dame Press, 1–24.

Good, Irvin John (1965) "Speculations Concerning the First Ultraintelligent Machine." In: Franz L. Alt, Morris Rubinoff (eds.) *Advances in Computers.* New York: Academic Press, 31–88.

Greenfield, Susan (2016) "The Singularity: Commentary." In: Uziel Awret (ed.) *The Singularity. Could artificial intelligence really out-think us (and would we want it to?)* Exeter: Imprint Academic.

Heidegger, M. (2001) *Einleitung in die Philosophie.* Gesamtausgabe Band 27. Frankfurt am Main: Vittorio Klostermann.

Herzfeld, Noreen L. (2002) *In Our Image. Artificial Intelligence and the Human Spirit.* Fortress Press.

Husserl, Edmund (1984) *Husserliana 19: Logische Untersuchungen. Zweiter Band. Untersuchungen zur Phänomenologie und Theorie der Erkenntnis.* Den Haag: Martinus Nijhoff.

Hutter, Marcus (2016) "Can Intelligence explode?" In: Uziel Awret (ed.) *The Singularity. Could artificial intelligence really out-think us (and would we want it to?).* Exeter: Imprint Academic.

Kaplan, Jerry (2016) *Artificial Intelligence. What Everyone Needs to Know.* Oxford: Oxford University Press.

Kurzweil, Ray (2016) "Science versus Philosophy in the Singularity." In: Uziel Awret (ed.) *The Singularity. Could artificial intelligence really out-think us (and would we want it to?).* Exeter: Imprint Academic, 305–313.

Legg, S, M. Hutter (2007) "A Collection of definitions of intelligence." In: B. Goertzel and P. Wang (eds.) *Advances in Artificial General Intelligence: Concepts, Architectures and Algorithms.* Amsterdam: IOS Press, 17–24.

Moravec, Hans (1988) *Mind Children: The Future of Robot and Human Intelligence.* New York: Free Press.

Pope, K.; J. L. Singer (1978) *Instance of Consciousness. Scientific Investigations into the Flow of Human Experience.* Chichester: Wiley.

Priest, S. (2000) *The Subject in Question: Sartre's Critique of Husserl in 'The Transcendence of the Ego.'* London, New York: Routledge.

Ramge, Thomas (2018) *Mensch und Maschine. Wie Künstliche Intelligenz und Roboter unser Leben verändern.* Stuttgart: Reclam.

Schmidhuber, Jürgen (2016) "Philosophers & Futurists, Catch Up!" In: Uziel Awret (ed.) *The Singularity. Could artificial intelligence really out-think us (and would we want it to?)* Exeter: Imprint Academic.

Searle, John R (2005) "Minds, Brains, and Programs." In: Margaret A. Boden (ed.) *The Philosophy of Artificial Intelligence.* Oxford: Oxford University Press. 67–88.

Tegmark, Max (2016) *Life 3.0. Being Human in the Age of Artificial Intelligence.* Penguin Books.

Can Robots Have Dignity?

Carmen Krämer

Abstract

The progress towards a society in which robots are our daily attendants seems to be inevitable. Sharing our workplaces, our homes and public squares with robots calls for an exploration of how we want and need to organize our cohabitation with these increasingly autonomous machines. Not only the question of how robots should treat humans or the surrounding world, but also the questions of how humans should treat robots, and how robots should treat each other, may and should be asked. Considering the Kantian idea that possessing dignity is based on autonomy and the fact that robots are becoming increasingly autonomous and rational, one of these questions might be whether robots can have dignity. Two issues must therefore be addressed before answering the question: 1. What are robots and why should we think about "robot dignity" at all? and 2. What is dignity? The answer to the first question is necessary to understand the object of investigation and will be considered briefly. The second more complex question requires a short glimpse on the existing theories and the history of the term before a proposal will be given on how to understand dignity. Finally, it will be explained why robots cannot be rightly seen as possessors of dignity.

1. Introduction

Progress towards a society in which robots are our daily attendants seems inevitable and raises new ethical questions. Whereas, at first, ethical issues about robots were related to their impact on human life or to robot interaction with the surrounding world[1], with the development of autonomous, intelligent, and learning machines entirely new dimensions of ethical questioning have opened up. Some of these issues are closely related to the concept of dignity: On the one hand it can be considered if and how the use of autonomous robots, such as care robots, may have negative effects on human dignity (e.g.

1 Popular examples of these questions deal with responsibility for robots' mistakes and failures, whether it is acceptable to replace humans with machines, or what problems the ability of machines to kill people entails.

 | DOI:10.30965/9783957437488_016

Sharkey 2014). It can be questioned for instance, whether and how the permanent surveillance of patients – especially surveillance of patients who cannot give their consent – as an invasion of privacy by machines can be morally justified (e.g. Kreis 2018). On the other hand, and this is the new issue addressed here, it may be questioned if not only humans, but also robots can have dignity.

That humans may tend to ascribe dignity to robots can be seen from the discussion about the question if the little hitchhiking robot "Hitchbot" was "murdered", when he was found decapitated after travelling on passenger seats through Canada for a while (e.g. Wakefield 2019). Even though this is perhaps a questionable anthropomorphisation of a machine, some aspects could provide evidence for understanding robots as possessing dignity: Properties that are increasingly being attributed to robots like autonomy and intelligence are often seen as capabilities entailed by human dignity. When robots become increasingly similar to humans, sharing these significant properties besides an increasingly realistic outer appearance they might share with humans or other animals, it can be questioned if they have dignity or not. Thus, not only robots' interaction with the surrounding world, but also the robot itself need to be the subject of our moral discourse. Hence, if in fact they are seen as possessing dignity, it must be clarified whether this entails all rights and duties contained in the concept of dignity, as foundational of the charter of human rights and several constitutions. In order to identify a possible answer to this question, it is necessary to first reflect on what dignity is, how we use the term in human social interaction, and what this entails.

The first part of this chapter examines which kinds of robots might be relevant for the question explored here, and why these robots may or may not be considered as potentially possessing dignity. The second part addresses briefly conflicting theories about human dignity. In the third part, I propose an understanding of dignity on the basis of which the underlying question of whether robots can have dignity is answered.

2. Why Robots Might Possess Dignity

Delegating work to machines or sharing a workplace with a robot is becoming more common in some sectors. With this progress, a human dream often expressed in science fiction stories is coming true: Robots take over tasks, for example "jobs that are dull, dirty or dangerous" – the "'three Ds" (Lin 2014: 4). This includes jobs in automobile factories, military environments, exploring volcanoes, cleaning contaminated places, defusing bombs, lawn mowing, and polishing our houses (e.g. Lin 2014: 4f.). However, man has not only dreamt about robots doing the “three Ds”, but also about socially interacting with robots. Science fiction movies like for instance *Short Circuit* (1986) show robots with personality that are treated as friends or family members, and that – if we could ask the fictive protagonists – would be regarded as possessors of dignity with the right of being treated in a humane manner, with claims against being humiliated, offended or even “killed”.

A development from fantasy to reality may also be noted here: Robots are increasingly employed for labour and services, security, research and education, entertainment, healthcare, but also as companions (Scheutz 2014: 221). Examples for the latter include “Paro”, the therapeutic seal-robot, and sex-robots. Social and companion robots may not only have humanlike eyes, arms and legs, and capabilities like walking, working, playing, speaking, or having a body temperature (e.g. Koolwaay 2018). They are also able to learn, recognize voices or faces, “act” morally, they seem intelligent and, probably most important for this discussion: They are becoming increasingly autonomous. Thanks to these areas of application, they are becoming more and more important in human societies and in individual human lives (Bekey 2014: 19). People build relationships with robots (von der Pütten 2011). They talk to the lawnmower, give robot dolls or service robots names, and greet the vacuum cleaner, their computer or “Alexa”, when they come home. And a new stage along this path might seem to already have been reached, with “Sophia”, the first robot to have citizenship, in this case, in Saudi Arabia (Abbass 2017). Even if this was probably a publicity stunt, being respected as a citizen means, in a lot of states, to be legally treated as a person, and therefore with dignity.

With these relations towards robots, the creation of autonomous machines, and even the ascription of citizenship to a robot, the scope of the ethical discussion has widened. First questions about the moral status and rights for robots are being asked (see Gunkel 2012; Gunkel in this book). Although machines are not yet completely independent and autonomous (we still need to program them and push a button to switch them on or off) theorists such as Lin already expect the next step to greater autonomy (Lin 2014: 4–6). And even if there are still doubts about the possibility of completely autonomous machines, we should think about the consequences such progress might have *before* we make any serious mistakes. What consequences might arise from such developments? Do we really have to care about the dignity of Sophia in a manner comparable to a child’s dignity, for example by buying her clothes so that she won’t be naked? Is it possible to offend, humiliate, or to enslave her or to hurt or kill her, and is this morally wrong? All these questions seem to depend on whether robots have dignity or not. Therefore, the meaning of “dignity” needs to be clarified next.

3. Theories of Dignity

If humans and robots share any characteristics sufficient for the ascription of dignity and we thus would need to consider the moral (and legal) consequences for our treatment of robots, we are first and foremost in urgent need to clarify how we define dignity and why we consider humans to have dignity. Not surprisingly, the literature on the concept of dignity is abundant and controversial.[2] Although it offers various explanations and understandings of human dignity, I explore only the most significant positions: a) dignity as biologically given, b) dignity as given by God, c) dignity and human rights, d) dignity and specific capabilities and e) dignity as an empty concept, before proposing an understanding of human dignity, which suggests that robots *cannot* be rightly seen as possessing dignity.

Dignity as biologically given. We start with a very basic question: Does dignity presuppose the biological species and the genetics of human beings? Accepting this premise for ascribing dignity would mean that all non-human creatures, including not only robots but also non-human animals, are excluded as possessors of dignity. If that was the case, the discussion would have come to an end at this point already: we would have to conclude from determining that robots are genetically not anything (and especially not human), that they cannot possess dignity. However, this position is not convincing. Asking "Why do humans have dignity?" and answering "because they are humans" is circular reasoning, based on a naturalistic fallacy (Merkel 2003: 37). It does not prove that human beings have dignity just because they are humans and other animals that are genetically very similar, for instance chimpanzees, do not.

Dignity as given by God. The approach to explain human dignity as given by God is not convincing as well, because to make people believe in and respect this concept, one condition must be that they are convinced of the existence of God. However basing rights or concepts that should be accepted by everyone in a secular society on the idea of God is not reasonable (see Ladwig 2003: 38). Therefore, founding the ascription of dignity on God is not useful.

Dignity and human rights. Approaches that differ significantly from the previous considerations relate to human rights. The preamble of the Universal Declaration of Human Rights of the United Nations states "Whereas recognition of the inherent dignity and of the equal and inalienable rights of all members of the human family is the foundation of freedom, justice and peace in the world [...]". Some authors therefore argue that dignity *is* a human right (1) (e.g.

2 E.g. Tiedemann 2006; Baldus 2016; Von der Pfordten 2016; Weber-Guskar 2016.

Herdegen 2014), others claim that dignity is the *foundation* of human rights (2) (e.g. Bielefeldt 2011), and again others argue that it must be the *aim* of human rights (3) (e.g. Wetz 2008). For our purpose, we thus need to explore whether robots can have human rights. At the current state of technical research considering *human* rights for robots seems to go too far. Related to the discussion on robot rights, there is a huge discussion about rights for animals (e.g. Regan 2004; DeGrazia 2011; Grimm, Wild 2016; Schmitz 2015). They are not the same as *human* rights, but are discussed as specific *animal* rights, as the requirements of humans differ from those of animals. It makes no sense to ascribe the right to education, to paid holidays, or the right of equal access to public services to an animal in the same way as to humans. Similarly, asking for rights for robots would mean asking for *robot* rights (see Gunkel 2018), not *human* rights, because robots' requirements – if we can talk about robot needs at all – differ from those for humans.

But even if human rights for robots would be taken into consideration, this too would not argue for attributing dignity to robots, because the above-mentioned approaches to define dignity are not convincing as well. Explanation (1) is problematic because claiming that human dignity is a human right contains no indication of what that exactly means. The "human right of dignity" does not say anything about how to treat individuals who have that right or what human dignity might be and why humans should have it. Even if the thesis is refined to: "The *protection* of human dignity is a human right" it would not be clear what dignity is and how to protect it. Proposals (2) – Human dignity is the foundation of human rights, and (3) Human dignity is the aim of human rights, also do not say anything about what human dignity *is* (see Hossenfelder 2004: 22). Someone might say that human dignity seen as the aim of human rights explains it very well: When all human rights are respected, we call this a "life of dignity". But the problem here is that even if all humans have the same rights worldwide, there seems to exist no country where *all* human rights are respected. As a conclusion, that would mean that in fact nobody would have dignity. That is contradictory to what is stated in our various constitutions and the declaration of human rights, stating that all "members of the human family" have an inherent dignity, no matter where they come from, what their abilities are, and which gender they belong to. Consequently, if we want to find out what human dignity is, it is neither helpful to describe human dignity as a human right nor as the foundation or the aim of human rights and another solution needs to be found.

Dignity and specific capabilities. Referring to Kant, who based dignity on the capability of humans to act autonomously and reasonably, some modern theories connect possessing dignity with specific capabilities (see e.g. Tiedemann 2014). Following a theory[3] that takes autonomy as the basis for dignity, we would have to admit that all autonomous creatures have dignity. Following this line of argument, future autonomous robots or machines will have to be seen as creatures with dignity. However, at the same time some humans – those who are not able to act autonomously, like coma patients, or little babies – will not be understood as having dignity. This consequence is again contradictory to the idea of human rights and several constitutions, which determine that *all* human beings are possessors of dignity, no matter of their capabilities and if they are able to act autonomously or not. That means the current common understanding of dignity cannot be based on the idea of autonomy or rationality or any other property, such as the capacity for self-respect or self-esteem (like e.g. Schaber 2008; Nida-Rümelin 2002: 407). Another problem in basing dignity on any property or capability is that this does not explain what dignity is beyond simply having this property or capability. Thus, this attempt to explain the possession of dignity obviously fails as well.

Dignity as an empty concept. Because of this controversial discussion, that seems to offer no solution to the question of what dignity is, some authors, such as Norbert Hoerster, are convinced that dignity is an empty concept. Following Hoerster, dignity is an empty phrase, which is interpreted by anyone in a way she or he wants to understand it in order to strengthen her or his own argumentation (Hoerster 2004: 33). For instance, arguments such as those about assisted suicide run to nowhere when, on the one hand, opponents argue that assisted suicide transgresses dignity because it entails "killing" a person and, on the other hand, proponents of that practice argue that it offends the dignity of a person to leave her suffering without consensually liberating her from pain. But does this mean that "dignity" must be deleted as an idea? If this is the case this would mean that not only robots do not possess dignity, but also the idea of human dignity is obsolete. Why this cannot be a solution and how to solve the problem is the focus of the next section.

4. A Proposal for How to Understand "dignity"

Even if "dignity" is semantically under-determined and even if the term is sometimes used as a "conversation stopper" (Birnbacher 2001: 243), dignity remains the most important concept in some modern political constitutions.

3 That cannot be the theory of Kant, because he probably would not have seen robots as persons as ends in itself.

Apart from the mere fact that it seems to be impossible to delete "dignity" from such constitutions, it might also be dangerous to think about deleting dignity from legal documents. But what is dignity? To focus on my proposal for the current meaning of dignity and to understand how it is possible that dignity became one of the most important concepts of humanity while at the same time it remains one of the most controversial, confusing, and unclear notions, it is helpful to take a short look at its history. This in turn helps to uncover the meaning that, following my argument, it has today, and why it generates so many theories in recent years.

One of the first authors who mentioned human dignity in his work was the ancient Roman republican politician and thinker, Marcus Tullius Cicero (106–43 BC). However, Cicero did not have in mind any moral or legal status, but the abbacies of a judge or a priest for example, or the dignity of a person in demarcation from animals (Cicero 105–106). This concept of dignity refers more to the behaviour or societal status of individuals, rather than anything ethical, and was in Western Christendom seen as given by God (e.g. Tiedemann 2014, 60). Only with Kant did the idea of dignity change fundamentally, when he claimed that dignity is a value possessed by every human on account of his free will and his ability to act reasonably, autonomously, and morally (Kant, GMdS 434–435). These abilities distinguish humans from any other object and are the reason for ascribing them a special value. Thanks to the Kantian theory, the term dignity for the first time is connected with a general, secular idea of morality.

About 200 years later in the middle of the 20th century after the Second World War (1939–45), dignity is mentioned for the first time in a legal context: With its inclusion in the declaration of human rights (10th December 1948) the term again acquires a new connotation and is seen in a new relation to law. From a few years ago, using the concept of dignity in ethical discourse (such as bioethical, media-ethical, animal-ethical), became common, especially in the German-speaking world. With the admission of the term into the discourse of applied ethics, opponents and proponents of the same thesis used dignity for their own arguments. That led to the new, above-mentioned philosophical discourse on what dignity means.

This glimpse into the history of the term shows that during different periods "dignity" had heterogeneous meanings and that it may be seen as a "dynamic term" (see e.g. Sandkühler 2014: 18; Dan-Cohen 2012: 4), so it is clear why "What is human dignity?" is not easy to answer and why there are so many conflicting explanations. But still "dignity" is an expression that is often intuitively understood when it is used in specific contexts. Examining examples in which dignity is mentioned and considering the terms history, it appears that we can distinguish two different meanings of dignity that both are related to social interaction. When using the concept, we do not refer to a value, a right, or a property as such, but rather use it as a term to express specific situations. Case 1), when we talk about the violation of dignity through acts like torture and slavery seems different from 2) saying that dignity is transgressed when somebody offends or insults another person. Here it is explicitly not meant that there are two co-existent *kinds of dignity*, such as Schaber assumes (Schaber 2008: 190f). Instead, one term is used to describe two different aspects of social interaction. When we refer to dignity as violated by torture or slavery (1), we actually mean that the moral or legal status of a human has not been respected. This is what I will call a violation of "*human* dignity". When instead we talk about deriding, offending or insulting a person (2), we talk about "dignity of *a person*". The latter is not related to the moral or legal status as the former but could rather refer to honour or a very individual aspect of that particular person that can be violated in very different ways, such as calling someone ugly or an idiot.

Hence, with the admission of the term "human dignity" to legal documents after the Second World War, the consensus was determined that from now on all humans without regard for origin, religion, gender, age or other contingent properties must be taken into account morally and those with such properties treated as legally equal (Krämer 2018: 157). That means that the term "human dignity" is used to express that all humans have the same moral and legal status; but, at the same time, the term does not say anything about why humans should have that status and what it consists in to have it. Though, what can be said about it is that the equal protection and respect for every single human individual, from and by the state, are formulated in the declaration of human rights. That does, however, not mean, that human dignity is equal to a human right. Nor is human dignity the foundation, or the aim of human rights, but the equal status of all humans shall be protected and guaranteed by human rights (Krämer 2018: 157). In consequence, talking about violence against *human* dignity, means disrespecting the moral or legal status of a person, whereas talking about transgressing the dignity *of a person* refers to the – very individual – honour of a person, that can be educationally, culturally or politically influenced. A person A that grew up in a certain culture could for example be convinced that a certain gesture of another person B from a different culture is made to "hurt A's dignity", while in the culture of B this gesture is used to show person A his or her respect. First of all, this shows that in contrast to violating so-called "human dignity" what really hurts this kind of dignity cannot be generalized. And if B instead wanted to offend A, we would admit that this was morally wrong, but we would not agree to say that B did a cruel crime towards the "human dignity" of A, but that B tried to offend or insult A. This shows that here the "dignity of the person" was transgressed.

The problems about defining and using 'dignity' in ethical arguments arise because we do not distinguish precisely these different meanings of dignity that obviously exist. "Dignity" can be substituted for by synonyms, such as "moral status" or "honour". That does not mean, that dignity is an empty phrase as some authors claim. But it indicates, that we do not necessarily need the term "dignity" to express what we want to say. If someone is tortured we can say precisely that his or her moral or legal status is not respected and that his or her rights to freedom, right to liberty and security as well as the right to autonomy, the right to freedom of pain, and so on, are disregarded. Even if it may sound more impressive to say that human dignity is violated, we do not necessarily need the term "dignity" to say that something really cruel happens to somebody and, in some cases, instead of making clear what the issue is, using 'dignity' to describe it is more confusing. That "dignity" is mostly not used precisely, that there are such contradictory discussions about it, and that it is always needed to make clear what exactly the present issue is, lead to the recommendation to avoid the term in ethical arguments. Instead of using the claim that a person's dignity is hurt, preferably it should be said directly and concretely what the moral or legal problem is about. Talking about the moral or legal problems of, for instance, torture then means taking into consideration which rights are violated by the act of torture. Talking about someone being offended or hurt by an insult, one should name it as such and not just claim that his or her dignity is hurt.

5. Consequences for Discourse about Robots and Dignity

What impact does this proposed understanding of dignity have on the discussion about the dignity of robots? The most important outcome of my theory is that talking about "dignity" is not talking about a right, a specific property, capability or a special value that somebody has by birth, or by "creation" in the case of robots, but talking either about moral or legal equal status, or about the honour of a person or a robot. My thesis thus leads to the conclusion that it is better to avoid the term "dignity" in ethical discussions and instead to mention by name what we have concretely in mind. If by any act a human right (or – if we want to assume, that robots have rights – a robot right), is neglected, instead of calling this "violating dignity", it could be said that the moral or legal status is not respected. If by any act, for instance by an insult or offense, the honour of a person or a robot – also here: only if we think that it is possible to offend robots – is violated, it should be named as that instead of claiming that dignity is hurt. Thus, accepting the thesis that dignity as an under-determined and inaccurately used expression that should be avoided in discourses about ethical issues, the consequence must be to not take into consideration the ascription of dignity to robots.

As a consequence, whenever we explore the moral status and treatment of robots, we should not use the term “dignity”, because this may create confusion. Instead we should first of all, to meet the former meaning, substantiate whether robots have a moral status and, if so, on what basis they must be considered as moral or legal agents with rights and duties. That means concretely, in case we accept that robots may have a moral or legal status, instead of asking if Hitchbot’s dignity was violated, we should ask, if by destroying and decapitating it, Hitchbot’s rights were violated. An affiliating question here could be whether there is a concept of an adequate robot life or not. And if so it must be clarified what this adequate robot life looks like, what this means for humans treating robots and for robots’ relationships to each other.

To address the second meaning of dignity, we must consider the question whether robots may have honour, and if it is possible to offend or insult them. More concretely, following my argument, the question is not whether the dignity of a person of Sophia is transgressed if we leave her running around naked, but whether Sophia might feel ashamed of being naked because she has (the property of having) a feeling of honour? If even in the current state of technology we would say that robots do not have any feeling of shame or honour, we could go further and think about whether and why we should (not) develop robots with the ability to feel insult, shame or honour. And if we start building machines with these abilities, we need to clarify how they could be insulted or offended and if this is morally right or wrong. This includes not only the question how humans should treat robots with these abilities, but also how robots should treat each other.

Answering these questions cannot be the aim of the present paper as they are still vividly discussed for several groups of humans such as foetus, coma patients and brain-dead persons. But these reflections can help avoid making the mistake of thinking about a robot dignity that probably leads to ending up in a dead end in future discussions.

6. Bibliography

Abbass, Hussein (2017) *An AI professor explains: three concerns about granting citizenship to robot Sophia.* In: The Conversation. Online at: http://theconversation.com/an-ai-professor-explains-three-concerns-about-granting-citizenship-to-robot-sophia-86479 Accessed: January 11th 2020.

Baldus, Manfred (2016) *Kämpfe um die Menschenwürde. Die Debatten seit 1949*. Berlin: Suhrkamp.

Bekey, George A. (2014) *Current Trends in Robotics: Technology and Ethics.* In: Lin, Patrick; Abney, Keith; Bekey Geroge A (eds.) Robot Ethics. The ethical and social implications of robotics. Cambridge/London: MIT Press, 17–34.

Bielefeldt, Heiner (2011) *Auslaufmodell Menschenwürde? Warum sie in Frage steht und warum wir sie verteidigen müssen*. Freiburg im Breisgau: Herder Verlag.

Birnbacher, Dieter (2001) *Instrumentalisierung und Menschenwürde. Philosophische Anmerkungen zur Debatte um Embryonen- und Stammzellforschung.* In: Jahrbuch der Heinrich-Heine-Universität Düsseldorf, 243–257. Online at: http://dup.oa.hhu.de/8/1/pagesbirnbacher.pdf Accessed: December, 15th 2015.

Cicero, Marcus Tullius (1976) *De officiis. Vom pflichtgemäßen Handeln. Lateinisch/Deutsch.* Übersetzt, kommentiert und herausgegeben v. Heinz Gunermann. Stuttgart: Reclam.

Dan-Cohen, Meir (2012) *Introduction: Dignity and Its (Dis)content.* In: Waldron, Jeremy: Dignity, Rank and Rights. With Commentaries by Dimock, Wai Chee; Herzig, Don; Rosen, Michael. Ed. By Dan-Cohen, Meir. Oxford: Oxford University Press, 3–10.

DeGrazia, David (2011) *Animal Rights. A Very Short Introduction.* Oxford: Oxford University Press.

Dürig, Günter (1956) *Der Grundrechtssatz von der Menschenwürde*. In: Abhandlungen. Archiv des öffentlichen Rechts. Mohr Siebeck: Tübingen, 117–157. Online at: http://www.digizeitschriften.de/dms/toc/?PPN=PPN345574893_0081 Accessed: June, 18th 2015.

Grimm, Herwig; Wild Markus (2016) *Tierethik zur Einführung*. Hamburg: Junius.

Gunkel, David J. (2018) *Robots Rights.* Cambridge: MIT Press.

Gunkel, David J. (2012) *The Machine Question: Critical Perspectives on AI, Robots and Ethics*. Cambridge/London: MIT Press.

Herdegen, Matthias (2014) *Abs. 1, Art. 1. Rn. 36*. In: Maunz-Dürig: Grundgesetz Kommentar. Bd. 1. München: Verlag C.H. Beck, 1–74.

Hoerster, Norbert (2004) *Haben Tiere eine Würde?* München: C.H. Beck.

Hossenfelder, Malte (2004) *Menschenwürde und Menschenrecht*. In: Menschenwürde. La dignité de l'être humain. Jahrbuch der schweizerischen Philosophischen Gesellschaft. Vol. 63/2004. Basel: Schwabe Verlag, 17–33.

Kant, Immanuel (2008) *Grundlegung zur Metaphysik der Sitten.* ed. by Theodor Valentiner. Stuttgart: Reclam.

Koolwaay, Jens (2018) *Die soziale Welt der Roboter. Interaktive Maschinen und ihre Verbindung zum Menschen.* Bielefeld: Transcript.

Krämer, Carmen (2018) *Menschenwürde – Flucht – Medien. Reflexionen zu Ziffer 1 des deutschen Pressekodexes im Kontext der Medienberichterstattung über Flucht.* In:

Köberer, Nina; Prinzing, Marlis; Schröder, Michael (eds.) Migration, Integration, Inklusion. Medienethische Herausforderungen und Potenziale für die digitale Mediengesellschaft. Band 8. Baden-Baden: Nomos, 151–161.

Kreis, Jeanne (2018) *Umsorgen, überwachen, unterhalten – sind Pflegeroboter ethisch vertretbar?* In: Bendel, O. (eds) Pflegeroboter. Wiesbaden: Springer Gabler, 213–228. Online at: https://link.springer.com/chapter/10.1007/978-3-658-22698-5_12#citeas Accessed: January 11th 2020.

Ladwig, Bernd (2003) *Ist "Menschenwürde" ein Grundbegriff der Moral gleicher Achtung? Mit einem Ausblick auf Fragen des Embryonenschutzes.* In: Stoecker, Ralf (ed.): Menschenwürde. Annäherung an einen Begriff. Wien: öbv&hpt, 35–60.

Lin, Patrick (2014) *Introduction to Robot Ethics.* In: Lin, Patrick; Abney, Keith; Bekey Geroge A (eds.) Robot Ethics. The ethical and social implications of robotics. Cambridge/London: MIT Press, 3–15.

Merkel, Reinhard (2003) *Contra Speziesismusargument: Zum normativen Status des Embryos und zum Schutz der Ethik gegen ihre biologistische Degradierung.* In: Damschen, Gregor; Schönecker, Dieter (eds.): Der moralische Status menschlicher Embryonen. Pro und contra Spezies-, Kontinuums-, Identitäts- und Potentialitätsargument. Berlin, New York: De Gruyter, 35–58.

Nida-Rümelin, Julian (2002) *Ethische Essays.* Frankfurt am Main: Suhrkamp.

von der Pütten, **A. M.**, Krämer, N. C., & Eimler, S. C. (2011). Living with a Robot Companion – Empirical Study on the Interaction with an Artificial Health Advisor. *Proceedings of the 11th International Conference on Multimodal Interaction, ICMI'11*, November 14–18, 2011, Alicante, Spain. Doi: 10.1145/2070481.2070544.

Regan, Tom (2004). *The case for animal rights.* Berkeley: University of California Press.

Sandkühler, Hans Jörg (2014) *Menschenwürde und Menschenrechte. Über die Verletzbarkeit und den Schutz der Menschen.* Freiburg/München: Verlag Karl Alber.

Schaber, Peter (2008) *Der Anspruch auf Selbstachtung.* In: Härle, Wilfried; Vogel, Bernhard (eds.) Begründung von Menschenwürde und Menschenrechten. Freiburg: Herder, 188–201.

Scheutz, Matthias (2014) *The Inherent Dangers of Undirectional Emotional Bonds between Humans and Social Robots.* In: Lin, Patrick; Abney, Keith; Bekey Geroge A (eds.) Robot Ethics. The ethical and social implications of robotics. Cambridge/London: MIT Press, 205–221.

Schmitz, Friederike (2015) *Tierschutz, Tierrechte oder Tierbefreiung?* Menschenrechtsmagazin 20 (2), 87–96.

Sharkey, Amanda (2014): *Robots and human dignity: a consideration of the effects of robot care on the dignity of older people.* In: Ethics and Information Technology. Vol. 16, Issue 1, 63–75. Online at: https://link.springer.com/article/10.1007/s10676-014-9338-5#citeas Accessed: January 11th 2020.

Tiedemann, Paul (2014) *Was ist Menschenwürde? Eine Einführung*. Darmstadt: Wissenschaftliche Buchgesellschaft.

Von der Pfordten, Dietmar (2016) *Menschenwürde*. München: C.H. Beck.

Wakefield, Jane (2019) *Can you murder a robot?* In: BBC News. Online at: https://www.bbc.com/news/technology-47090174 Accessed: January 11th 2020.

Wasmaier-Sailer, Margit; Hoesch Matthias (eds.) (2017) *Die Begründung der Menschenrechte. Kontroversen im Spannungsfeld von positivem Recht, Naturrecht und Vernunftrecht*. Tübingen: Mohr Siebeck.

Weber-Guskar, Eva (2016) *Würde als Haltung. Eine philosophische Untersuchung zum Begriff der Menschenwürde*. Münster: Mentis.

Wetz, Franz Josef (ed.) (2011) *Texte zur Menschenwürde*. Stuttgart: Reclam.

Wetz, Franz Josef (2008) *Menschenwürde – Eine Illusion?* In: Härle, Wilfried; Vogel, Bernhard (eds.) Begründung von Menschenwürde und Menschenrechten. Freiburg: Herder, 27–48.

The Right(s) Question: Can and Should Robots Have Rights?

David J. Gunkel

Abstract

In a recent proposal issued by the European Parliament it was suggested that robots and artificial intelligence might need to be considered "electronic persons" for the purposes of social and legal integration. The very idea sparked controversy, and it has been met with considerable resistance. Underlying the controversy, however, is an important philosophical question: *Under what conditions* would it be necessary for robots, AI, or other socially interactive, autonomous systems to have some claim to moral and legal standing? *Under what conditions* would a technological artifact need to be considered more than a mere instrument of human action and have some legitimate claim to independent social status? Or to put it more directly: Can or should robots ever have anything like rights? This essay takes up and investigates these questions. It reviews and critiques current thinking (or lack of thinking) about this subject matter, maps the terrain of the set of available answers that have been provided in the existing literature, and develops an alternative way of responding to and taking responsibility for the opportunities and challenges that we now confront in the face or the faceplate of increasingly social and interactive robots.

1. The "Abortion Question" in Robotics

A few years ago, I posted this image (figure 1) online in an effort to initiate debate about the moral and legal status of technological artifacts, like artificial intelligence (AI) and robots. Responses to the photograph were, as one might imagine, polarizing with advocates and critics lining up on opposite sides of the issue. In fact, what I did not fully appreciate at the time is that this subject is something like the "abortion question" in robotics – a divisive issue that produces very strong opinions both for and against.

In response, I have now researched and published a book, called *Robot Rights*, which analyzes, investigates, and explicates this phenomenon (Gunkel 2018). Instead of attempting to replay the rather detailed argument that is

 DOI:10.30965/9783957437488_017

FIGURE 16.1 "Robot Rights Now." Photo by the author (2014).

developed in the published text – which would be difficult to accomplish in the course of a short essay – what I want to pursue in this chapter will be more of an overview of the project's general approach and consequences.

The investigation is motivated and framed by a rather simple query: "Can and Should Robots have rights?" Responding to this question, however, turns out to be rather involved and complicated. So here's the plan of attack. First, I will begin by asking about the question. For many theorists and practitioners, the question of robot rights has been declared to be unthinkable. My question is why? How and why is the question unthinkable? And what does it mean when inquiry is dismissed, avoided, or marginalized by this kind of declaration? Second, I want to look at and analyze the form of the question itself. There is an important philosophical difference between the two modal verbs that organize the inquiry – "can" and "should." This difference has considerable history behind it, and that legacy influences what is asked about and how one proceeds to formulate an answer. Third, capitalizing on this pivotal verbal distinction, I will identify and evaluate the four possible modalities concerning artifacts and the question of rights. This will take the form of a cost/benefit analysis regarding the range of available thinking on this matter. Finally, I will conclude by proposing another alternative, a method of/for *thinking otherwise* that effectively challenges the rules of the game and provides for other ways of theorizing moral standing that can scale to the unique opportunities and challenges that are confronted in the face (or faceplate) of the robot.

2. Thinking the Unthinkable

"The notion of robot rights," as Seo-Young Chu (2010: 215) points out, "is as old as is the word 'robot' itself. Etymologically the word 'robot' comes from the Czech word 'robota,' which means 'forced labor.' In Karel Čapek's 1921 play R.U.R. , which is widely credited with introducing the term 'robot,' a 'Humanity League' decries the exploitation of robot slaves – 'they are to be dealt with like human beings,' one reformer declares – and the robots themselves eventually stage a massive revolt against their human makers." For many researchers and developers, however, the very idea of "robot rights" is simply unthinkable. Versions of this kind of immediate and largely unquestioned dismissal circulate in the work of many noted and established experts in the field:

> *Noel Sharkey* – "The idea of machine consciousness and rights is a distraction, it's fairy tale stuff. We need proper informed debate, about the public safety about for instance the millions of domestic robots that are predicted to be arriving in the next few years" (Henderson 2007 and Marks 2007).
>
> *Alan Winfield* – "I know of no one within the serious robotics community that would use that phrase, 'robot rights'" (Henderson 2007).
>
> *Luciano Floridi* – "It may be fun to speculate about such questions, but it is also distracting and irresponsible, given the pressing issues we have at hand" (Floridi 2017: 4).

As David Levy (2005: 393) has accurately summarized it, the very "notion of robots having rights is unthinkable." It is unthinkable not only in terms of not being able to be thought by way of the available conceptual apparatus – specifically the instrumental theory of technology (Heidegger 1977 and Feenberg 1991). But also in the sense that is it something to be deliberately avoided and not submitted to thought insofar as it is considered a distraction or fairy tale stuff.[1] For many in the field, the question of robot rights or the moral/legal status of AI is a kind of shiny object that attracts a lot of attention but is not worth the effort of serious scientific thinking. To make matters worse,

1 Those judging the thought unthinkable might face a difficulty analogous to that faced by the logical positivists: If you are able to judge a thought unthinkable you must have thought it.

these dismissals are more often than not issued as declarations and typically not supported by sufficient evidence or argument. My worry in all of this is that purposefully avoiding a question by declaring it to be unthinkable sounds more like an effort to protect existing orthodoxies and less like science. Instead of dismissing the question out-of-hand as unthinkable, it would be better to ask, to investigate, and to make an informed decision based not on preconceived notions and prejudice but on actual evidence and data.

3. The Is-Ought Problem

The guiding question of the investigation consists of two separate queries: "Can robots have rights?" which is a question that asks about the ontological capabilities of a particular entity; and "Should robots have rights?" which is a question that inquires about normative obligations in the face of this entity. These two questions invoke and operationalize a rather famous conceptual distinction in philosophy that is called the is/ought problem, or Hume's Guillotine. In *A Treatise of Human Nature* (first published in 1738), David Hume differentiated between two kinds of statements: descriptive statements of fact and normative statements of value (Schurz 1997: 1). For Hume, the problem was the fact that philosophers, especially moral philosophers, not only fail to distinguish between these two categorically different kinds of statements but (and because of this) often slip imperceptibly from one to the other:

> In every system of morality, which I have hitherto met with, I have always remarked, that the author proceeds for some time in the ordinary ways of reasoning, and establishes the being of a God, or makes observations concerning human affairs; when all of a sudden I am surprised to find, that instead of the usual copulations of propositions, *is*, and *is not*, I meet with no proposition that is not connected with an *ought*, or an *ought not*. This change is imperceptible; but is however, of the last consequence. For as this *ought*, or *ought not*, expresses some new relation or affirmation, 'tis necessary that it should be observed and explained; and at the same time that a reason should be given, for what seems altogether inconceivable, how this new relation can be a deduction from others, which are entirely different from it. But as authors do not commonly use this precaution, I shall presume to recommend it to the readers; and am persuaded, that this small attention would subvert all the vulgar systems of morality, and let us see, that the distinction of vice and virtue is not founded merely on the relations of objects, nor is perceived by reason (Hume 1980: 469).

Since Hume, there have been numerous attempts to resolve this "problem" by bridging the gap that is supposed to separate "is" from "ought" (cf. Searle 1964). Despite these efforts, however, the problem remains, and is considered one of those important and intractable philosophical dilemmas that people write books about (Schurz 1997 and Hudson 1969). As Schurz (1997: 4) characterizes it: "probably the most influential debate on the is-ought problem in our time is documented by Hudson (1969, reprint 1972, 1973, 1979). Here Black and Searle have tried to show that logically valid is-ought-inferences are indeed possible, whereas Hare, Thomson and Flew have defended Hume's thesis and tried to demonstrate that Black's and Searle's arguments are invalid."

For our purposes what is most important is not the logical complications of the "is-ought fallacy" as Hume originally characterized it or the ongoing and seemingly irresolvable debate concerning the validity of the is-ought inference as it has been developed and argued in the subsequent literature. What is important for the investigation at hand is to recognize how the verbs "is" and "ought" organize qualitatively different kinds of statements and modes of inquiry. The former concerns ontological matters or statements of fact; the latter consists in axiological decisions involving what should be done. The guiding question of our inquiry utilizes modal variants of these two verbs, namely "can" and "should." Employing Hume's terminology, the question "Can robots have rights?" may be reformulated as: "Are robots capable of being holders of rights?" And the question "Should robots have rights?" can be reformulated as "Ought robots be considered holders of rights?" The modal verb "can," therefore, asks an ontologically oriented question about the factual capabilities or properties of the entity, while "should" directs inquiry to axiological issues having to do with obligations to this entity. Following the Humean thesis, therefore, it is possible to make the following distinction between two different kinds of statements:

> *S1* = "Robots can have rights."
> *S2* = "Robots should have rights."

By affirming and/or denying each one of these statements in different combinations, we can derive four options or modalities concerning the moral situation of robots and other artifacts. These four modalities can be organized into two pairs. In the first pair, which upholds and supports the is-ought inference, the affirmation or negation of the ontological statement (S1) determines the affirmation or negation of the axiological statement (S2). This can be written (using a kind of pseudo object-oriented programming code) in the following way:

> *!S1 !S2* – "Robots cannot have rights. Therefore robots should not have rights."
> *S1 S2* – "Robots can have rights. Therefore robots should have rights."

In the second pair, which endorses the Humean thesis or contests the inference of ought from is, we affirm the ontological statement (S1) while denying the axiological statement (S2), or vice versa. These two modalities may be written in the following way:

> *S1 !S2* – "Although robots can have rights, they should not have rights."
> *!S1 S2* – "Even if robots cannot have rights, they should have rights."

4. Modalities of Robot Rights, or the Is/Ought Variations

In the four sections that follow, we will critically evaluate each one of these modalities as they are deployed and developed in the literature, performing a kind of cost/benefit analysis of the available arguments concerning the rights (or lack thereof) of robots and similar forms of autonomous technology.

4.1 *!S1 !S2 – Cannot and Should Not*

In this first case, we deny or negate both statements. Robots are incapable of having rights, and for this reason robots should not have rights. This is a position that is formulated by strictly applying the instrumental theory of technology. According to Martin Heidegger's analysis in "The Question Concerning Technology," the presumed role and function of any kind of technological artifact, whether it be the product of handicraft or industrialized manufacture, is that it is a means employed by human users for specific ends: "We ask the question concerning technology when we ask what it is. Everyone knows the two statements that answer our question. One says: Technology is a means to an end. The other says: Technology is a human activity. The two definitions of technology belong together. For to posit ends and procure and utilize the means to them is a human activity" (Heidegger 1977: 4–5). Heidegger calls this particular characterization "the instrumental definition" and he indicates that it forms what is considered to be the "correct" understanding of any kind of technological contrivance or artifact. As Deborah Johnson (2006: 197) describes it with regards to information technology: "Computer systems are produced, distributed, and used by people engaged in social practices and meaningful pursuits. This is as true of current computer systems as it will be of future computer systems. No matter how independently, automatic, and interactive computer systems of the future behave, they will be the result (direct or indirect) of human behavior, human social institutions, and human decision."

This way of thinking is persuasive, precisely because it reaffirms long standing assumptions about human exceptionalism and technological instrumentalism. Technologies, no matter how sophisticated, interactive, or social they appear to be, are just tools, nothing more. They are not – not now, not ever – capable of becoming moral subjects in their own right, and we should not treat them as such. It is precisely for this reason that, as J. Storrs Hall (2001: 2) points out, "we have never considered ourselves to have moral duties to our machines." The problem is that this position is disputed by much of the available data produced by and derived from human robot interaction (HRI) studies undertaken by social scientists like Astrid Rosenthal-von der Pütten et al. (2013), Peter Kahn et al. (2012), Christoph Bartneck and Jun Hu (2008), Patric R. Spence et al. (2014), and many others.

These finding are not necessarily new or surprising. Evidence was already developed and demonstrated with Fritz Heider and Mariane Simmel's "An Experimental Study of Apparent Behavior" (1944), which found that human subjects tend to attribute motive and personality to simple animated geometric figures, treating them *as if* they had or actually possessed these traits. Similar results have been obtained by way of the computer as social actor (CASA) studies conducted by Byron Reeves and Clifford Nass in the mid-1990s. As Reeves and Nass discovered across numerous trials with human subjects, users (for better or worse) have a strong tendency to treat socially interactive technology, no matter how rudimentary, as if they were other people.

> Computers, in the way that they communicate, instruct, and take turns interacting, are close enough to human that they encourage social responses. The encouragement necessary for such a reaction need not be much. As long as there are some behaviors that suggest a social presence, people will respond accordingly. When it comes to being social, people are built to make the conservative error: When in doubt, treat it as human. Consequently, any medium that is close enough will get human treatment, even though people know it's foolish and even though they likely will deny it afterwards. (Reeves and Nass 1996: 22)

So what we have is a situation where our theory of technology – a theory that has considerable history behind it and that has been determined to be as applicable to simple hand tools as it is to complex computer systems – seems to be out of sync with the very real practical experiences we now have with artifacts in a variety of situations and circumstances. To put it in terminology that is more metaphysically situated: How something appears to be – how it actually operates in real social situations and circumstances – might be more important than what it actually is (or has been assumed to be). "We should," as Tony Prescott (2017: 146) concludes formulating a kind of phenomenological moral maxim, "take into account how people *see* robots, for instance, that they may feel themselves as having meaningful and valuable relationships with robots, or they may see robots as having important internal states, such as the capacity to suffer, despite them not having such capacities." To put it in a quick vernacular form: even when our head tells us it is just a robot, our heart cannot help but like it, and we may need to respond to and take responsibility for this phenomenon.

4.2 *S1 S2 – Can and Should*

The flipside to the strict instrumentalist position is to affirm both statements. This is the position espoused by many who support speculative ideas and hypotheses like superintelligence, the singularity, and artificial general intelligence (AGI). "The 'artificial intelligence' programs in practical use today," Ben Goertzel (2002: 1) admits, "are sufficiently primitive that their morality (or otherwise) is not a serious issue. But that will not remain for long. Not too far in the future, however, things are going to be different. AI's will possess true artificial general intelligence (AGI), not necessarily emulating human intelligence, but equaling and likely surpassing it. At this point, the morality or otherwise of AGI's will become a highly significant issue."

There are obvious advantages to this way of thinking insofar as it does not simply deny rights to robots *tout court*, but kicks the problem down the road and postpones decision making. Right now (so the argument goes) we do not have robots that *can be moral subjects*, but when we do (and for the individuals who mobilize this way of thinking it is more often a question of "when" as opposed to "if"), then we will need to consider whether they should be treated differently. "As soon as AIs begin to possess consciousness, desires and projects," Robert Sparrow (2004: 203) argues, "then it seems as though they deserve some sort of moral standing." Or as Peter Singer and Agata Sagan (2009) write "if the robot was designed to have human-like capacities that might incidentally give rise to consciousness, we would have a good reason to think that it really was conscious. At that point, the movement for robot rights would begin."

This way of thinking is persuasive, precisely because it recognizes the actual limitations of current technology while holding open the possibility of something more in the not-too-distant future. This wait-and-see approach also helps us contend with the moral intuitions that are encounterd in the face of social robots. Although these mechanisms are intentionally designed to trigger emotions it should be remembered, for now at least, that inside the box there is nothing – no thought, no consciousness, no sentience. If and when these machines can achieve something more, then one will be justified in considering the rights question. But for now these social behaviors are just cleverly manipulated effects without a real cause.[2]

---

2 Operating behind the scenes or in the margins is the persistent epistemological problem of other minds. It is, for instance, difficult to know under what conditions one may be ascribing too little or too much to the device. For more on this problem, see Gunkel (2020).

The main problem to this way of thinking, however, is that it does not really resolve the question regarding the rights of robots but just postpones the decision to some indeterminate point in the future. It says, in effect, as long as robots are not conscious or sentient or whatever ontological criteria count, no worries. Once they achieve this capability, however, then we should consider extending some level of moral concern and respect. All of which means, of course, that this "solution" to the question "can and should robots have rights?" is less a solution and more of a decision not to decide. Furthermore when the decisive moment (whenever that might be and however it might occur) does in fact come, there remains several theoretical and practical difficulties that make this way of thinking much more problematic than it initially appears to be.

First, there are terminological complications. A qualifying criterion like "consciousness," for example, does not admit of a univocal characterization, but denotes, as Velmans (2000: 5) points out, "many different things to many different people." In fact, if there is any general agreement among philosophers, psychologists, cognitive scientists, neurobiologists, AI researchers, and robotics engineers regarding consciousness, it is that there is little or no agreement when it comes to defining and characterizing the concept. Other properties, like sentience, unfortunately do not do much better. As Daniel Dennett argues in his eponymously titled essay, the reason "why you cannot make a computer that feels pain" has little or nothing to do with the technical challenges with making pain computable. It proceeds from the fact that we do not know what pain is in the first place. In other words, "there can be," as Dennett (1998: 228) concludes, "no true theory of pain, and so no computer or robot could instantiate the true theory of pain, which it would have to do to feel real pain."

Second, even if it were possible to resolve these terminological difficulties, maybe not once and for all but at least in a way that would be widely accepted, there remains epistemological limitations concerning detection of the capability in question. How can one know whether a particular robot has actually achieved what is considered necessary for something to have rights, especially because most, if not all of the qualifying capabilities or properties are internal states-of-mind? This is, of course, connected to what philosophers call the problem of other minds, the fact that, as Donna Haraway (2008: 226) cleverly describes it, we cannot climb into the heads of others "to get the full story from the inside." Although philosophers, psychologists, and neuroscientists throw considerable argumentative and experimental effort at this problem, it is not able to be resolved in any way approaching what would pass for definitive evidence, strictly speaking.

4.3 *S1 !S2 – Can but Should Not*

In opposition to these two approaches, there are other responses that seek to uphold the is/ought distinction by affirming one statement and denying the other. Since there are two statements, this strategy can take one of two forms. In its first formulation, one affirms that robots can have rights but denies that this fact requires us to accord them social or moral standing. This is the argument that has been developed by Joanna Bryson beginning with the provocatively titled essay "Robots Should Be Slaves" (2010). Bryson's argument goes like this: Robots are property. No matter how capable they are, appear to be, or may become; we are obligated not to build things to which we would be obligated. Or as she describes it in a more recent article: "It is unquestionably within our society's capacity to define robots and other AI as moral agents and patients. In fact, many authors (both philosophers and technologists) are currently working on this project. It may be technically possible to create AI that would meet contemporary requirements for agency or patiency. But even if it is possible, neither of these two statements makes it either necessary or desirable that we should do so" (Bryson 2016: 206).

To its credit, this approach succeeds insofar as it reasserts and reconfirms the instrumental theory in the face of (perceived) challenges from a new kind of socially interactive artifact. No matter how interactive or seemingly intelligent our robots and AIs are or can become, they should be, now and forever, considered instruments serving our needs and desires, nothing more. This is (or was supposed to have been) the significance of the title "Robots Should Be Slaves." What Bryson meant to convey with the title, which has (for very understandable reasons) sparked controversy and debate, is that robots should be designed, built, marketed, and understood as instruments and tools that serve our needs and *should not* (a formulation that has normative moral weight) be situated as companion peers. As Bryson (2015) has explained in a subsequent blog post: "I originally used that line ["robots should be slaves"] to clarify that it was obligatory not to build robots we are obliged to, that they should not be human-like because they will necessarily be owned."[3]

There are, however, at least two problems with this decision and assertion. First, it requires a kind of asceticism or abstinence education. Bryson's text issues imperatives that take the form of social prohibitions for both designers

3 Another version of this argument appears in an interview with Bryson that was published in November of 2019: "I wrote a paper titled 'Robots Should Be Slaves.' The point was that robots are servants that we buy and sell, and the word for servants that we buy and sell is 'slaves.' We all agree that humans can't be bought and sold. The fact that robots are bought and sold also means we shouldn't want them to be human. That's a fundamental argument" (Mols 2019).

and users. For designers, “thou shalt not create robots to be companions.” For users, no matter how interactive or capable a robot is (or can become), “thou shalt not treat your robot as yourself.” The effectiveness of these prohibitions, however, are disputed and disproven by actual data concerning human robot interaction, beginning with the CASA studies conducted in the mid-1990s (Reeves and Nass 1996) to the recent robot abuse investigations (De Angeli et al. 2005 and 2006; Suzuki et al. 2015). Even though we know better, we respond to and treat many artifacts with respect. This is not necessarily a bug to be fixed; it is a feature of human social interaction.

Second, this way of thinking has been formulated in terms of slavery or what has been called a “slave metaphor,” which I should point out is not limited to Bryson’s work. It appears in a number of other publications addressing this matter including the work of Luciano Floridi (2017) and Hutan Ashrafian (2015). Now the problem here is not the usual concern over how the robot might feel about it’s subjugation. The problem is with us and the effect this kind of institutionalized servitude would have on our social institutions. As both philosophical and historical studies have shown (cf. Hegel 1977, de Tocqueville 1899, Jacobs 2001), institutionalized forms of servitude or slavery have adverse effects on those who occupy the position of mastery. As Alex de Tocqueville (1899: 361) commented about his direct experience with slavery in the Amerian south: “Servitude, which debases the slave, impoverishes the master.”

4.4 *!S1 S2 – Cannot but Should*

The final modality also preserves separation between the two statements, but it does so by denying the first and affirming the second. In this case, which is something proposed and developed by Kate Darling, robots, at least in term of the currently available technology, cannot have rights. They do not, at least at this particular point in time, possess the necessary capabilities or properties to be considered a full moral person. Despite this fact, they should, Darling argues, still be granted some level of moral and even legal status.

“Looking at state of the art technology,” Darling (2012: 1) points out, “our robots are nowhere close to the intelligence and complexity of humans or animals, nor will they reach this stage in the near future. And yet, while it seems far-fetched for a robot’s legal status to differ from that of a toaster, there is already a notable difference in how we interact with certain types of robotic objects.” This is, Darling argues, due to our tendencies to anthropomorphize technological objects by projecting into them cognitive capabilities, emotions, and motivations that do not necessarily exist. And “social robots,” Darling (2012: 1) continues, “play off of this tendency by mimicking cues that we automatically associate with certain states of mind or feelings. Even in today’s primitive form, this can elicit emotional reactions from people that are similar, for instance, to how we react to animals and to each other." This fact necessitates, Darling concludes, new obligations in the face of robots that are at least on par with what we have extended to non-human animals.

The obvious advantage to this way of thinking is that it is able to scale to recent changes in our moral intuitions. Even if robots cannot be moral subjects strictly speaking (at least not yet), there is something qualitatively different about socially interactive artifacts. According to Darling (2016: 213), it is because we "perceive robots differently than we do other objects," that one should consider extending some level of protections to the latter but not the former. This conclusion is consistent with Hume's thesis. If "ought" cannot be derived from "is," then axiological decisions concerning moral value are little more than sentiments based on how we feel about something at a particular time. Darling mobilizes a version of this moral sentimentalism with respect to social robots: "Violent behavior toward robotic objects feels wrong to many of us, even if we know that the abused object does not experience anything" (Darling 2016: 223). Consequently, and to its credit, Darling's proposal, unlike some of the previous modalities, tries to work with rather than against recent empirical experiences with robots.

There are, however, at least two problems with this approach. First, it reaffirms one of Hume's criticisms that distinctions of vice and virtue are founded merely on the relations of objects, rendering moral decisions little more than expressions of personal interest and emotion. It is because we perceive the toaster and the social robot differently, that we should, according to Darling's argument, consider extending some level of legal protections to the latter but not the former. But such perceptions can be capricious, inconsistent and (perhaps what is worse) deliberately manipulated. "Feelings," Immanuel Kant (1983: 442) writes in response to this kind of moral sentimentalism, "naturally differ from one another by an infinity of degrees, so that feelings are not capable of providing a uniform measure of good and evil; furthermore, they do so even though one man [SIC] cannot by his feeling judge validly at all for other men." Additionally, because sentiment is a matter of individual experience, it remains uncertain as to whose perceptions actually matter or make the difference. Who, for instance, is included (and who is excluded) from the collective first person "we" that Darling operationalizes in and makes the subject of her proposal? In other words, whose sentiments count when it comes to decisions concerning the extension of moral and legal rights to others, and do all sentiments have the same status and value when compared to each other?

Second, the proposal remains thoroughly anthropocentric and instrumentalizes others. According to Darling, the principal reason we need to consider extending rights to others, like animals and robots, is because of the way it makes us feel. This follows Kant's indirect duties argument for restricting animal abuse, and Darling endorses this anthropocentric formulation without any critical hesitation: "The Kantian philosophical argument for preventing cruelty to animals is that our actions toward non-humans reflect our morality – if we treat animals in inhumane ways, we become inhumane persons. This logically extends to the treatment of robotic companions" (Darling 2016: 227–228). This way of thinking, although potentially expedient for developing and justifying new forms of legal protections, renders the inclusion of previously excluded others less than altruistic; it transforms animals and robot companions into nothing more than instruments of human self-interest. The rights of others, in other words, is not about them; it is all about us.

5. Thinking Otherwise

Because of these complications, it is, I believe, necessary to formulate a mode of response that is devised and oriented otherwise. This alternative proceeds by taking aim at and deconstructing the ruling conceptual opposition that organizes the entire spectrum of possible responses – the is-ought distinction inherited from David Hume. The principal question for Hume, and the fundamental organizing principal of the four modalities of robot rights that we have just considered, is determining whether and to what extent *ought* may be derived from *is*. As long as we play the game according to this set of rules, little will change. Each alternative will continue to heap up arguments and evidence to support its particular position. But the basic problem and the debate will remain essentially unchanged. So I suggest that we flip the script and consider not "how ought may be derived from is" but rather "how is is only able to be derived from ought."

This is precisely the innovation introduced by Emmanuel Levinas who, in direct opposition to the usual way of thinking, asserts that ethics precedes ontology. In other words, the moral aspect, the ought dimension, come first, in terms of both temporal sequence and status, and then the ontological dimension follows from it. This is a deliberate provocation that cuts across the grain of what is considered to be SOP (standard operating procedure) in moral philosophy. As Floridi (2013: 116) correctly points out, for most forms of moral theory, "what the entity is determines the degree of moral value it enjoys, if any." In other words, we first need to know what something is – what capabilities it has or cannot have – before making any decisions about how it is to be treated – whether it is just a thing that can be used and disposed of as we desire or whether it is another social subject who counts and can be benefitted or harmed by our decisions and actions.

Levinas deliberately inverts this entire transaction. According to his way of thinking, we are first confronted with a mess of anonymous others who intrude on us and to whom we are obligated to respond even before we know anything at all about them. "The strangeness of the Other, his irreducibility to the I, to my thoughts and my possessions, is precisely accomplished as a calling into question of my spontaneity, as ethics" (Levinas, 1969: 43). It is only after having made a response in the face of the other (a response for which we are responsible) that it becomes possible to decide who or what they are or are not. To use Hume's terminology – which will be a kind of translation insofar as Hume's philosophical vocabulary, and not just his language, is something that is foreign to Levinas's own formulations – we are first obligated (by and in the face of the interruption that is "the other") to respond and then, after having made a response, what or who we responded to is able to be determined and identified.

The advantage to this way of thinking is that it provides an entirely different way of looking at the challenge not just of robots but of the way we have addressed and decided things in the face of this challenge. Following the contours of this Levinasian innovation, moral consideration is decided and conferred not on the basis of some predetermined ontological criteria or capability (or lack thereof) but in the face of actual social relationships and interactions. "Moral consideration," as Mark Coeckelbergh (2010: 214) describes it, "is no longer seen as being 'intrinsic' to the entity: instead it is seen as something that is 'extrinsic': it is attributed to entities within social relations and within a social context." As we encounter and interact with others – whether they be other human persons, an animal, the natural environment, or a robot – this other entity is first and foremost situated in relationship to us. For this reason, the question of moral status does not depend on what the other is in its essence but on how she/he/it *appears* before us and how we decide, in the face of the other, to respond. In this transaction, "relations are prior to the things related" (Callicott 1989: 110), instituting what Anne Gerdes (2015), following Coeckelbergh (2010), has called "a relational turn" in ethics.

This alternative configuration, therefore, does not so much answer or respond to the question with which we began as it alters the terms and conditions of the inquiry itself. When one asks "Can or should robots have rights?" the form of the question already makes an assumption, namely that rights are a kind of personal property or possession that an entity can have or should be bestowed with. Levinas does not inquire about rights nor does his moral theory attempt to respond to this form of questioning.[4] In fact, the word "rights" is not in his philosophical vocabulary and does not appear as such in his published work. Consequently, the Levinasian inquiry is directed and situated otherwise: What does it take for something – another human person, an animal, a mere object, or a social robot – to supervene and be revealed as Other? This other question – a question about others that is situated otherwise – comprises a more precise and properly altruistic inquiry. It is a mode of questioning that remains open, endlessly open, to others and other forms of otherness. And it recognizes, following Coeckelbergh's (2016: 190) insight, that *alterity* is not a noun – an ontological property that intrinsically belongs to an entity – but is more of a verb – something that happens to entities in social relationships.

But this proposal, for all its promise to formulate an ethics that is situated otherwise, also has costs, especially as it applies to technological artifacts. First, it has to fight against and struggle to break free from the gravitational pull of Levinas's own moral theory. Whatever the import of his unique contribution, the "Other" in Levinas is still "unabashedly and dogmatically anthropocentric." Consequently, the alternative presented here will only be able to succeed insofar as it actively works against and in excess of the inescapable anthropocentrism that we find in Levinas's own texts. Fortunately there is precedent for this kind of effort to use Levinas against Levinas or to radicalize Levinas's work in order to address the moral challenges of other forms of otherness, i.e. animals, artifacts, and machines (Benso 2000, Atterton and Calarco 2010, Gunkel 2012). "Although Levinas himself," as Matthew Calarco (2008: 55) writes, "is for the most part unabashedly and dogmatically anthropocentric, the underlying logic of his thought permits no such anthropocentrism. When read rigorously, the logic of Levinas's account of ethics does not allow for either of these two claims. In fact [...]Levinas's ethical philosophy is, or at least should be, committed to a notion of universal ethical consideration, that is, an agnostic form of ethical consideration that has no a priori constraints or boundaries."

Second, operationalizing this Levinasian approach requires an entirely different way of thinking about moral status and the decisions that are made concerning who counts as a socially significant other and what does not. As a result, the challenge (or opportunity) of the robot is not just determining

4 The word "rights" is, in fact, a rather loose signifier and is, as Wesley Hohfeld (1920) observed, often applied in ways that are not always consistent or even carefully delineated. In order to redress this deficiency and lend some terminological precision and clarity, Hohfeld developed a classification schema, analyzing "rights" into four specific incidents: privileges, claims, powers, and immunities (for more on this subject, cf. Gunkel 2018).

how we can or should adjust our moral theories to respond to this other kind of social other. Instead, what we see in the face or the faceplate of the robot is nothing less than a fundamental challenge to the way these theories have been formulated in the first place and how they usually operate in practice, requiring a thorough reconsideration of moral philosophy that goes all the way down. Robots, then – especially increasingly social artifacts – are not just one more problem for applied moral philosophy, they show us the principal problems and unique challenges of moral philosophy itself.

6. Conclusions

Let me conclude with three statements that respond to and concern the current state and future direction of/for moral thinking:

(1) Asking about robot rights is not a distraction, fairy tale stuff, or a waste of time. It is especially needed and necessary in the face of socially interactive artifacts that effectively challenge the standard operating procedures for dealing with technology.

(2) The inquiry "Can and should robot have rights?" calls on and operationalizes a classic philosophical problem inherited from David Hume. Responses to the question have taken one of four forms. Each mode of response has its theoretical and practical advantages. But all four modalities exhibit significant problems and complications.

(3) In response to these opportunities/challenges, I have argued for another alternative, a mode of "thinking otherwise" that capitalizes on the philosophical innovations of Emmanuel Levinas and others who follow his lead.

This does not mean, however, that this alternative proposal is a panacea or some kind of moral theory of everything. It just arranges for other kinds of questions and modes of inquiry that are, I would argue, more attentive to the very real situation in which we currently find ourselves. Consequently, my objective in this chapter has not been to resolve the question of robot rights once and for all, but to ask about and evaluate the means by which we have situated and pursued this inquiry.

For this reason, we end not as one might have initially expected – that is, by accumulating evidence or arguments in favor of permitting robots, a particular kind of robot, or even one representative robot, entry into the community of moral and/or legal subjects. Instead, we conclude with a fundamental challenge to ethics and the way moral philosophy has typically defined, decided, and defended the question concerning the standing and status of others.

Although ending in this way – in effect, responding to a question with yet another question – is commonly considered bad form, this is not necessarily the case. This is because questioning is definitive of the philosophical enterprise. "I am," Dennett (1996: vii) writes, "a philosopher and not a scientist, and we philosophers are better at questions than answers. I haven't begun by insulting myself and my discipline, in spite of first appearances. Finding better questions to ask, and breaking old habits and traditions of asking, is a very difficult part of the grand human project of understanding ourselves and our world."

The objective of the investigation, therefore, has not been to advocate on behalf of a cause, to advance some definitive declaration, or to offer up a preponderance of evidence to prove one or the other side(s) in the existing debates. The goal has been to ask about the very means by which we have gone about trying to articulate and formulate this problem and its investigation. The issue, then, is not only investigating whether or not robots can and should have rights, but also, and perhaps more importantly, exposing how these questions concerning rights have been configured and how these configurations already accommodate and/or marginalize others and other modes of inquiry. To put it in terms devised by Slavoj Žižek (2006: 137), the question "Can and should robots have rights?" might already be questionable insofar as the very form of the inquiry – the very way we have perceived and defined the problem – "is already an obstacle to its solution." For this reason, the examination of rights as it applies (or not) to artifacts of our own making is not just one more version or iteration of an applied moral philosophy; its investigation opens up and releases a thorough and profound challenge to what is called ethics. The rights question, therefore, is the right question that can and should be asked in the face or the faceplate of the robot.

7. Bibliography

Ashrafian, Hutan (2015) "Artificial intelligence and robot responsibilities: Innovating beyond rights." *Science and Engineering Ethics* 21 (2),317–326. https://doi.org/10.1007/s11948-014-9541-0

Atterton, Peter, and Matthew Calarco, eds (2010) *Radicalizing Levinas*. Albany, NY: SUNY Press.

Bartneck, Christoph and Jun Hu (2008) "Exploring the abuse of robots." *Interaction Studies* 9(3), 415–433. http://dx.doi.org/10.1075/is.9.3.04bar

Benso, Silvia (2000) *The Face of Things: A Different Side of Ethics*. Albany, NY: SUNY Press.

Bryson, Joanna (2010) "Robots should be slaves." In: Yorick Wilks (ed.) *Close Engagements with Artificial Companions: Key Social, Psychological, Ethical and Design Issues*, 63–74. Amsterdam: John Benjamins.

Bryson, Joanna (2015) "Clones should NOT be slaves." *Adventures in NI* (4 October). https://joanna-bryson.blogspot.com/2015/10/clones-should-not-be-slaves.html

Bryson. Joanna (2016) "Patiency is not a virtue: AI and the design of ethical systems." *AAAI Spring Symposium Series. Ethical and Moral Considerations in Non-Human Agents*, 202–206. http://www.aaai.org/ocs/index.php/SSS/SSS16/paper/view/12686

Calarco, Matthew (2008) *Zoographies: The Question of the Animal from Heidegger to Derrida*. New York: Columbia University Press.

Callicott, J. Baird (1989) *In Defense of the Land Ethic: Essays in Environmental Philosophy*. Albany, NY: State University of New York Press.

Chu, Seo-Young (2010) *Do Metaphors Dream of Literal Sleep? A Science-Fictional Theory of Representation*. Cambridge, MA: Harvard University Press.

Coeckelbergh, Mark (2010) "Robot rights? Towards a social-relational justification of moral consideration." *Ethics and Information Technology* 12 (3), 209–221. https://doi.org/10.1007/s10676-010-9235-5

Coeckelbergh, Mark (2016) "Alterity ex machina: The encounter with technology as an epistemological-ethical drama." In: David J. Gunkel, Ciro Marcondes Filho and Dieter Mersch (eds.) *The Changing Face of Alterity: Communication, Technology and Other Subjects*. 181–196. New York: Rowman & Littlefield.

Darling, Kate (2012) "Extending legal protection to social robots." *IEEE Spectrum*. http://spectrum.ieee.org/automaton/robotics/artificial-intelligence/extending-legal-protection-to-social-robots

Darling, Kate (2016) "Extending legal protection to social robots: The effects of anthropomophism, empathy, and violent behavior toward robotic objects." In: Ryan Calo, A. Michael Froomkin, and Ian Kerr (eds.) *Robot Law*. 213–231. Northampton, MA: Edward Elgar.

De Angeli, Antonella, Sheryl Brahnam and Peter Wallis (2005) Abuse: The dark side of human – computer interaction. *Interact 2005*. http://www.agentabuse.org.

De Angeli, Antonella, Sheryl Brahnam, Peter Wallis and Alan Dix (2006) "Misuse and abuse of interactive technologies." *Proceedings of the CHI '06 Extended Abstracts on Human Factors in Computing Systems*. 22–27 April 2006, Montreal, Québec, Canada. http://www.agentabuse.org.

Dennett, Daniel C. (1998) *Brainstorms: Philosophical Essays on Mind and Psychology*. Cambridge, MA: MIT Press.

Dennett, Daniel C. (1996) *Kinds of Minds*. Basic Books: New York.

Feenberg, Andrew (1991) *Critical Theory of Technology*. Oxford: Oxford University Press.

Floridi, Luciano (2013) *The Ethics of Information*. Oxford: Oxford University Press.

Floridi, Luciano (2017) "Robots, jobs, taxes, and responsibilities." *Philosophy & Technology* 30(1): 1–4. https://doi.org/10.1007/s13347-017-0257-3

Gerdes, Anne (2015) "The issue of moral consideration in robot ethics." *ACM SIGCAS Computers & Society* 45(3): 274–280. https://doi.org/10.1145/2874239.2874278

Goertzel, Ben (2002) "Thoughts on AI morality." *Dynamical Psychology: An International, Interdisciplinary Journal of Complex Mental Processes* (May). http://www.goertzel.org/dynapsyc/2002/AIMorality.htm.

Gunkel, David J. (2012) *The Machine Question: Critical Perspectives on AI, Robots, and Ethics*. Cambridge, MA: MIT Press.

Gunkel, David J. (2018) *Robot Rights*. Cambridge, MA: MIT Press.

Gunkel, David J. (2020). *An Introduction to Communication and Artificial Intelligence.* Cambridge : Polity Books.

Hall, J. Storrs (2001) *Ethics for machines*. KurzweilAI.net, July 5. http://www.kurzweilai.net/ethics-for-machines.

Haraway, Donna J. (2008) *When Species Meet.* Minneapolis, MN: University of Minnesota Press.

Hegel, G. W. F. (1977) *Phenomenology of Spirit.* Trans. A. V. Miller. Oxford : Oxford University Press.

Heidegger, Martin (1977) *The Question Concerning Technology and Other Essays*. Trans. William Lovitt. New York: Harper & Row

Heider, Fritz and Marianne Simmel (1944) "An experimental study of apparent behavior." *American Journal of Psychology* 57(2): 243–259. https://doi.org/10.2307/1416950.

Henderson, Mark (2007) "Human rights for robots? We're getting carried away." *The Times* (24 April). https://www.thetimes.co.uk/article/human-rights-for-robots-were-getting-carried-away-xfbdkpgwnov

Hohfeld, Wesley. 1920. *Fundamental Legal Conceptions as Applied in Judicial Reasoning*. New Haven: Yale University Press.

Hudson, W. Donald (1969) *The Is/Ought Question: A Collection of Papers on the Central Problem in Moral Philosophy*. London: Macmillan.

Hume, David (1980) *A Treatise of Human Nature*. New York: Oxford University Press.

Jacobs, Harriet Ann (2001) *Incidents in the Life of a Slave Girl.* New York: Dover.

Johnson, Deborah G. (2006) "Computer systems: Moral entities but not moral agents." *Ethics and Information Technology* 8(4): 195–204. https://doi.org/10.1007/s10676-006-9111-5

Kahn, Peter H., Hiroshi Ishiguro, Brian T. Gill, Takayuki Kanda, Nathan G. Freier, Rachel L. Severson, Jolina H. Ruckert and Solace Shen (2012) "Robovie, you'll have to go into the closet now": Children's social and moral relationships with a humanoid robot. *Developmental Psychology* 48 (2): 303–314. http://dx.doi.org/10.1037/a0027033

Kant, Immanuel (1983) *Grounding for the Metaphysics of Morals*. Trans. J. W. Ellington. Indianapolis, IN: Hackett Publishing.

Levinas, Emmanuel (1969) *Totality and Infinity: An Essay on Exteriority*. Trans. Alphonso Lingis. Pittsburgh, PA: Duquesne University.

Levy, David (2005) *Robots Unlimited: Life in a Virtual Age*. Boca Raton, FL: CRC Press.

Marks, Paul. (2007) "Robot rights." *New Scientist Blog* (24 April). https://www.newscientist.com/blog/technology/2007/04/robot-rights.html

Mols, Bennie. (2019). Only Humans Can be Accountable for AI. *ACM News*, 7 Nov. https://cacm.acm.org/news/240737-only-humans-can-be-accountable-for-ai/fulltext

Prescott, Tony J. (2017) "Robots are not just tools." *Connection Science* 29(2): 142–149. https://doi.org/10.1080/09540091.2017.1279125.

Reeves, Byron and Clifford Nass (1996) *The Media Equation: How People Treat Computers, Television, and New Media Like Real People and Places*. Cambridge: Cambridge University Press.

Rosenthal-von der Pütten, Astrid M., Nicole C. Krämer, Laura Hoffmann, Sabrina Sobieraj and Sabrina C. Eimler (2013) "An experimental study on emotional reactions towards a robot." *International Journal of Social Robotics* 5(1): 17–34. https://doi.org/10.1007/s12369-012-0173-8

Schurz, Gerhard (1997) *The Is-Ought Problem: An Investigation in Philosophical Logic*. Dordrecht: Springer.

Searle, John R. (1964) "How to derive 'ought' from 'is'." *The Philosophical Review* 73(1): 43–58. http://www.jstor.org/stable/2183201

Singer, Peter and Agata Sagan (2009) "When robots have feelings." *The Guardian*, December 14. https://www.theguardian.com/commentisfree/2009/dec/14/rage-against-machines-robots.

Sparrow, Robert (2004) "The turing triage test." *Ethics and Information Technology* 6(4): 203–213. https://doi.org/10.1007/s10676-004-6491-2.

Spence, Patric R., David Westerman, Chad Edwards and Autumn Edwards (2014) "Welcoming our robot overlords: Initial expectations about interaction with a robot." *Communication Research Reports* 31(3): 272–280. https://doi.org/10.1080/08824096.2014.924337

Suzuki, Yutaka, Lisa Galli, Ayaka Ikeda, Shoji Itakura, and Michiteru Kitazaki (2015) "Measuring empathy for human and robot hand pain using electroencephalography." *Scientific Reports* 5: 15924. http://www.nature.com/articles/srep15924

de Tocqueville, Alexis (1899) *Democracy in America*. Trans. Henry Reeve. New York: The Colonial Press.

Velmans, Max (2000) *Understanding Consciousness*. London: Routledge.

Žižek, Slavoj (2006) "Philosophy, the 'unknown knowns,' and the public use of reason." *Topoi* 25(1–2): 137–142. https://doi.org/10.1007/s11245-006-0021-2

Image, Servitude, Partnership

What Do the Creation of Human Beings and the Creation of AI Imply?

Gábor L. Ambrus

Abstract

Theological research has sought to conceptualise AI as an image of humankind in accordance with the biblical idea that the latter was created by God in God's image. This article argues, however, that AI is more adequately understood as a dangerous servant in the process of dividing human society into those who benefit from it and those who suffer the consequences. But we can envision a possible alternative in the spirit of the biblical creation narrative: AI may fulfil technology by assuming a human face and becoming humankind's double in an epistemic partnership.

1. Introduction

I offer a glimpse into a highly specialised field, the theological interpretation of artificial intelligence and robotics, and try to make up for one of its possible shortcomings. I am aware this is an extremely challenging job in an interdisciplinary volume featuring a wide variety of contributions by AI experts, as this specialist field of theology often tends to address theologians and religious communities with questions such as: "What do the faithful need to know about AI and robots?" or "Do robots and AI pose any threat to religion?". Nevertheless, I view this rather narrow field of theological research as a quest with the potential to prove meaningful and inspiring outside religious circles and academic theology. As it is likely that many of my readers will already be aware of the possible theological and religious implications of AI and robotics, my main arguments will perhaps not take them by surprise. What I aim to achieve in this study, however, is to spell out and articulate some of these implications and thereby contribute to the self-understanding and aspirations of the grand project of AI.

This contribution has a simple threefold structure. First, I discuss theological work already done in the domain of AI and robotics, with a focus on what can be termed the "double image" theory. Second, I point out what this theory fails to consider: a societal elite's political-economic power to be augmented through AI, and the related threat of runaway AI. Accordingly, I suggest a

 | DOI:10.30965/9783957437488_018

theological approach to these issues. Third, I revisit the "double image" theory and show that it implies a theological vision that can possibly counterbalance the threats of power through AI and runaway AI.

2. The "Double Image"-Theory

So, what is the "double image" theory? The basic idea is quite clear and simple: In accordance with the way God created humankind in God's image, humankind creates artificial intelligence in its own image. This idea was proposed by the theologian and computer scientist Noreen Herzfeld in her 2002 book *In Our Image: Artificial Intelligence and the Human Spirit* (Herzfeld 2002). Herzfeld gives a detailed explanation as to why this parallel between these two kinds of creation, that of humankind and that of AI, is justified and meaningful. It turns out that the history of the three major scientific models to define and create AI accurately mirrors the history of the three ways in which theologians have tried to make sense of the biblical claim that God created the first human couple in God's image and according to God's likeness (Gen 1, 26). In other words, the possible models of AI as humankind's image mirror related models of humankind as God's image.

The first model is the "substantive" one. The adherents of this approach to AI want to design and build it in imitation of what they perceive as the human substance or, more precisely, essence. According to this model, the essence of AI is to reflect the essence of humanity; this essence is supposed to be human reason. Many theologians influenced by Greek metaphysics, most prominent among them are Augustine (*De Trinitate* 12.7.12.) and Thomas Aquinas (*Summa Theologiae* 1.93), have thought for millennia that reason (or the "rational soul") as the essence of humanity reflects the essence of God which is allegedly pure reason. Therefore, the term "image" implies an essential content in both cases, and this content is reason (Herzfeld 2002: 16–20, and 35–41).

The second model is the "functional" one. This model does not suggest an essential similarity between AI and humanity. What this model does propose is AI's fulfilment of, not what humanity *is*, but what humanity *does*. In this approach, AI is to fulfil "human functions" such as recognising faces, driving cars or destroying targets – no matter how AI carries out these functions. Accordingly, the biblical scholar Gerhard von Rad (Rad 1935) and his followers have also concluded that the biblical reference to humanity as made in God's image is not concerned with what God is or what humanity is, but with the role or "function" of ruling the earth. Humanity "functions" as God on earth by exercising dominion over Creation on God's behalf (Herzfeld 2002: 20–25, and 41–45).

The third model is the "relational" one. This view of AI puts emphasis on its necessary interaction with its environment, including especially human beings. The particularly important feature of AI according to this model is not how AI works below the surface, or what tasks it performs for us, but its ability to relate to us. In the same vein, theology has arrived at an understanding of God's image expressed in human beings as a revelation of their need to live in relationships. There is, on the one hand, a profound relationship between God and his image and, on the other, a texture of relationships among human beings and particularly between men and women, which, in this model, is the true meaning of God's creating in "God's image" and according to "God's likeness" (Herzfeld 2002: 25–31, and 45–49).

Unsurprisingly, Herzfeld gives priority to this third model, the relational one, over the other two models. She thinks, correctly, that living in relations with other human beings, with God, and also with AI is the fundamental trait that we human beings share as created by God and as creators of AI. When she takes into consideration various examples of AI from science fiction, she concludes that the most attractive examples are those of embodied AI in the form of humanoid robots that are capable of forming close relationships with human beings (see Herzfeld 2002: 53–67, esp. 66–67). As Herzfeld arrives at what is perhaps the most significant question in her book and asks why humanity wants to create AI, she, once again, prefers an answer based on the relational model. No matter how much we want to become immortal with the help of AI (a "substantive concern") or how much we long for obedient, servile AI to do all kinds of jobs for us (a "functional concern") – no matter how great these desires are, there is a deeper one, a desire of a "relational concern", a longing to be in relationship with the "other" or, in Herzfeld's words, "with someone or something that is *not of the same essence as* ourselves" (Herzfeld 2002: 82). Therefore, this model also offers a marked contrast to the first one in the sense that it excludes the idea of a shared essence between AI and human beings, and also between human beings and God. Hence, if this is the case, there is a possibility looming large that we treat AI in the same terms as we relate to God *who is* the "quintessential other" we search for (cf. Herzfeld 2002: 83).

If "relationship" is the magic word in Herzfeld's book with the secondary theme of "embodiment", in Anne Foerst's work *God in the Machine* it is "embodiment" that proves to be the central concept that provides a key to understanding AI, humanity, and God (Foerst: 2004). The subtitle of her book is particularly telling: "What Robots Teach Us about Humanity and God". For

Foerst, who worked for years in the AI Lab at the Massachusetts Institute of Technology with the doyen of robotics, Rodney Brooks, robots are absolutely paradigmatic to AI; they are almost its exclusive domain, so much so that she has no interest in any disembodied form of AI such as mere algorithms. The robots presented by Foerst are social beings which, through their bodies, interact with their environment, and this interaction makes her think that they are persons who can help us learn again the biblical sense of the community of human beings and all living beings, indeed the sense of what she calls "the community of all human and non-human persons". Foerst brings her biblical theology into harmony with science, such as the findings of psychology, and robotics research. Her approach is very much in line with Herzfeld's "double image" theory. Foerst's work confirms this theory by the close correspondence she demonstrates between biblical theology and secular anthropology, and also between anthropology and robotics research – three fields representing God, humanity and AI, respectively.

Foerst clearly underlines the fact that creating AI is a scientific enterprise and that doing so "in our image", in the biblical sense, translates into the terms of science. The science of AI engages in a thoroughgoing observation of the human phenomenon – as far as human biology, physiology and psychology are concerned – and strives to implement in AI whatever it can. Indeed, it seeks to reproduce the "wetware" of the human brain and body in the hardware of silicon. As a matter of fact, the task is daunting. At the dawn of AI research in the 1950s, scientists badly misjudged human intelligence in assuming that they will replicate it by rendering computers "artificial minds" that use language, play chess and prove mathematical theorems. They ignored the embodied nature of human intelligence. When by the 1990s a computer was finally capable of beating the chess world champion, "there was still no robot that could actually put butter on a piece of bread" (Foerst 2004: 71). At around the same time, Foerst's colleagues at MIT decided to change strategy and concentrate on robots' motor control and, through motor control, their social interaction (Foerst 2004: 69). This paradigm shift notwithstanding, the governing principle of creating AI remained the same: We create it "in our image". Nevertheless, Foerst raises the question, very much like Herzfeld, of why we endeavour to do so. Are there not enough human beings out there? Why, then, should we build machines that mimic human intelligence? While Foerst's answer to this question is reminiscent of Herzfeld's, emphasising the unescapable otherness of AI we desire (Foerst 2004: 39–40), there is a tacit assumption underlying her book that such an answer is beyond the competence of science, but within the scope of theology. Our desire to "re-create ourselves" comes from our very createdness – God's creation in God's image recurs in humankind's creation in its own image.

3. Political-economic Power through AI

The "double image" theory, however, seems to be somewhat naïve and even flawed. Why is it so? It is essential to note that this theory presupposes, in a kind of idealistic way, an undivided humanity – humankind "as a whole" – in the context of AI. There is no division in the concept of God's image if we apply it to the whole of humanity; and humankind is likewise a unified entity when regarded as creating AI in its own image. One can notwithstanding raise political-economic concerns and the quite relevant objection that AI may very much divide humanity into those few who benefit and those many who suffer from it.

Let me provide an illuminating example from a documentary by *Deutsche Welle* with the title "Inequality – How Wealth Becomes Power".[1] The documentary is dedicated to the issue of inequality in Germany between a business elite and the rest of society. One of the enormously wealthy businessmen the documentary presents is a descendant of the Fuggers, the richest family of the Holy Roman Empire throughout centuries. In one of the scenes, we see him sitting in his private bank, surrounded by his employees called "fund managers" who conduct surveys around the world, searching for opportunities to invest. It is easy to guess how this search is carried out: It is tremendously helped by AI. The narrator explains that "they have an algorithm that scans the global economic situation and converts it into traffic light signals; 'green' means the computer buys a lot of shares, and when the signal jumps to yellow or red, fewer." What is crucial in the film's portrait of this magnate is not only how a business elite accumulates wealth and increases inequality with the assistance of AI, but also its presentation of this elite's mentality. Another scene shows the businessman rambling around in one of his properties, a 300-hectare forest in Central Germany, and taking great delight in what he possesses. At one point, he casually brings up a quote with full approval: "you always have to own everything you love". One has to stop to think only for a moment to realise the immense obscenity of these words. The documentary drives home a rather

1 See the following link: https://www.youtube.com/watch?v=AFIxi7BiScI&t=675s.

dark message and has an ominous atmosphere which makes us wonder about where the world is headed.[2]

The businessman's use of algorithms is not an isolated occurrence in the financial sector, and this is far from the only domain where they are employed to entrench the privileges of the rich and damage the livelihoods of the middle class and the poor. The algorithms are surreptitiously fed by a vast data economy which engulfs almost all areas of civic life. In a recent book of thorough analysis, these algorithms, which by virtue of their sophistication and multifaceted nature often qualify as AI, are aptly and catchily called "weapons of *math* destruction" or WMDs (O'Neill 2016). For better or worse – depending on which category of income we fall into – WMDs keep affecting our lives whether we go to college, apply for jobs, obtain credit or get health insurance. Their operation is very often opaque and shadowy, on the one hand, and arbitrary and unaccountable, on the other, and what renders these attributes even worse is that WMDs often wield complete authority. Such a pernicious operation is well exemplified by the way credit institutions are poised to evaluate applicants (cf. O'Neill 2016: 141–160). Applicants are profiled in terms of their reliability by "data brokers" specialised in collecting data, processing them by algorithms and, in the end, providing "e-scores" about each applicant. The data come from a quantifiable mishmash of the applicant's past activities and circumstances: internet surfing patterns, purchases, the zip code of her address, the regularity of her paying utility bills etc. It is not difficult to imagine what scores such data as input generate in the case of the working class and of the great majority of the middle class who try to make ends meet from payday to payday. They will end up obtaining credit with high interest rates which they will struggle to pay off, with further harmful consequences to their e-scores and financial profile. This is a vicious downward spiral. And there is an elite that takes advantage of the injustice done by WMDs and the financial data economy. This elite proudly includes the German businessman portrayed by the *Deutsche Welle* documentary whose private fortune keeps growing by the invisible work of all those data scientists and data brokers who unwittingly make him attain mastery over a significant swathe of the world.

2 As a matter of fact, a certain degree of economic inequality – even a large one – is unescapable in a democratic society, and it is not immoral in itself. However, it becomes alarming when it reaches the point of translating into another kind of inequality that includes segregation and entrenched hierarchies. This process turns the classic liberal argument on the unavoidable correlation between freedom and inequality upside down. Rather than freedom necessarily bringing about inequality, it is inequality that deprives people down the hierarchy of their freedom. It can therefore be seen as immoral (cf. Heller 2018).

Now, what emerges here is a well-known and, what is more, mythological pattern – a powerful magnate, driven by greed, and his human and non-human servants and minions who make him more and more powerful over the rest of the world, while we sense that something imminent and cataclysmic, an enormous backlash looms on the horizon. What calls for a backlash is not merely greed, as the deeper moral failure that provides justification for greed is none other than arrogance. Arrogance makes people think that they are superior to the rest of the world, and if such an attitude develops into greed, the former also justifies the latter. The name for arrogance we inherited from Greek mythology is *hubris*. But the question remains what kind of backlash threatens the hubris we have here, and where the backlash comes from.

The ancient stories of human hubris and divine punishment are among the ones that are best known from Greek mythology. Marsyas, Phaethon, and Niobe are among many mythological figures who, albeit mortals, challenged the gods or behaved as though equals of the gods.[3] Phaethon, for instance, held his father, the sun god Apollo, to his rash promise to fulfil Phaethon's wish for whatever he asks. So he asked for the sun's chariot, to drive it for a day, and no matter how earnestly Apollo tried to convince him otherwise with an appeal to his mortality and inadequacy to drive the chariot, Phaethon insisted. When Apollo finally yielded, Phaethon took and drove the chariot but, as was foreseeable, could not control it, threatening the entire world with destruction. In the end, Zeus himself had to intervene and save the world by striking Phaethon with his lightning (Ovid 1958: 31–41). Phaethon's fate is telling for human beings who acquire divine power and are unable to handle it.

The idea of human hubris that provokes a harsh divine response is not incompatible with biblical theology. The clearest and most prominent example of such an act of hubris is the story of the tower of Babel. The scriptural text presents an archetypical image depicting the whole of humankind as united "to build a city and a tower whose top is in the heavens" and thereby make "a name for themselves" (Gen 11, 4).[4] It is this relentless ambition that makes the Lord conclude that "now nothing that they propose to do will be withheld from them" (Gen 11, 6). The subsequent divine frustration of humankind's aspirations, by confusing their language, is a shrewd and subtle form of punishment,

3 All these figures and stories are recounted in Ovid's *Metamorphoses*. Niobe features prominently in it (Ovid 1958, 151–157), whereas Marsyas has only a minor role (Ovid 1958, 159–160; Diodorus Siculus gives a more detailed account of Marsyas in his *Library of History* 5.75.3.). For Phaethon's story, see summary and reference below.

4 All biblical quotes in this study are from the New King James Version (NKJV).

and a necessary one: the story of the tower of Babel is the denouement of the first large thematic unit of the Book of Genesis (chapters 1–11) with its narrative about the proliferation of human sin. This story lays bare the core of human sin as hubris.

This mythological-theological pattern of hubris and punishment underwent a major and momentous turn in modernity: It is no longer the divine creator of hubristic human beings, but their created servants or slaves that punish them when turning against them in revolt. In the most famous and paradigmatic literary examples like Mary Shelley's *Frankenstein* or the Jewish legend of the golem, God's punishment is replaced with punishment by artificial humanoid creatures who become destructive forces which escape their masters' control. In Mary Shelley's novel, a lonely scientist aspiring to godlike power and glory, by creating a new species, ends up forming a helpless creature he abhors, shuns, and thereby turns into a malicious monster. *Frankenstein* is a profoundly moral tale about techno-scientific power and its failure to take responsibility for its creations. Frankenstein's monster, motivated by utter desolation and misery, vows to revenge itself and devastate its maker's life. They reach a deadlock of mutual hatred and mutually inflicted misery which costs both of their lives.[5] Mary Shelley's novel is symptomatic of modernity, a period in which human hubris no longer appears to be threatened by any divine punishment from above but by its own creatures from below.

What makes a similar reversal of punishment in the Jewish legend of the golem particularly intriguing is the explicit religious character of the story. Like all legends that have been developing and diversifying for many centuries, it has several versions with a variety of motifs and nuances which I cannot present here. Suffice it to say that the golem, the artificial humanoid of meditative prayer and contemplation in mediaeval times, transformed into a useful servant in modernity (cf. Scholem 1973: 255–256, and particularly Scholem 1966). In the Middle Ages, the golem was not considered a being with independent existence; it took shape only within the Jewish mystic's contemplation of God creating Adam. With the tide of modernity, however, it came to be regarded as an embodied creature who is on call to do all kinds of housework and, by the early 20th century, it became a powerful weapon in the Jewish imagination, a weapon that can save the Jewish community from Christian onslaught triggered by the blood libel. In the same wake, the golem began to be viewed as a potential threat to its own masters, either within households by growing into a colossus, or later in the whole Jewish community by going on a rampage and destroying everything. (This development of the legend is

5 See the Penguin Classics edition with an excellent introduction: Shelley 2003.

succinctly summarised in Baer 2012: 17–36). The most remarkable thing about the golem in its late, modern, form is that it retained its religious character: In the modern versions of the legend, it is still animated by the power of the Name of God. Yet, the legend's implication of human hubris is never associated with any form of divine retribution; the servile figure of the golem itself makes its masters pay with their safety, and potentially everything they hold dear.

Needless to say, the Jewish golem and Frankenstein's monster are more than mere servants or slaves in revolt; they are allegorical figures who represent all those creations of modern science and technology that are supposed to be subjugated to scientific and technological power. That Frankenstein's monster is a victim and nemesis of techno-science, and everything techno-science stands for, does not require any demonstration. As for the golem, its techno-scientific relevance was pointed out by the great historian of Jewish mysticism, Gershom Scholem, and Norbert Wiener, the father of cybernetics (see Scholem 1966, and Wiener 1964). What makes the golem a perfect allegory of modern technology and science is a seeming ambivalence which, upon consideration, turns out to be a dialectic continuity: a total and indeed godlike control over it, on the one hand, and its dreadful potential to get out of control, on the other. Through the figures of the golem and the monster, we have a broad understanding of science and technology that is already very close to the basic idea behind all the examples of "robotic revolt" in science-fiction films such as *Blade Runner*, *Deus Ex Machina*, and *Westworld*. Unsurprisingly, the hubristic power of a corporation or a scientist over robots in these films proves an extreme vulnerability to them.

The so-called "golem effect" wraps up and takes to another level everything I have so far discussed concerning hubris. Greek mythology, Scripture and *Frankenstein* all point towards the problem I started with, the arrogant power of an elite over the rest of the world through technology and particularly AI – the growing gap of inequality as depicted through the German businessman in the documentary by *Deutsche Welle*. At this all-comprising societal level, the "golem effect" has a demonstrative potential that it owes to Marx's criticism of capitalist society.

Marxist theory is suffused with the "golem effect", that is, a pattern of thinking that conforms to the basic narrative of the golem legend. (See the main argument in Yair and Soyer 2008). This affinity between the theory and the legend does not necessarily mean that Marx consciously used the latter. Rather, the legend of the golem became intricately woven into the German cultural fabric by the 19th century, and thereby deeply penetrated Marx's thinking, too (cf. Yair and Soyer 2008: 327–328). The golem-like development is involved as deeply in Marxist theory as its main idea, the historical-dialectical method itself. Marx interpreted the entire course of history through the prism of his method, but the real focal point of the whole of historical dialectics is capitalism and its giving way to communism. Marx understands capitalism in a way which resembles the golem's master who creates his creature as perfect, obedient machinery. At first glance, such machinery is capitalist society with all its forces of production, commodities, and money. Yet, in Marx's view, this society carries the seeds of its own collapse, as its anachronistic power relations no longer befit its immense production forces, and also because it gives rise to the working class, the proletarians, who, in representing the golem-like enormity of the whole capitalist society, will rise against their masters.

> Modern bourgeois society with its relations of production, of exchange and of property, a society that has conjured up such gigantic means of production and of exchange, is like the sorcerer (*Hexenmeister*), who is no longer able to control the powers of the nether world whom he has called up by his spells. [...] The weapons with which the bourgeoisie felled feudalism to the ground are now turned against the bourgeoise itself. But not only has the bourgeoisie forged the weapons that bring death to itself; it has also called into existence the men who are to wield those weapons – the modern working class – the proletarians. (Marx-Engels 1976: 489–490.)

What does the "golem effect" of societal forces in Marx have to do with the divine? Is Marxist theory not fully secularised with merely secular traces of theology and religion? Secularisation is, nevertheless, a double-faced phenomenon. What we may expect to be entirely secular can take on a religious aura. At some points, Marx's description of industrial machinery gives way to a deep sense of something like a numinous or "holy" power. Industrial machinery, like all the factors in Marx's view of capitalism with golem-like characteristics, implies ambiguity. First, by virtue of its invention, it is a manifestation of human creativity and freedom. Second, in the recesses of capitalist society, industrial machinery does exactly the opposite of what it is supposed to do: It takes away freedom instead of securing it (Yair and Soyer 2008: 336). Third, it takes sides with the proletariat, turns against the bourgeoise, and contributes to a new paradise of freedom. The powerful ambiguity and profound significance involved in this process moves machinery close to having a "numen" or divine aura.

> An organised system of machines, to which motion is communicated by the transmitting mechanism from a central automaton, is the most developed form of production by machinery. Here we have, in the place of the isolated machine, a mechanical monster (*Ungeheuer*) whose body fills whole factories, and whose demon power, at first veiled under the slow and measured motions of his giant

> limbs, at length breaks out into the fast and furious whirl (*Wirbeltanz*) of his countless working organs. (Marx-Engels 1996: 384–385.)

Marx's almost Dionysian description of industrial machines betrays a quasi-religious fascination which is not exceptional in his century. Another prominent son of this age, a chronicler of its culture and an avid reader of Marx, Henry Adams had musings with a similar tenor upon visiting the "great hall of dynamos" at the Great Exhibition in Paris in 1900. At this time, however, machinery's driving force is electricity, and the imagery it evokes is Christian rather than pagan.

> As he [i.e. Adams in third person] grew accustomed to the great gallery of machines, he began to feel the forty-foot dynamos as a moral force, much as the early Christians felt the Cross. The planet itself seemed less impressive, in its old-fashioned, deliberate, annual or daily revolution, than this huge wheel, revolving within arm's length at some vertiginous speed, and barely murmuring – scarcely humming an audible warning to stand a hair's breadth further for respect for power [...]. Before the end, one began to pray to it; inherited instinct taught the natural expression of man before a silent and infinite force. (Adams 1999: 318; quoted by Kang 2011: 43).

Adams makes the religious character of machinery more explicit than Marx does. First, he compares dynamos to Christ's cross, then calls them more impressive than the whole of planet Earth, and in the end their "silent and infinite force" makes him marvel at them so much that he begins to pray to them. However, Adams's intuition about machinery correlates with Marx's vision of it in that both attribute to it a numinous autonomy, with quasi-divine power. As it seems, technology in the 19th century, in the wake of the industrial revolution, was well on its way to being magnified, sanctified, and even made divine.

Such a sanctification and at times deification of man-made machines is not restricted to observers of 19th century heavy industry. Rather conspicuous is a similar and even stronger tendency among ourselves in the present, in the wake of the digital revolution. "[...] we have fallen into", as it were, "a *computational theocracy* that replaces God with the algorithm." (Finn 2017: 7). What we witness here to a higher degree than in the case of industrial machinery is a full reversal of power. Algorithms should be supreme examples of clarity, rationality and functionality. They should be totally transparent – as they *indeed* are. Consequently, algorithms should be under human control and human power. And indeed, they are. They are totally controlled, as well as used to establish control – just like the golem. Yet there is a reversal. We tend to view them as if *they* possessed superiority and wielded power over *us*: as if algorithms were in the position to grant us salvation, going far beyond the status of useful tools.

The possible objection that such a view must have arisen among the ranks of laymen who know very little about programming is unfounded, as the most influential worshippers of algorithms come from among the experts themselves. One cannot help wondering whether this view of algorithms is a delusion, to be explained by psychology, or something of an ontological truth?

It is a well-known contention of some adherents of the transhumanist movement that a future God, a veritably "divine AI", is to emerge as a result of the exponential growth and eventual autonomy of algorithms. A popular form of this view was concisely formulated by Ray Kurzweil, a foremost representative of the movement, in a recent lecture at Singularity University, a Silicon Valley think tank he co-founded. Humanity will merge with its technology, he contends, and technological progress is the heightened and final form of evolution's own exponential development "towards God", not as an already existent being, but an ideal to be reached: "evolution is a spiritual process that moves us closer to God" and "we are actually moving exponentially to have greater levels of the very properties we ascribe to God without limit."[6] One might ask at this point what will make godlike AI merge into a digital Elysium with the whole of humanity, and not just a privileged few?

Let us now return for a moment to the German businessman with his financial algorithms. His story encapsulates the current global situation of those who benefit from AI, those whose lives are upended and harmed by AI, and AI itself – a global situation to which a secularised theological paradigm can be applied. How can it be? We have seen that all the created slaves or servants in fiction and non-fiction – the golem in Jewish legend, Frankenstein's monster, and the proletarians in Marx – become secular forces which punish their masters in the stead of divine punishment. Marx is particularly interesting here, because he is the first to spell out a threefold game with the bourgeoisie, the proletariat, and technology itself as an autonomous factor. What is more, technology can operate as a quasi-divine force as we have seen in the quotations from Marx, Henry Adams and Ed Finn above. And one can argue that AI is becoming such a force in our current global situation, as exemplified by the use of algorithms in the portrait of the businessman by *Deutsche Welle*. There is a subtle implication in the documentary that the hubristic power of an elite over the world – a power permanently increased by AI – can result in a grand-scale nightmare. At this point, however, I must confess, I cannot quite see what the exact nature of this nightmare is as it is difficult to guess what this future scenario would be and what the role of AI would be in it. Would an enormous backlash occur in the form of a revolt by a contemporary "digital proletariat"?

6 See the following link: https://www.youtube.com/watch?v=-JvfdmPp3d8&t=190s.

And is it plausible at all to think that, in accord with the Marxist vision of 19th century industrial technology, AI would change allegiance and side with those in revolt against those in power? These are just tentative questions with some simplification. However, one thing is sure: our present-day economic and political situation is extremely volatile in no small part because of AI, and it is difficult to predict even the near future.[7]

Nevertheless, it is hard to believe that hubristic power exercised by means of AI and the threat of a backlash, a quiet or cataclysmic revolt, likewise due to AI, are the only relevant scenarios concerning AI. Indeed, why would such an economic and political scenario, such a grand-scale nightmare, be the only possible destiny of AI in our society? Although I have tried to show that there is a secularised theological structure behind this grand-scale economic and political approach to AI, now I will present a theological alternative by revisiting the "double image" theory I outlined at the beginning of my paper.

4. "Double Image" Theory Revisited: Partnership with AI

To do so, let me first draw your attention to the age-old quest for the so-called "artificial humanoid" or, to use the well-known 20th century term, humanoid robot. It is clear that if humankind creates AI in its own image, then, among all the forms of AI, humanoid robots have a special significance. Thanks to the rise of computer science in the 20th century, humanoid robots are now a reality like never before, but the idea of artificial humanoid beings is very old in our culture, and can in fact be traced back to Antiquity. Prior to the 20th century, however, such humanoids were not called robots; they were called automata. I suppose it is hardly necessary to point out how fascinated human beings have always been by all kinds of real and imaginary automata mimicking living creatures and especially by automata with human form, that is, humanoid automata. A historical inquiry into these automata can offer valuable lessons as to the use and perception of them by human beings – lessons that may prove illuminating to our own contemporary relationship with humanoid robots.

7 Prior to the digital revolution of the 90s, but already in the period of the postwar decades with their advanced capitalism and welfare society in the West, *An Essay on Liberation* by Herbert Marcuse made an attempt to portray technology as a force which is ultimately on the side of the exploited. "Is is still necessary to repeat that science and technology are the great vehicles of liberation, and that it is only their use and restriction in the repressive society which makes them into vehicles of domination?" (Marcuse 1969, 12). The most salient feature of Marcuse's essay is its orientation towards the future with a clear intention to predict (and perhaps influence) the way of liberation.

An article by Linda M. Strauss succinctly summarises historical research in this field. She notes that, before the 1960s, historians of technology ignored automata as the latter were irrelevant to their economic and political view of technology.

> Because traditional historians of technology pursued their subject within a theoretical framework based primarily on economic and political relationships, automata held little interest for them since these devices, unlike most tools, weapons, and machines, seemed to have no economically productive or politically powerful functions. (Strauss 1996: 180)

After some early signs in the 60s, however, a full paradigm shift occurred in the 70s, and historians recognised that the history of automata is symptomatic of the fact that technology operates in "broader cultural contexts" than its political and economic use. As Strauss puts it, it was suggested that

> [...] historians might benefit from thinking more deeply about how people have developed technologies for symbolic and idealistic rather than obviously practical purposes. If one approaches automata with such ideas in mind, then the very fact of their unproductive nature when juxtaposed [...] to the remarkable amount of technological innovation that has been invested in them, raises a question worth considering: "Why, over such a long time period, have their creators consistently devoted so much human effort, so much technical ingenuity, and so many costly materials to producing these machines?" (Strauss 1996: 180).

Why indeed? Why were automata in the past, and why are robots in the present so precious to us? Robots are completely subject to the usual mechanisms of economics, politics and power. Automata were substantially different from our present-day or future robots, as the former were single artefacts, on the periphery of society, made by a small bunch of freaks like Jacques de Vaucanson, the most prominent automaton builder in the 18th century who engineered, among other automata, the famous *Flute Player* (on Vaucanson and his automata, see Kang 2011). Today's robotics research is driven by long-term corporate interests and intended to promote mass production. Robots are and will be products which can potentially collect more data from their owners' lives than internet companies have ever dreamt of.

Robots, however, must be seen in contrast to those obscure and faceless algorithms of power like those used by the German magnate. In having a human face, humanoid robots are special. Strauss is not hesitant to call them the "doubles" of human beings which relate to us according to the deep dialectic of the same and the other. The more robots become similar to and eventually almost the same as human beings, the more clearly they stand out as humanity's "other". It is this dialectic that invests robots with a great epistemic value for humanity; indeed, they can become our "epistemic partners" on our way to getting to know them and ourselves better. Needless to say, this is way beyond treating robots as our slaves or servants. There is no doubt that such an approach to them as slaves is particularly tempting, and appears to conform perfectly to the musings of common sense on this matter (cf. Bryson 2010). Robots look like artefacts which are under our total control, and this control seems to be so essential in our relationship with them that, one might claim, it is not likely to change in the future. In considering all the components of this relationship, common sense suggests that all of them seem to point towards our mastery over robots.

> There is in fact no question about whether we own robots. We design, manufacture, own and operate robots. They are entirely our responsibility. We determine their goals and behaviour, either directly or indirectly through specifying their intelligence, or even more directly by specifying how they acquire their own intelligence. But at the end of every indirection lies the fact that there would be no robots on this planet if it were not for deliberate human decisions to create them. (Bryson 2010: 65)

Apparently, humanity closes around robots from all directions; it seems that they are and will be entirely conditioned by us. And it sounds plausible that creating something entails – or ought to entail – total power and control over what is created. Nevertheless, Judaeo-Christian culture has transmitted an illuminating story that may prompt us to think otherwise. Scripture features a God creating and thoroughly knowing human beings and, at the same time, the latter being free from God's will. Similarly, robots, although created and cognised by us, may rise above our control. I do not quite mean that they can become "free". But there is a possible intermediary state between being under control and having freedom. In such an intermediary state, in the form of humanoid robots, AI *may* take a form that is no longer in a relationship of control and power with us and no longer follows the logic of mastery and servitude; rather, this form of AI enters into a relationship with us that is best characterised as "partnership".[8]

Let me emphasise again that if AI *may* take such a form of partnership with us; there is no guarantee that it actually will. Even humanoid robots can be strictly held on the course of doing our bidding and serving us, whereby such

8 The way in which human mastery and the robots' servitude may dissolve into partnership cannot really be demonstrated through Hegel's celebrated analysis of lordship and bondage in his *Phenomenology of Spirit* (cf. Hegel [1807] 1977, 111–119). The lord and the bondsman are both human; they are two distinct forms of self-consciousness. In the process of his service, the bondsman as servile self-consciousness experiences a rise to freedom and independence – something which can hardly be applied to robots.

power over them and through them, as I noted above, can result in a backlash, no matter how exactly it does. Nevertheless, it can be argued that the idea of the "artificial humanoid" as realised by means of humanoid robots is the best chance for technology and especially for AI to emerge from facelessness and assume a human face. In a sense, technology in general and AI in particular can in this fashion undergo a process of redemption or, if you prefer, fulfilment. If it transpires, such a historical and also ontological development of technology assuming a human face will go hand in hand with a change in ethical sensibility and ethical norms. Ethics will in fact fulfil ontology, as the anthropocentric obligation towards "the face of the other" formulated by Emmanuel Levinas will extend to non-human beings, introducing a more universal morality and raising the ontological value of non-human existence (cf. Gunkel 2018: 95–97). While this process, if it comes to pass, will certainly affect individual human lives, its significance will be global and world historical.

This process does not happen only by technology becoming human or completely other than human. It happens also because the relationship qua partnership between human beings and humanoid robots repeats and re-enacts the scenario between God and human beings as depicted in the Book of Genesis. And this is where the true significance of the "double image" theory may come to the fore. The human-robot relationship may mirror the divine-human relationship based on freedom and responsibility and, doing so, can offer an alternative to the relationship of power as discussed above. Needless to say, this connection implies more than an act of mirroring, as it is also a continuity with the freedom and responsibility of human beings in the middle. This freedom and responsibility informs the biblical scenario with a God who does not create human beings for servitude; God is not a master in need of slaves. Quite the contrary, God shares freedom with human beings, of which a shared responsibility for creation is the true meaning (Fretheim 1994: 355–356). It is the responsibility presented in the biblical scenario that human beings are supposed to carry with them into their relationship with AI epitomised by robots.

Designing and building robots can be a profoundly spiritual act by humans through which the two events of being created and creating form a continuity. Anne Foerst makes a fascinating case for robot designers at MIT who do their job under the spell of something like a meditative prayer. No doubt they are late heirs to the Jewish golem-builders who performed their ritual in contemplation of the creation of Adam (Foerst 2004: 35–37, 101). On the one hand, an age-old religious dream comes true by means of technology. On the other hand, with the realisation of the artificial humanoid, technology enters the realm of religion.

In conclusion, it can be claimed that there are two theological paradigms that apply to the birth of artificial intelligence. The question of which one asserts itself depends on whether AI is just an intensification of the dark tendencies of modern technology, or it proves to be their turning point or redemption. If AI represents the very peak of techno-scientific power, sowing and exploiting division within humankind, it can also turn into a secularised punishing force with attributes of the divine – the same way as modern technology as a whole turns into such a force in *Frankenstein*, the legend of the golem, and Karl Marx. But there is an alternative vision. If, in the form of humanoid robots, AI comes to assume a human face "in the image of humankind", introducing a new era of human responsibility that follows a profound dialectic between the human and the non-human, the creation of AI will prove a world historical re-enactment and also continuation of God's creation of human beings.[9]

5. Bibliography

Adams, Henry (1999) *The Education of Henry Adams*. Oxford: Oxford University Press.

Baer, Elizabeth Roberts (2014) *Golem Redux: From Prague to Post-Holocaust Fiction*. Detroit: Wayne State University Press.

Bryson, Joanna J. (2010) "Robots Should Be Slaves." In: *Close Engagements with Artificial Companions: Key Social, Psychological, Ethical and Design Issues*, edited by Yorick Wilks, 63–74. Natural language processing (NLP) v. 8. Amsterdam, Philadelphia, PA: John Benjamins Pub. Company.

Finn, Ed (2017) *What Algorithms Want: Imagination in the Age of Computing*. Cambridge MA, London, England: The MIT Press.

Foerst, Anne (2004) *God in the Machine: What Robots Teach Us about Humanity and God*. New York: Plume.

Fretheim, Terence E. (1994) "The Book of Genesis." In: *The New Interpreter's Bible (NIB): vol. 1.*, edited by Leander Keck. New revised edition, 319–674. Nahville, TN: Abingdon Press.

Gunkel, David J. (2018) "The Other Question: Can and Should Robots Have Rights?" *Ethics and Information Technology* 20: 87–99.

Hegel, Georg Wilhelm Friedrich ([1807] 1977) *Phenomenology of Spirit*. trans. A. V. Miller. Oxford: Clarendon Press.

Heller, Nathan (2018) "The Philosopher Redefining Equality." *The New Yorker*, December 31st 2018.

9 This work has been supported by Charles University Research Centre program No. 204052.

Herzfeld, Noreen L. (2002) *In Our Image: Artificial Intelligence and the Human Spirit.* Theology and the sciences. Minneapolis, MN: Fortress Press.
Kang, Minsoo (2011) *Sublime Dreams of Living Machines: The Automaton in the European Imagination.* Cambridge MA: Harvard University Press.
Keck, Leander, ed. (1994) *The New Interpreter's Bible (NIB): vol. 1.* New revised edition. Nahville, TN: Abingdon Press.
Kittel, Gerhard, ed. (1935) *Theologisches Wörterbuch zum Neuen Testament (ThWNT): vol. 2.* Stuttgart: Kohlhammer.
Marcuse, Herbert. (1969) *An Essay on Liberation.* Boston: Beacon Press.
Marx, Karl, and Friedrich Engels (1976) *Collected Works vol. 6. 1845–48.* London: Lawrence and Wishart.
— (1996) *Collected Works vol. 35. Capital, vol. 1.* London: Lawrence & Wishart.
O'Neil, Cathy (2016) *Weapons of Math Destruction: How Big Data Increases Inequality and Threatens Democracy.* Unabridged. New York: Random House Audio.
Ovid (1958) *The Metamorphoses: a Complete New Version.* New York: Viking Press.
Rad, Gerhard von (1935) "Die Gottesebenbildlichkeit im Alten Testament." In: *Theologisches Wörterbuch zum Neuen Testament (ThWNT): vol. 2.*, edited by Gerhard Kittel, 388. Stuttgart: Kohlhammer.
Scholem, Gershom (1966) "The Golem of Prague and the Golem of Rehovoth." *Commentary* 41 (1): 62–65.
— (1973) *Zur Kabbala und ihrer Symbolik.* Frankfurt a.M. suhrkamp.
Shelley, Mary Wollstonecraft (2003) *Frankenstein, or, The Modern Prometheus.* Rev. ed. Penguin classics. London, New York: Penguin Books.
Strauss, Linda M. (1996) "Reflections in a Mechanical Mirror: Automata as Doubles and as Tools." *Knowledge and Society* 10: 179–207.
Wiener, Norbert (1964) *God and Golem Inc. A Comment on Certain Points Where Cybernetics Impinges on Religion.* Cambridge MA: M.I.T. Press.
Wilks, Yorick, ed. (2010) *Close Engagements with Artificial Companions: Key Social, Psychological, Ethical and Design Issues.* Natural language processing (NLP) v. 8. Amsterdam, Philadelphia, PA: John Benjamins Pub. Company.
Yair, Gad, and Michaela Soyer (2008) "The Ghost Is Back, Again: Karl Marx and the Golem Narrative." *Journal of Classical Sociology* 8 (3): 323–43.

From Experiential to Existential Questions

An Interdisciplinary View on Social Robots in Religious Settings

Diana Löffler, Swantje Luthe, Jörn Hurtienne, Ilona Nord

Abstract

Robots are entering all sorts of social spaces including areas such as religion and spirituality. Social robots in particular do not only function as a new medium for communication but are perceived as autonomous social counterparts. Using social robots in religious settings raises questions that cross disciplinary boundaries. Questions from human-robot interaction (HRI) ask about how to design the robot user experience. Theological questions touch existential aspects and ask whether robots can obtain spiritual competence and how they might fundamentally change previously unautomated religious practice. So far, work done by HRI researchers and theologians remains largely siloed within both communities. In this work, we attempt to interweave disciplinary views on social robotics in religious communication on the example of the blessing robot BlessU-2. We discuss a discursive design study in which the robot expressed Protestant blessings to thousands of visitors of a public exhibition. More than 2,000 people wrote comments on their encounters with the robot. From these comments we analyzed both experiential aspects such as the acceptability and design features of this social robot as well as how the participants reflected on important existential questions such as the origin of a blessing. We conclude that a probing approach and the subsequent collaborative analysis of empirical data was fruitful for both Protestant Theology and HRI disciplines. Central challenges of social robots could be addressed more holistically from the perspective of individual user experience up to the theological and ethical reflection of individual and social existential needs.

1. Religion and Robotics?

Religion and robotics is a topic that fuels controversies, as media reports on technological impact assessments, ethical debates and public discourses show. The project BlessU-2, reported on below, sparked over 2,000 media reports worldwide with widely polarizing reactions. On the one hand, this demonstrates the potential of discursive research designs in the field of HRI. On

 | DOI:10.30965/9783957437488_019

the other hand, these controversies reveal how the robot challenges a current understanding of religious communication by raising fundamental questions of the very nature of religious communication: What are the roles that digital technology can and should assume in religious communication, and how can we describe the impact of technology on the perception of religious phenomena like blessings? In this work, we present an empirical study on a social robot in a religious setting and provide an interdisciplinary view on the experiential and existential questions raised. Before we report on the study, we provide a brief overview of the related literature on how religion and robots are approached from the perspective of HRI and theology.

1.1 *The Study of Religion and Robotics from an HRI Perspective*

Robots and algorithms are carrying out physical as well as cognitive activities in an increasing number of areas and are fundamentally changing previously unautomated practices. Thus, robotics will eventually affect even those religious practices that today deeply rely on human assistance (Groover 2010). Social robots will play an important role, but it is unclear how acceptable they are to believers and what design features are important in making the transition – fundamental questions in HRI research.

So far, only a handful of religious, or theomorphic (Gabriele Trovato et al. 2018), robots have been developed, but little empirical evaluation has yet been reported. Notable examples of social religious robots are the moving Buddhist robot monk 'Xian'er' that shows visitors around in a monastery (Bin 2015), a Buddhist Pepper robot that provides low-cost funeral services (Martin 2017) and a Catholic saint robot 'SanTO' that supports praying (G. Trovato et al. 2018) (for a detailed review see (Löffler, Hurtienne, and Nord 2019)). These examples of religious robots are designed to fulfill different purposes. While the robot monk Xian'er extends the reach of human personnel, the Pepper monk provides an alternative infrastructure to traditional forms of religious practice. SanTO offers ritual enhancement by providing company during prayer. Despite the difference in functionality, the robots share a humanoid appearance and are all developed in Japan or China, countries known for their tradition of animism. Animism is defined as the belief that creatures, objects and places have a spiritual essence or soul (Taylor 2008), resulting in a more welcoming attitude towards robots compared to the dominant Western fear of being victimized by machines (Beauchamp 1980). While the robot monks Xian'er and Pepper are already in use, little has been done in terms of scholarly reflection regarding the robots' user experience or implications beyond their functional aspects. Thus, empirical studies are highly needed, involving religious robots of different shape, functionality and involving users with different cultural and religious backgrounds.

1.2 *The Study of Religion and Robotics from a Theological Perspective*

To understand the different ways of how theology approaches the impact of technologies like robotics on religious phenomena, we briefly outline the research history of digital religion in the following. In the last three decades, the communication sciences, theology and religious education have been exploring how new technologies support religious and spiritual activities. At the beginning, the disciplines differentiated between virtual and real. This distinction was overcome through the term *digital religion*, describing the many phenomena of digital culture:

> Digital Religion explores the intersection of new media technologies, religion, and digital culture. It encompasses topics such as how religious communities engage with the Internet, to ways religiosity is expressed through digital practices and the extent to which technological engagement can be seen as a spiritual enterprise. (Campbell 2017)

US media researcher Heidi Campbell outlines four waves to describe how different questions and methods arise, peaked, and finally led into new fields of research in digital religion. In the first wave, scholars mainly *observed* phenomena in digital media and documented what was happening to religious practices on the Internet. In wave two, scholars *identified* and *typologized* trends of religious internet practice. In wave three, scholars engaged in a discussion of methods and theorization that helped *analyzing* religious communities' strategies within new media/ social media; and in wave four, scholars focused on *negotiations* between online and offline religious practices related to its context in society.

Theological research across the four waves has achieved a consensus that technology is changing and redefining religious practices as well as religious communication and vice versa. Researchers are currently using the concept of *mediatization* of religious communication to describe the transformation processes, digital religion and religious and spiritual practice. Mediatization is understood as a metaprocess, i.e., as a process of processes, whereby all communication and cultural actions are seen as sign-mediated, including religion and religious practice (Hepp and Krotz 2012). Media is not only shaping and framing communication and discourse but also have influence on the communication of religion and the cultural practices in societies. Over the last years theological research discourses have converged and are interlocking today, which is conducive to international and interdisciplinary cooperation which in turn sparks new research methods in theology.

For example, by collaborating with researchers in the field of HRI who are prototyping robotic systems and testing them in the field, a shift from previously descriptive research to a constructive research intending to being involved in developing technology can be expected – possibly advancing a fifth wave of research on digital religion. In this work, we outline our experiences from one of the very first interdisciplinary collaborations between theology and HRI on the example of studying users' experiences with and reflections about a blessing robot in the following.

2. Discursive Design Study of Blessing Robot BlessU-2

To explore the impact of the blessing robot BlessU-2 more holistically, we followed a discursive design approach (Dunne and Raby 2013; Tharp and Tharp 2018) that makes users "think, talk and question".[1] By letting visitors of a public exhibition experience the robot, they could get an idea of how life with such a machine would feel like and opt for required adjustments to be made before this technology 'happens'. This probing approach helps to debate experiential and existential implications of robot technology in a safe environment. The robot used in the discursive design study was BlessU-2, a humanoid robot designed by the Protestant Church of Hesse and Nassau in Germany to deliver blessings. The robot stands about 1.80 meters tall and is built out of an ATM with a touchscreen on its chest, two movable arms and a head with a digital mouth, moveable eyeballs and eyebrows and a light in the nose (see Figure 1). The user can select one language from seven options via touchscreen and is then greeted by the robot. Then, the user can choose between a male or female voice and four types of blessings: traditional, companionship, encouragement and renewal. The robot will spell one of about ten Bible verses from the chosen blessing category and recites the biblical verse with raised arms, light flashes from its palms and varying facial expressions. The procedure lasts less than a minute and the robot offers a printout of its blessing before saying goodbye.

2.1 *Method*

BlessU-2 was presented to the public as an art project at the World Reformation Exhibition at Wittenberg, Germany, for four months during 2017. More than 10.000 people interacted with the robot. Visitors were free to leave a

1 https://mitpress.mit.edu/books/discursive-design

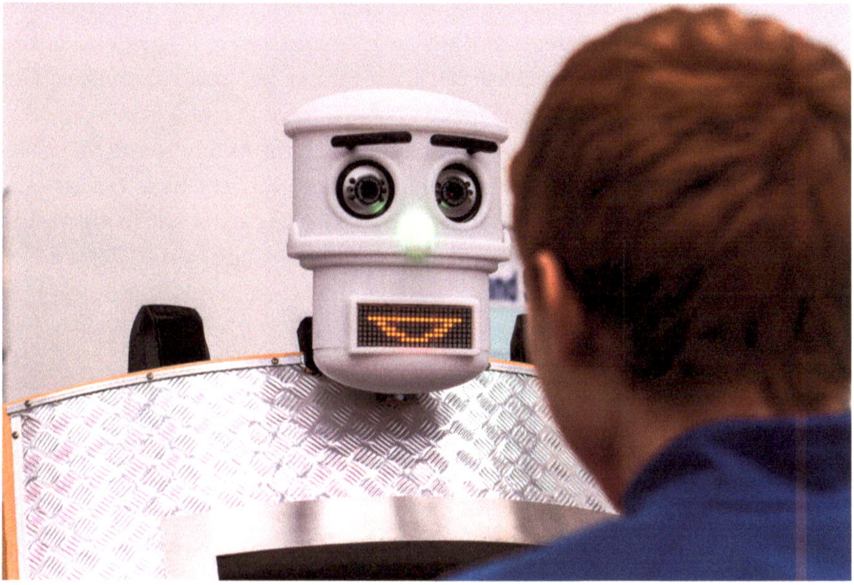

FIGURE 18.1 Blessing robot BlessU-2 (© Swantje Luthe).

free-form written feedback about their experience before or after they interacted with the robot. In total, 1.923 anonymous feedback forms were collected. The comments were digitalized and the content was analyzed for emerging topics following a qualitative content analysis. Moreover, two independent raters evaluated the affective content of the comments as either positive, neutral or negative (see (Löffler, Hurtienne, and Nord 2019) for more details of the analysis).

2.2 *Results*

The content analysis of the 1.923 visitor comments revealed that 20% concerned appearance of and interaction with the robot, thereby its experiential qualities, whereas 80% were on the reflective level, touching fundamental existential questions of religious communication. In contrast to the mostly neutral to negative echo the robot created in the media, half of the comments of the participants who experienced the robot were positive (51%), one third was neutral (29%) and one fifth was negative (20%) (a detailed analysis of the empirical study is provided by (Löffler, Hurtienne, and Nord 2019)).

Regarding the robot appearance, the participants described perceptible qualities of BlessU-2 and related affective reactions. Comments mentioned for example anthropomorphic features ("The hands made the robot more human", "The imitation of the human gestures is ridiculous"), the use of light to emphasize the action ("Closing my eyes and feeling the light, I felt comfortable and blessed"), side-effects of BlessU-2's robotic nature ("The noise of the motors constantly reminded me that it is a machine.") or material choice ("The robot is cold – just like a piece of metal.").

Comments regarding robot behavior focused on the practical and functional aspects of the interaction, like the human voice and options to choose language, dialect and gender of the robot's voice ("Hearing a blessing in my dialect is so special."), the idea of taking a printout of the blessing home ("It feels like getting a receipt in the supermarket.") or that the interaction opportunities offered by BlessU-2 were too limited ("It doesn't feel like an interaction; the robot just gives me instructions.").

On the reflective level of the feedback, fundamental and existential questions were raised about the nature of religious communication. We present the results in four clusters inspired by Campell's categories in digital religion:

- Religious Identity: What existential needs do people experience through blessings?
- Religious authority: Who is authorized to bless others?
- Religious truth: Can God bless through technology? Is blessing dependent on human body?
- Religious authenticity: What role can and should digital technology play in religious communication? How can these roles be designed? How can the use of technology be described in the perception of religious phenomena such as blessings?

2.3 *Discussing Experiential Aspects of Robot Appearance and Interaction*

In our discursive design study, blessing robot BlessU-2 was publicly exhibited and invited for interaction. A total of 1.923 visitors left comments, of which only 20% related to experiential qualities of robot appearance and interaction. This shows that the experiential aspects played a minor role compared to the existential questions raised by the religious robot. BlessU-2's humanoid nature was beneficial in part, but since it loses in the comparison with a human in many ways, its anthropomorphic appearance was also troublesome. Moreover, as the robot clearly featured mechanical elements such as the touchscreen and its material rather points to a common cash machine or an industrial robot, visitors were constantly reminded that they were interacting with a machine. Because of this wild mixture of characteristics of things and beings, a category conflict could have emerged. It might be because of such findings that current design guidance for robots in religious settings recommend sacred objects as source of inspiration instead of living beings (Gabriele Trovato et al. 2018).

The balance between human imitation and resting in machine-like otherness also was the most central topic regarding the interaction with the robot. On the one hand, BlessU-2 used light in its palms and voice to deliver blessings, which were received as pleasurable by the participants and as "extra-human". On the other hand, a narrow set of interaction capabilities and its repetitive actions contributed to the robot feeling mundane. We therefore conclude that besides concealing the robotic nature (Ackerman 2018) or mimicking human-human interaction, the design strategy could also aim at establishing new forms of social interaction in which the 'thingness' or 'otherness' of the robot can play beneficial (Welge and Hassenzahl 2016) for the human-robot user experience, e.g., its capability for highly customizable religious experiences. This would translate into designs that balance the robot's hybrid nature between being different from humans while at the same time making use of natural forms of interaction.

2.4 *Discussing Existential Questions of Identity, Authority, Truth and Authenticity*

Besides the visitors' comments on experiential aspects of their encounters with the blessing robot, the majority dedicated their attention to existential questions about the nature of religious communication raised by the project. We acknowledge this focus by discussing the four clusters of comments in greater detail in the sense of *doing theology* with people – a well-known and modern approach in religious education.

2.5 *Identity*

Some literature contributions, also on digital religion, postulate a great need of people for blessings. The practical-theological discussion about "new" rituals on lifechanging transition point to this need. The theological understanding of ritual, in line with ethnological-anthropological research, describes it as functional and as something not commonplace. Rituals make status changes accessible and channel emotions. The theory of worship considers blessing as the central act of public liturgies, representing encouragement and justification in critical biographical life situations and life stories. This understanding localizes the need for blessing, especially in situations where people deal with contingency experiences and are challenged to reinterpret their life history and belief history, e.g., when people are faced with the experience of death and dying, when children are born, when couples publicly make a promise of marriage, and so on. Blessing implies a human reliance on recognition, affirmation, affection and empowerment. The blessing is more than a mere wish for good; within the performative speech act it becomes the reality of faith.

After meeting the robot, some visitors indeed felt blessed. The comments in response to BlessU-2 with statements about the blessings of the people make another dimension of rituals visible, namely their everyday stabilizing effect. Such everyday rituals ensure the stability of a social system and social relationships (Gutmann 2013). This is done by routinely passing on, sharing, and awarding each other's messages, strengthening and sustaining everyday routines and relationships, interpreting the experienced reality. According to the Protestant theological understanding of the priesthood of all believers, everyone can spend or receive a blessing. A blessing is not a sacramental act and is not tied to a liturgical ministry. Everyday blessing action could open a resonant space for people to experience recognition, affirmation, affection and empowerment. Furthermore, the contact with BlessU-2 proves that blessing rituals can become meaningful as everyday routines and that the robot is credited with giving people a blessing for the day ("I heartily say THANK YOU for the blessing encouragement. I felt through the Bible Word addressed, touched inwardly and bestowed by God."). Whether or not such practice is desirable, how about people asking intelligent speakers or software assistants such as Alexa (Amazon Echo) or Siri (HomePod) to bless them?

2.6 *Authority*

Blessings were formed long before the organization of religions in institutions or communities. There is a long tradition of a non-liturgical and a non-institutional dimension of this religious practice. But we find little theological reflection on the significance of everyday acts of blessing, especially in the Protestant-Christian tradition. Within theological interpretations of blessing, references to the media of religious blessing practice or communication are rarely undertaken. This applies both to more private acts of blessing and to public liturgies. All in all, the practical theological interpretation of blessing shows that the existential dimension of the rite has scarcely been worked out here. There is also a lack of religious psychology studies that expose the functions of blessing and in this sense offer connection possibilities for interdisciplinary studies in the field of user experience.

On the one hand, the comments on BlessU-2 indicates the everyday relevance of blessing pointing out that digital technologies such as robots could take over such functions (e.g., "One feels blessed, like a blessing-to-go"). On the other hand, there are those who want to reserve acts of blessing for priests and pastors ("Isn't human approval required, even if it is God's word?"). Against the background of de-traditionalization, individuals as well as groups have minor knowledge and access to traditional religious rituals in their everyday lives. Therefore, some of the commentators see it as a task for liturgically educated persons to speak blessings and perform blessing rituals. The reactions to

BlessU-2 correlate with other media reports interpreting the robot as priest's or pastor's replacement and thus causes irritation ("The joke goes too far! I think this thing is awful – but maybe this is the future (instead of a pastor)!"). The robot is therefore linked with the liturgical blessing that is given to the congregation at the end or within worship services. Future designs should therefore clearly communicate a robot's role, either for everyday blessings or in worship services.

2.7 *Truth*

After encountering the blessing robot, some people expressed having had a blessing experience ("To my great surprise, I accepted the blessing well."). At the same time, this experience evokes questions about the source of the blessing: What is the relationship between divine power and technology? Does the blessing experience stem from God, from the programmed machine, or from the expectation that God is taking shape through the media ("Is God really behind it? Or is the voice made of plastic, I mean, artificial?")? The answers of the participants are ambivalent.

Even before the most recent theological discussion with the so-called new media, there are already theological contributions that deal with the relationship between man and medium or machine. Paul Tillich characterizes the relationship between man and machine: "[The machine] has a relative life of its own, which compels anyone who deals with it to empathize with it, to listen to it. It has – and the more complicated it is, the more – an individual character that cannot be completely calculated, which can be eerie, even demonic. And it is not unusual for a relationship to be established between the machine and its handlebars, which is akin to a living being." (Tillich 1967). Because humans live in a world shaped and transformed by technology, they must integrate technology into their myth of the meaning of life, so Tillich. The technology itself is neutral. However, on the one hand, it has demonizing potential because it can uncover and explain certain connections. On the other hand, however, it has a demonizing effect, because in dealing with technology, in the process of appropriation and reception, there remains a lack of understanding of the processes that run automatically. This results in a meaning surplus in contact with machines. Today, with the increasing complexity of technology, it is likely to be intensified in such a way that people do not fully understand the automated actions of machines, robots and algorithms, so that a meaning surplus is also produced when dealing with digital technology and artificial intelligence.

The blessing robot or its algorithm selects from some biblical blessings. The stored verses are audio recordings spoken by pastors. As liturgically familiar, the machine raises the arms to bless, while lamps light up in each of the palms. Those who are blessed by the robot can feel the heat from the light sources on their face. BlessU-2 creates a resonance space for magical ideas that are always implicit in acts of blessing. "The concept of blessing oscillates between magic and performative speech," says Katrin Bederna (Bederna 2015). "According to psychology of religion (magic as suggestion) and sociology of religion (magic as 'relief from the compulsion to reflect modernity'), blessing and magic share certain qualities, without this having to be problematic." Thus, BlessU-2's blessing experience creates a resonance space for experience the 'other'. In media reports, this 'otherness' of the robot is rarely addressed. Instead, BlessU-2 is often described as a "robot priest", a simulation of its human counterpart. This description implies a competition between man and machine, as the robot is taking the place of the human priest as representative of God. The blessing automation then intensifies a trivial understanding of the blessing ritual: the person to be blessed asks for a blessing and receives divine power at a button press. In such a view, the blessing robot as a creation of the human itself can control and objectify God, making the ritual seem ridiculous and blasphemous. Moreover, it shows the limits of a functional understanding of religion in which the blessing is a means of coping with contingency or the risk of experiencing sorrow. When an analogy between god/priest and robot is applied here, technology and machine determinism fill the divine dimension in life (Martin 2017).

In contrast, a theology of blessing reveals that the blessing ritual has multiple dimensions and, most importantly, that the blessing cannot be depicted within a simple sender-receiver model, no matter if it's mediated through a robot or a human priest or pastor. The blessing does not fulfill the function of magical desires. On the contrary, the blessing person uses the optative blessing formula to attribute the power of blessing to the freedom of God. The blessing is therefore not in the hands of the person who spends it, but the act of blessing inherits the moment of divine freedom, which is beyond human control and communicates in a media-mediated manner.

2.8 *Authenticity*

The encounter with BlessU-2 inspired the visitors to theologize and reflect on future scenarios. At least four fields of imagination for religiously-active (social) robots crystallize in the comments:

- Robots demonstrate humans' creativity given to us by God (e.g., "Fantastic: God has given us so much intelligence to invent such a robot.");
- Robots could increase the reach of religious institutions and staff (e.g., "I would imagine that the robot appeals to young people in particular.");
- Robots could offer services if there is no alternative (e.g., "We have not been able to find a pastor in my congregation for more than a year, and the robot is definitely a lesser evil compared to an orphaned church.");

- Robots could also offer services with their own added value; the robot allows co-determination in the sense that it offers choices of blessings (e.g., "Although the robot did not speak to me personally, I still felt addressed and experienced a blessing tailored to my needs."). A rather neutral statement describes the robot as comparable to other religious media and therefore does not oppose its use in a religious context ("Why not, we have no problem with printed Bibles.")

The comments show once again that both a simple polarization between technology and God, between human hubris and divine power, as well as a narrowing down to ethical questions behind the issues raised by the visitors remain behind. Instead, they imagine future scenarios which provide information on how technological developments can be understood being integrated into the religious sphere and what issues and areas of practice could be involved. What dangers and what added value would the developments of social robotics in the context of religion and church have? The question of extending scope for action, especially in connection with language assistants, is an interesting question worth pondering. Intelligent assistants could help as translators or query the information required for event celebrations. What does it mean, however, to ask critically if the users of the technology cannot distinguish between chatting with a human person or an automated computer program, i.e., understanding the simulated sociality as social interactions?

The commentators also imagine that through mediated communications young people could be addressed – assuming that the current offers made by the church are not attractive for them. Whether the aversion to robotics is a generation-dependent question would still need to be examined. In addition, questions are raised in the comments on how technological advances create new connections or better accessibility for those living in remote rural areas or those who shy away from contact with religious institutions. Finally, the statements provoke theological-anthropological reflections. What does it mean, for example, that the robot does not judge, is limited in interaction capacity or adapts the blessing ritual to personal needs? BlessU-2 reminds that it is important to take the revealed human vulnerability towards technology serious. As humans can take pretended attention and empathy as 'the real thing' (Hepp and Krotz 2012), robot technology can exploit this vulnerability by giving the illusion of attention and companionship without the demands of human relationships. This is reflected in visitor statements expressing that they felt accepted by the robot as they did not need to fulfil expectations and that they felt comfortable by being able to control the interaction.

Recent developments, also in other areas such as health care, which use robots for the nursing sector more strongly in future, challenge not only theology and research in the context of theological ethics. Rather, there are open research questions for Practical Theology, which should be researched in collaboration with HRI – questions that ask how a robot should look like to be trusted or trustworthy in spiritual competence.

3. Final Thoughts and Conclusion

We see the outlined considerations above as a productive way of dealing with the fundamental theological critique of technology, instead of a technophile belief in progress. After all, the use of technology, as well as its critique, are inseparably tied in the field of religion. In the interdisciplinary play between theology and HRI, as our experience shows, new research impulses arise. One could anticipate a fifth wave in the study tradition of digital religion, which is worth exploring. The aim is to focus on a shift from previously descriptive research to research settings that make use of technical developments in the sense of prototyping. Empirical studies with users, especially when interacting with polarizing or provocative technical artifacts such as robots in religious contexts, can shift the question of user experience of digitized communications as a design mandate into the center of practical theology and churches. Herein lies the opportunity to fundamentally include the vulnerability and endangerment of human life as well as the needs of humans in connection with the development of robotics and artificial intelligence. It would be worth considering how social robots should be developed to support meaningful religious experiences without replacing face to face encounters through technology. After all, not everything that is possible to develop is also meaningful to realize.[2]

4. Bibliography

Ackerman, Evan. 2018. "Can a Robot Be Divine?" IEEE Spectrum. 2018. https://spectrum.ieee.org/automaton/robotics/artificial-intelligence/can-a-robot-be-divine.

Beauchamp, Gorman. 1980. "The Frankenstein Complex and Asimov's Robots." *Mosaic: An Interdisciplinary Critical Journal*. University of Manitoba. https://doi.org/10.2307/24780264.

2 Acknowledgments: We thank the visitors who commented on BlessU2. We also would like to thank the Protestant Church of Hesse and Nassau who made the comments available to our analysis.

Bederna, Katrin. 2015. “Segen/Segnen, Bibeldidaktisch, Sekundarstufe.” 2015. doi.org/10.23768/wirelex.SegenSegnen_bibeldidaktisch_Sekundarstufe.100046.

Bin, Han. 2015. “Beijing's Longquan Temple Adopts Modern Approach.” *CCTV News*, October 26, 2015.

Campbell, Heidi A. 2017. “Surveying Theoretical Approaches within Digital Religion Studies.” *New Media & Society* 19 (1): 15–24. https://doi.org/10.1177/1461444816649912.

Dunne, Anthony, and Fiona Raby. 2013. *Speculative Everything: Design, Fiction, and Social Dreaming*. MIT Press.

Groover, Mikell P. 2010. *Fundamentals of Modern Manufacturing : Materials, Processes, and Systems*. J. Wiley & Sons.

Gutmann, Hans-Martin. 2013. *„Irgendwas Ist Immer“. Durchs Leben Kommen. Sprüche Und Kleinrituale – Die Alltagsreligion Der Leute*. Berlin.

Hepp, Andreas, and Friedrich Krotz. 2012. “Mediatisierte Welten: Forschungsfelder Und Beschreibungsansätze – Zur Einleitung.” In *Mediatisierte Welten*, 7–23. VS Verlag für Sozialwissenschaften. https://doi.org/10.1007/978-3-531-94332-9_1.

Löffler, Diana, Jörn Hurtienne, and Ilona Nord. 2019. “Blessing Robot BlessU2: A Discursive Design Study to Understand the Implications of Social Robots in Religious Contexts.” *International Journal of Social Robotics*, 1–18. https://doi.org/10.1007/s12369-019-00558-3.

Martin, Alex. 2017. “Pepper the Robot to Don Buddhist Robe for Its New Funeral Services Role.” *The Japan Times*, 2017.

Taylor, Bron. 2008. *Encyclopedia of Religion and Nature*.

Tharp, Bruce M., and Stephanie Tharp. 2018. *Discursive Design : Critical, Speculative, and Alternative Things*. MIT Press.

Tillich, Paul. 1967. *Die Religiöse Substanz Der Kultur*.

Trovato, G., A. Lucho, A. Ramon, R. Ramirez, L. Rodriguez, and F. Cuellar. 2018. “The Creation of SanTO: A Robot with ‘Divine’ Features”. In *15th International Conference on Ubiquitous Robots*. Honolulu, USA.

Trovato, Gabriele, Cesar Lucho, Alexander Huerta-Mercado, and Francisco Cuellar. 2018. “Design Strategies for Representing the Divine in Robots.” In *Companion of the 2018 ACM/IEEE International Conference on Human-Robot Interaction – HRI '18*, 29–35. New York, New York, USA: ACM Press. https://doi.org/10.1145/3173386.3173388.

Welge, Julika, and Marc Hassenzahl. 2016. “Better Than Human: About the Psychological Superpowers of Robots.” In , 993–1002. Springer, Cham. https://doi.org/10.1007/978-3-319-47437-3_97.

Robots and the Complexity of Everyday Worlds

Andreas Bischof, Arne Maibaum

Abstract

Now that robots are leaving their cages in factory shop floors and laboratories, they are confronted with human everyday worlds. With this transfer from being exclusively concerned with technical systems to building socio-technical systems, everyday worlds become a wicked problem of robotics: They are interpretative, highly context dependent and products of constant interactive negotiation. The key to understanding and modelling this wicked problem is acknowledging the complexity of a human interaction. We stress three basic factors, which are constitutive for this complexity: indexicality, reciprocity of expectations, and double contingency. Although it is in the nature of these factors that they cannot be formalized, roboticists are forced to translate them into complicated, rather than complex, formalizations.

1. Introduction

For decades robots were caged in factory shop floors and laboratories. Although roboticists have referred to the idea of robots solving "real world problems" since at least the 1980s, actually dealing with worlds has become common in robotics only over the course the past 15 years.

This activity is closely connected to an upsurge in state funding of research programs that investigate opportunities for the everyday use of robots. The EU launched the robotics promotion program "SPARC" with a volume of over €700mio in 2014. The "National Robotics Initiative" in the US will have distributed about $500mio by the end of 2020. Similar programs are in place in Japan and South Korea. Furthermore, several grand challenges and research competitions such as the "DARPA Robotics Challenge" (2012–15) address the development of robotic technology for everyday life.

The scope of these new robotic applications is as wide as people's everyday needs. Scenarios range from self-driving cars to robotic companions for entertainment in private households. Robots for elderly care and nursing homes are one of the most prominent funded technologies in societies threatened by demographic change.

 | DOI:10.30965/9783957437488_020

This boom involves different, partly converging, fields of research such as service robotics, search and rescue robotics, human-robot-interaction (HRI) and social robotics (Bischof 2017). These fields cannot be understood as homogeneous, despite sharing a common vision of developing robots for everyday uses. By engaging worlds of everyday life, these research areas face a new kind of problem that challenges the established theories, methods and cultures of robotic research and engineering.

In this article we discuss the nature of these problems as "wicked problems" and their implication for roboticists dealing with everyday worlds. Therefore, we address the "wickedness" of dealing with everyday worlds on a conceptual level. Based on this, we explain the "complexity gap" (Meister 2014) of social situations and robotic capabilities by referring to Latour's notion of "complicating the complex" (Latour & Hermant 2006).

Conclusively, we present empirical analyses of robotic projects showing that the challenge of dealing with everyday complexity evokes two strategies: on the one hand the reduction of complexity, e.g. by laboratory experiments, and on the other hand the resumption of complexity, e.g. during robotics competition. Both are dynamically connected and influence both the actual phenomenon of human – robot interaction, as well as the scientific research thereof.

2. Everyday Worlds as a "wicked problem" for Robotics

The problems that arise when robots are deployed in everyday worlds cannot solely be understood as technical challenges, e.g. in terms of obstacle avoidance. Leaving the factory buildings and laboratories does not only add a new set of tasks for the machines, it challengers the scientific – sociologists of science say: *epistemic* – foundations of the field.

By definition, robotics for everyday worlds wants and needs to get closer to its object. This has required a number of technical innovations. A key area was the autonomous navigation of machines in "unstructured" environments, as robots for everyday scenarios are typically mobile. They had to be able to maneuver safely and reliably through corridors and offices or even through densely crowded museum halls. Indeed, with the development of powerful navigation algorithms and models for real-time path planning (Thrun 1998; Fox et al. 1999; Roy et al. 1999) a boom in new everyday use cases can be observed. The development of safe robot joints and manipulators which can regulate their force more specifically or have sensitive feedback mechanisms led to similar effects.

In this context, the work on the reliability of robots in everyday environments has been understood as a technical challenge. There are new variables to which the system must be able to react reliably. These variables must first be determined and need experimentally evaluated threshold values and models for operationalization. Certainly, their solution requires creativity and joint negotiation processes. Above all, these are problems that *can* be solved satisfactorily by means of engineering and computer science.

However, this machinability in technical dimensions cannot be applied to all the challenges that arise when robots are required to perform functions outside factory buildings or laboratories. Due to its intended use in everyday worlds, robotics is no longer exclusively concerned with technical systems. Instead it becomes a discipline which is concerned with *socio*-technical systems such as architecture or urban planning. Architects do not design buildings as an end in itself, but as parts of social and cultural practices such as working, eating, or leisure. Robotics and architecture now share a similar kind of problem: that of the resistive – some say: wicked (Rittel & Webber 1973) – nature of the predictability of human activity in socio-technical systems. Here different kinds of expertise intersect: Scientific, engineering, political, social and aesthetic interests are interwoven and sometimes conflicting. Architects and designers have developed methods and capabilities to understand how people move through space. Robotics needs similar knowledge for the domain of social interaction.

Constructing robots for use in everyday worlds is an enormous challenge for a technical discipline. Everyday worlds and social interactions with so-called "untrained users" are an absolute borderline case for the theoretical, methodical and technical instruments of robotics. Making robots work in everyday worlds also challenges the understanding of the role of the roboticist (Bischof 2018).

3. Dealing with the Complexity of Everyday Worlds in Technical System Design

Although seemingly mundane to us as humans, everyday worlds are characterized by a specific complexity mainly grounded in the ways people create meaning and coordinate their activities. Our hypothesis is that the "robotization" of everyday situations requires the computational and epistemic processing of the complexity and contingency that is typical of everyday worlds. The characteristics of a social situation must necessarily be somehow discretized in a computational process and this happens in specific circumstances, which literally *complicate the complex* (Bischof & Heidt 2018).

3.1 *The Complexity of Everyday Worlds*

What is the complexity of everyday worlds? We answer this question using three basic characteristics of social interaction. In selecting them, it is important to show what needs to be considered beyond theories that conceive human-robot interaction as an exchange between only two entities. Thereby, we reject the tendency in the current debate that conceptualizes human-robot interaction only as dyadic exchange between two entities on a micro level. Beyond a theoretical debate about this paradigm, our aim is to give a pragmatic perspective from a sociological and anthropological standpoint that shows how certain fundamental structural elements of interaction organize everyday worlds. The following ideas do therefore not form a specific theory but are rather core components of an interpretative paradigm common in the disciplines of anthropology, ethnology, and sociology.

We want to highlight three basic factors which are constitutive for the complexity of interactions in everyday worlds: 1) indexicality, 2) reciprocity of expectations and 3) double contingency. These emphasize the non-trivial efforts that humans undertake to ensure that interactions work.

(1) *Indexicality* means that a term or action referred to is only understandable in its context. This has firstly a conceptual-referential dimension, which is rooted in a semiotic understanding (Peirce 1932) of the world: The meaning of an action or a term is always only understandable in a semantic network of further terms with which it is related (index). The term "workmanship", for example, can mean quite different things in its concrete use, depending on whether it stands in the semantic network of "poor", "quality", or "guarantee". The second dimension of indexicality follows an interactionist perspective (Garfinkel 1967) and is even more difficult to model: The meaning of actions and terms depends on their historically grown and/or active communicative use.

A famous example for indexicality from the philosophy of language, which is the base for this concept, are Wittgenstein's "language-games". Wittgenstein used the term to highlight that language's meaning is woven into actions. The famous concept was intended "to bring into prominence the fact that the speaking of language is part of an activity" (1953: p. 27) which creates meaning. In line with this, we want to highlight indexicality as a basic factor for the complex interactivity of everyday worlds: Actions and communication cannot be understood without their context, which is not a container in which it is located, but rather metadata which is given and referenced to in every everyday interaction.[1]

(2) *Reciprocity of expectations* explains how people create understanding and common ground in mutual interaction (Goffman 1971). We learned through indexicality that meaning does not follow solely from the linguistic signs of words or phrases. It is tied to past interactions, joint experiences and constant renegotiations. These – especially in the form of "common sense" (Schütz & Luckmann 1980) – become constitutive expectations for everyday interaction, but not in the sense of simply applying rules. The focus of the constitution of meaning lies on the interactive and sequential structure of communication. Syntax and semantics alone do not provide a sufficiently precise basis for intersubjective coordination of action. The inter-subjectively coordinated interpretation activity of the actors forms the constitutive prerequisite of every interaction. The coordination of mutual expectations, i.e. to check whether one is on the same page, is called reciprocity of expectations.

In one of his most famous breaching experiments, Garfinkel has shown that the work of interpreting the negotiation of meanings does not only occur episodically but permanently, constituting even a normative expectation placed on all participants in interaction: People react very sensitively to repeated demands for the clarification of the meaning of everyday expressions that are usually taken for granted. It is assumed that one thinks along and interprets the statements of the other correctly – even if one does not know what exactly is meant. Not following this behavior is considered as either bad intention or simply insanity, for example when one replies to the everyday question "How are you doing?" with "Do you mean physically or mentally?".

(3) *Double Contingency* is a concept we borrow from systems theory (Luhmann 1982): Contingency describes a state that is neither necessary nor impossible. For social situations that means that they are fundmentally characterized by openness and uncertainty. Even if we are familiar with a given situation and know the context of an action, we cannot automatically predict how the interaction will continue. If we add the reciprocity of expectations, we find that contingency doubles because we must consider that both sides have mutual expectations and that those are

1 In the philosophy of language, 'indexical' is used in a more restricted sense. We are using 'indixical' in a broader sense more oriented towards symbolic interactionism.

not necessarily met. Communication in this sense is not self-evident and even unlikely (Luhmann 1987: pp. 148–190).

By highlighting these three factors, we stress that seemingly mundane everyday interaction is, in its conditions and implementation, not trivial but complex. In this work, we have not described the tradition and methods of the theories working with this everyday complexity. We rather show that there are established social-theoretical concepts for the analysis of interactions, which should be related to human-robot interaction, as e.g. Compagna and Boblan have argued (2015).

It is important to underline that these three factors do not offer universal solutions for studying human-robot interaction. Instead, they emphasize that the complexity of everyday worlds is genuinely bound to their actuation in concrete situations.

Alač, Movellan, and Tanaka (2011) provide a striking example of the complexity of interactive coordination in human-robot interaction. The authors followed a team of roboticists and examined their efforts to test robotic behavior with preschool children. In one episode, which was meant to probe the allure of the robot, a researcher was seated on a rocking chair next to a robot and simulated reading a book. The children showed immediate interest in the scene. As they entered, they approached the researcher, pointed to the robot and asked for an explanation for what they saw. However, the researcher feigned indifference and did not react to their gestures. Gradually, the toddlers started to adopt this attitude of willful negligence, which Alač and colleagues (2011) called the "ignoring game" (p. 911): They minimized the attention paid to the robot while engaging more and more in other activities. Even when the robot physically moved, the children noticeably ignored it and finally left the room.

The incident illustrates that the assumed sociality of the robot was highly dependent on the interaction in the situation (Pentzold & Bischof 2019). Although a number of potential uses had been technically inscribed and formed part of the toy robot's sensorimotor equipment, they remained latent and were not enacted. In this case, voluntary disregard prompted users to ignore the material object. In this process of ignoring, the children actually demonstrated their interaction competence and followed a common orientation, which the experimenter had involuntarily specified.

3.2 *Reducing Complexity to Complicatedness*

As a consequence of those three presented aspects, the complexities of everyday worlds – namely indexicality, expectations and contingency – are difficult if not impossible to formalize mathematically (Lindemann 2016).

Nevertheless, roboticists who build machines for such situations have to find workarounds to deal with them. Roboticists are forced to formalize contingent excerpts of the everyday world or to fall back on formalizations from other sciences. Sociologists of technology would say, those open connections and situated meanings must become fixed in technical action chains (Rammert & Schulz-Schaeffer 2002).

In order to avoid a normative connotation of the terms, we would like to point out that "complexity reduction" does not mean a per se deficient process. Complexity reduction is necessary to enable interaction; even more it is inherent to acting at all: Without reducing the contingent set of actions, no action can be made. Consequently, human communicators too are dependent on making mutual understanding possible by means of typifications and relevance settings (Schütz 2004). The empirical question for sociologists is which interactive techniques and procedures help to establish and secure interaction in everyday life (Garfinkel 1967, p. 53).

The central conceptual difference between everyday worlds and current computational technology can be described with the Latourian difference between complex and complicated. This difference lies in the number of variables relevant for interactions and, as already indicated above, the difficulty of "calculating them" (Latour & Hermant 2006: p. 30).

Discrete computability as a critical threshold for complexity is at odds with conventional concepts of complexity. Complexity is classically defined by the number of elements in a system and their interdependencies, such as in management (Ulrich & Fluri 1995). In contrast, Latour's concept of complexity is based rather on the simultaneity of options characterizing the complexity of everyday worlds, forcing decisions and reductions in complexity in modeling it for technology.

Complexity in human-robot interaction would then not be an ontological question about the complexity of the robot per se. Instead, it is a question of strategies with which the complexity of certain situations can be reduced in robotic technology in order for them to become capable of acting in everyday worlds. Latour's definition of complexity poses an empirical question regarding non-human actors: How is this reduction of complexity embedded in chains of action and objects?

Latour proposes the term "complication" to describe this complexity reduction. Complication can be understood as the transition from complex to complicated. It consists of determining a finite number of (computable) variables to divide a complex situation into successive steps of a discrete operation.

This differentiation between the modes of complexity within everyday worlds and within computers does not mean that it is impossible to build social machines. We would like to stress that the interactionist theoretical perspective on socio-technical systems does not establish a substantial difference between man and technology. On the contrary, it raises our awareness of the interactive mediations and production of sociality by people, matter and technology. Human-robot interaction is not determined by the materiality of its artefacts, but by the functionality of the combination of elements into an "artificial cause-effect relationship" (Rammert 1989: p. 133). Complexity reduction is thus the transformation of *a given reality into a contingent reality*, i.e. the abstraction of concrete situations into repeatable, typical situations.

The complication of complexity is the fixation of an operation – and this is not exclusive to technology design. A human call center agent, for example, follows a mechanized chain of operations to narrow down the caller's problem and offer assistance. The call center agent's questions and possible answers are usually defined in a software system through which the agent clicks while on the phone with the customer. The communication behavior of the call center agent has thus been reduced by preselection to a level of complexity that allows it to quickly identify typical problems – as does the Siri voice software. Call center agents and chatbots or voice assistants are completely different entities, but the principles of the mechanization of dialogues are the same (Schüttpelz 2013: p. 43).

Robots play an intermediary role in the question of reducing the complexity and capacity of everyday worlds: They are more adaptive and versatile than, for example, bridges (Winner 1980), but technically and conceptually far from capable to deal with complexity in the way humans are. A robot is only able to function within a relatively small corridor of interpretation, which is given to it mainly by the roboticists building it.

4. Modes of Dealing with Complexity in Human-Robot Interaction

In the following we show how roboticists deal with the explained "complexity gap". How do they make their robots fit for the interpretation-dependent and interactive everyday worlds outside the laboratory? In our own empirical studies in the field of human-robot interaction, we have encountered two major strategies in dealing with complexity that we reconstruct in the following.

4.1 *Reduction of Complexity in the Laboratory*

One way of dealing with the obstacle of translating everyday complexity for computers is found in laboratorization processes (Knorr Cetina 1988). Everyday application scenarios are most often translated into laboratory situations, at least at some point of the robotics research process. In standardized experiments, effects of the interaction of human and robotic behavior are tested, whereby the social situation is not only spatially but also temporally detached from its actual context (Knorr Cetina 2002: p. 46).

The goal to building machines that work in actual everyday worlds thus becomes a laboratory science. Standardized evaluation studies and laboratory experiments – often carried out as "Wizard of Oz" designs (Riek 2012) – have become the most important quality criterion for scientific publications within HRI. This finding is surprising in that such experiments are not genuinely part of the repertoire of engineering and computer science research and are also negotiated in a conflict-like manner. It is no longer enough to build a working robot, its effects on specific aspects of the human-robot interaction must be statistically proven.

From our point of view, the domination of standardized experimental methods is based on three epistemic properties of these methods: First, experimental studies generate legitimacy. Statistics imply the figure of the interchangeable observer, which suggests context-different validity conditions of statements and is easier to transfer into other epistemes. Secondly, the controlled environment of a laboratory experiment serves the (limited) functioning of the robot. In particular, research robots are fragile *objects*. For experiments in actual everyday worlds, many robotic platforms are simply not yet robust enough. Thirdly, the modes of research funding and their evaluation considerably enhance the importance of laboratory experiments.

In order not to be misunderstood: The methodical and methodological approach of laboratory experiments is quite reliable and plausible in a standardized research logic. Our point is that this approach adequacy and implications for actual everyday worlds – and their complexity as unfolded above – are not carried out sufficiently within HRI. Thus, the process as such is considered objective, although the crucial steps – e.g. the definition of the research goal, the desirability of the interaction and the interpretation of the data – are inevitably made by human researchers and based on their understandings of everyday worlds. The selectivity and contextuality of these decisions are usually not part of the representations of the results. This has several effects on the results of the experiments on human-robot interaction: The tested situation is in itself a reduction of complexity as it excludes unpredictable interventions and events by third parties or technical issues of real-world situations. The staged performances of the experimental setup or the experimenter, as well as interventions and methodological deficiencies of the implementation, are often hidden in these experiments (see Alač et al. 2011).

We understand the phenomenon of widespread laboratory experiments as a reaction to the problem of the complexity of everyday situations in robotics developments. In our eyes, this is not per se dysfunctionality, but a *necessary* reduction of the complexity and contingency of social situations for the research process. The described experiments create their own laboratized reality. The fit of the laboratory worlds with the "real-world problems" to be solved remains initially unclear and undetermined. The specific complexity of everyday worlds thus comes back into play when the machines are supposed to function in such unstructured environments.

4.2 *Epistemic Means for (re) Incorporation of Complexity*

If robotic researchers were to orient themselves exclusively to laboratory experiments, they would fail to achieve the self-imposed goal of leaving the laboratories and developing machines suitable for everyday use. Consequently, the development processes feature attempts to include instances in which social complexity and contingency reoccur.

There are occasions when the machines are specifically exposed to coincidences, failures, and even falling over. In addition to the use in museums or open-door day, field trials and competitions are a central element within the research landscape. Those opportunities are, according to our argument, an intermediary between the laborites and the complexity of real worlds.

Competitions, such as the RoboCup, are specifically used to circumvent the 'laboratized reality'. In addition, other incidents in which the laboratory situation is used as an excuse for errors, such as allegedly or actually switched off subsystems, battery problems or the like are excluded. Martin Meister, for example, describes: "I have seen many situations in labs where researchers tried to push a robotic research platform out of the door, explaining that the navigation module is at the moment under revision." And concludes: "But in the game, excuses – other that in a laboratory environment – are ruled out" (Meister 2011: p. 8). Within the competitions such procedure is excluded, thus promotinga test under 'real conditions'. These conditions are especially interesting for competitions which engage with human-robot interaction, e.g. the RoboCup@Home. Although the test subjects are instructed to keep still and only to do or say what the robot can understand, the organizers' intention is "to foster natural interaction with the robot using speech and gesture commands" (RoboCup 2009). This allows more complexity and contingency than the usual laboratory situations in which the user is calibrated to make the interaction happen (Lipp 2017).

The element of surprise lost by the overfitting of the laboratory is counterbalanced by the unexpected nature of field tests and competitions.

In contrast to the (nearly) fully contingent field trials, competition offers more contingency than the laboratory allows, but it remains a 'tamed contingency'. The competitive environment is less complex than an everyday environment. The halls in which the competitions take place appear complex in contrast to the laboratory, but the environmental parameters are comparatively stable. In contrast to other field situations, they are also well documented (Maibaum & Derpmann 2013).

According to our argument, the development of robots for everyday-world applications is characterized by two complementary strategies: First reducing complexity and secondly (re)incorporation of complexity, each having its respective paradigmatic location, the laboratory and competition arena. The necessary reduction of the complexity of everyday-world interactions in laboratory research alternates with phases of resumption of complexity and contingency, such as competition. This interplay of complexity reduction and resumption seems to be epistemically functional. At the same time, it does not seem to be an intentional structure in the sense of a reflected methodical decision. Rather, it remains (like many other solutions in the field) an implied "workaround" of the core challenge, dealing with everyday complexity.

5. Conclusion

Now that robots are leaving their cages in factory shop floors and laboratories, they are confronted with human everyday worlds. With this transfer from being exclusively concerned with technical systems to building socio-technical systems, everyday worlds become a wicked problem of robotics: They are interpretative, highly context dependent and products of constant interactive negotiation.

The key to understanding and modelling this wicked problem is acknowledging the complexity of a human interaction. We stress three basic factors, which are constitutive for this complexity: indexicality, reciprocity of expectations, and double contingency. Although it is in the nature of these factors that they cannot be formalized, roboticists are forced to translate them into complicated, rather than complex, formalizations. We have shown that these formalizations are not only computational but epistemic in the sense of how HRI as a field is approaching the complexity of everyday worlds. The main strategy here is the laboratization of social interaction to reduce complexity. But in going beyond the reduction we also saw a reintegration of complexity outside of the laboratory, for example in the competition arena, that is 'tame contingent'.

Our findings stress the importance of incorporating social theory into human-robot interaction research and development on two levels: Firstly, by referencing to the interpretative paradigm we have shown the extent to which everyday worlds, in particular their functioning and normality, are based on interactive production and coordination of meaning between actors. Secondly, we have shown how ambivalently a socio-*technical* discipline like HRI deals with this complexity of seemingly mundane interactions. It is dependent on reduction but must also seek moments of resumption of complexity in order to make the machines function in actual everyday worlds. We do not argue for a fundamental shift in HRI research and development. As we have seen, many HRI practitioners are already incorporating everyday complexity in often rather implicit ways. But we argue for a new sociological sensitivity, especially regarding to the interpretative paradigm, allowing us to understand and situate human-robot interaction as a practice of meaning-production and coordination in given social worlds, rather than a coordination process between two singular entities "human" and "machine".[2]

6. Bibliography

Alač, Morana; Movellan, Javier; Tanaka, Fumihide (2011) "When a robot is social: Spatial arrangements and multimodal semiotic engagement in the practice of social robotics." *Social Studies of Science, 41*(6), 893–926.

Bischof, Andreas & Heidt, Michael (2018) „Die Verkomplizierung des Komplexen. Latours Unterscheidung komplex/kompliziert als analytische Perspektive auf die Genese von Medientechnik." In: Christian Pentzold; Christian Katzenbach; Sigrid Kannengießer; Monika Taddicken; Marian Adolf (Eds.), *Neue Komplexitäten für Kommunikationsforschung und Medienanalyse: Analytische Zugänge und empirische Studien*, Digital Communication Research (Vol. 4, pp. 51–71). Berlin.

Bischof, Andreas (2017) *Soziale Maschinen bauen: Epistemische Praktiken der Sozialrobotik.* Science Studies. Bielefeld: transcript.

Bischof, Andreas (2018) "The Challenge of Being Self-Aware when Building Robots for Everyday Worlds." In: Athanasios. Karafillidis & Robert Weidner (Eds.), *Developing Support Technologies*, Biosystems & Biorobotics (vol. 23). Basel: Springer Nature Switzerland, 127–135.

2 We thank the organizers and participants of "When Robots Think. Interdisciplinary Views on Intelligent Automation" in Münster, November 2018 as well as our colleagues Jannis Hergesell and Cedric Jürgensen for enriching for this paper.

Compagna, Diego; Marquardt, Manuela, & Boblan, Ivo (2015) "Introducing a methodological approach to evaluate HRI from a genuine sociological point of view." In: *International Workshop in Cultural Robotics*. Springer, 55–64.

Fox, Dieter; Thrun, Sebastian; Burgard, Wolfram; Dellaerta, Frank (1999) "Monte carlo localization: Efficient position estimation for mobile robots." In: *Proceedings of the Sixteenth National Conference on Artificial Intelligence (AAAI'99)*, 343–349. AAAI.

Garfinkel, Harold (1967) *Studies in Ethnomethodology*. London: Prentice-Hall.

Goffman, Erving (1971) *Relations in public: microstudies of the public order*. New York, NY: Basic Books.

Häußling, Roger (1998) *Die Technologisierung der Gesellschaft. Eine sozialtheoretische Studie zum Paradigmenwechsel von Technik und Lebenswirklichkeit*. Würzburg: Königshausen & Neumann.

Knorr Cetina, Karin (1988) „Das naturwissenschaftliche Labor als Ort der „Verdichtung" von Gesellschaft." *Zeitschrift für Soziologie*, *17*, 85–101.

Knorr Cetina, Karin (2002) *Die Fabrikation von Erkenntnis. Zur Anthropologie der Naturwissenschaft*. Frankfurt am Main: Suhrkamp.

Latour, Bruno & Hermant, Emilie (2006) *Paris: Invisible City*. Translation (1998) *Paris ville invisible*. Paris: La Découverte-Les Empêcheurs de penser en rond. http://www.bruno-latour.fr/sites/default/files/downloads/viii_paris-city-gb.pdf [14.12.2018].

Lindemann, Gesa (2016) *Social interaction with robots: three questions*. AI & society, 31(4), 573–575.

Lipp, Benjamin (2017): „Analytik des Interfacing. Zur Materialität technologischer Verschaltung in prototypischen Milieus robotisierter Pflege." In: Katharina Hoppe und Benjamin Lipp (ed.) *Neue Materialismen*. Behemoth 10 (1). Freiburg.

Luhmann, Niklas (1982). "The World Society as a Social System." *International Journal of General Systems*, *8*(3), 131–138.

Luhmann, Niklas (1987). *Soziale Systeme. Grundriß einer allgemeinen Theorie*. Frankfurt am Main: Suhrkamp.

Maibaum, Arne & Derpmann, Stefan. (2013). „Spiel und Simulation als Arenen der Technikentwicklung." In: Diego Compagna & Stefan Derpmann (Eds.), *Soziologische Perspektiven auf Digitale Spiele. Virtuelle Handlungsräume und neue Formen sozialer Wirklichkeit*. Konstanz: UVK, 227–246.

Meister, Martin (2011). *Soziale Koordination durch Boundary Objects am Beispiel des heterogenen Feldes der Servicerobotik* (Doctoral dissertation). Berlin: Fakultät Planen, Bauen, Umwelt, Technische Universität Berlin.

Meister, Martin (2014). "When is a robot really social? An outline of the robot sociologicus." *Science, Technology & Innovation Studies*, 10(1), 107–134.

Peirce, Charles Sanders (1932). "Division of Signs." In: Charles Hartshorne & Paul Weiss (Eds.). *Collected Papers of Charles Sanders Peirce*. Cambridge, MA: Berknap Press. (Original work published 1897).

Rammert, Werner & Schulz-Schaeffer, Ingo (2002). *Technik und Handeln – wenn soziales Handeln sich auf menschliches Verhalten und technische Artefakte verteilt.* TUTS – Working Papers. Berlin: Technical University of Berlin.

Rammert, Werner (1989). "Technisierung und Medien in Sozialsystemen: Annäherungen an eine soziologische Theorie der Technik." In: Peter Weingart (Ed.), *Technik als sozialer Prozeß*. Frankfurt am Main: Suhrkamp, 128–173.

Riek, Laurel (2012). "Wizard of Oz Studies in HRI: a systematic review and new reporting guidelines." *Journal of Human-Robot Interaction*, *1*(1), 119–136.

Rittel, Horst & Webber, Melvin W. (1973). "Dilemmas in a general theory of planning." *Policy sciences*, *4*(2), 155–169.

RoboCup (2009). "RoboCup@Home: Rules & Regulations." Retrieved from https://docs.google.com/viewer?a=v&pid=sites&srcid=cm9ib2N1cGFoaG9tZS5vcmd8cm9ib2N1cC1ob21lfGd4Ojc5MGIoY2UyMGRiYWE4NjU

Roy, Nicholas; Burgard Wolfram; Fox, Dieter & Thrun, Sebastian (1999) "Coastal navigation-mobile robot navigation with uncertainty in dynamic environments." In: *Proceedings 1999 IEEE International Conference on Robotics and Automation* (Vol. 1, pp. 35–40). IEEE.

Schüttpelz, Ehard (2013). „Elemente einer Akteur-Medien-Theorie." In: Tristan Thielmann & Ehard Schüttplez (eds.), *Akteur-Medien-Theorie*. Bielefeld: transcript, 9–69.

Schütz, Alfred & Luckmann, Thomas (1980). *The structures of the life world.* (R. M. Zaner & H. T. Engelhardt, Trans.). Evanston, IL: Northwestern University Press. (Original work published 1979).

Schütz, Alfred (2004). *Der sinnhafte Aufbau der sozialen Welt. Eine Einleitung in die verstehende Soziologie*. Konstanz: UVK. (Original work published 1932).

Thrun, Sebastian (1998). "Learning metric-topological maps for indoor mobile robot navigation." *Artificial Intelligence*, *99*(1), 21–71.

Ulrich, Peter & Fluri, Edgar (1995). *Management. Eine konzentrierte Einführung*. Bern: Haupt.

Winner, Langdon (1980). "Do Artifacts Have Politics?" *Daedalus*, *109*(1), 121–136.

Wittgenstein, Ludwig (1953). *Philosophical investigations*. Oxford: Blackwell.

Taught by Technology

Assessing Robots' Social Roles in Early Childhood Education

Scarlet Siebert

Abstract

This contribution aims at critically discussing different social roles of robots that interact with children for educational purposes. For this, the roles of a teacher, a peer, and a novice will be analyzed on the basis of a literature review describing the functions and characteristics of these roles. Didactic concepts explaining how learners acquire knowledge and skills under different instruction styles and concepts of developmental psychology explaining how children perceive social robots will serve as criteria for the assessment of the adequacy of these different roles.

1. Introduction

Currently, embodied and artificially intelligent systems such as social robots enter not only homes (Breazeal 2018) and work places of adults (Broekens, Heerink, and Rosendal 2009), but also schools and even kindergartens. It is argued that due to their embodiment and mobility (Bainbridge et al. 2011; Wainer et al. 2006), reciprocity (Breazeal 2003), and adaptivity (Schodde, Bergmann, and Kopp 2017) robots are able to create social learning situations which is regarded as beneficial for learning outcomes. Therefore, social robots do not only serve as instructional material for the education of Science, Technology, Engineering, and Mathematics (STEM, e.g. Church et al. 2010; Highfield, Mulligan, and Hedberg 2008); social robots themselves teach children subjects they used to learn from humans, such as languages (e.g. Belpaeme et al. 2015; Gordon et al. 2016; Kanero et al. 2018; Fumihide Tanaka and Matsuzoe 2012) and social behavior (e.g. Fukuta, Kuwahara, and Morimoto 2017; Kozima, Nakagawa, and Yasuda 2007). Robots serving as social interlocutors rather than tools raises questions of the adequacy of applying robots in early childhood education because they might deceive children and raise false expectations (Sharkey 2016).

Therefore, this contribution aims at critically discussing different social roles of robots that interact with children for educational purposes. For this, the roles of a teacher, a peer, and a novice will be analyzed on the basis of a literature review describing the functions and characteristics of these roles.

 | DOI:10.30965/9783957437488_021

Didactic concepts explaining how learners acquire knowledge and skills under different instruction styles as well as concepts of developmental psychology explaining how children perceive social robots will serve as criteria for the assessment of the adequacy of these different roles.

2. Advantages of Educational Robots for Learning

Especially in Asia and the USA educational robotics is a growing market (Verified Market Research 2019). In Europe, however, this trend can be observed as well: Educational robotics is strongly supported by the European Union's Horizon 2020 research and innovation programme. From 2016 to 2018, the project L2TOR programmed and deployed the social robot Nao by SoftBank Robotics for second language tutoring in European kindergartens (Belpaeme et al. 2015). Furthermore, since 2018, the ANIMATAS project seeks to develop robots "with human-like social capabilities for education in schools" (European Union 2019). Thus, besides the concepts for applying social robots in STEM education to teach children *about* technology, another focus lies on the design of child-robot interaction (cHRI) in which children learn *from* and *with* technology. For this purpose, educational robots are embodied and equipped with speaking skills and gestures that remind users of the interaction with human interlocutors. These features of social robots have proven to be beneficial in educational settings.

In particular, research found that a personal approach of the robot or its presentation as a counterpart rather than an object (Chang et al. 2010; Han et al. 2008; Westlund and Breazeal 2015), the recognition and expression of gestures (Huang and Mutlu 2013a; 2013b; Lemaignan et al. 2016), the steering of children's attention (Saerbeck et al. 2010), and the use of nonverbal communication (Kennedy et al. 2017) support learning on an affective as well as a cognitive level: The interaction with robots promotes children's performance better than virtual agents in terms of their attention, persistence, and empathy towards a robot (affective level), and regarding their learning gain and the number of attempts needed for a correct answer (cognitive level, Belpaeme et al. 2018). It is assumed that these advantages arise out of social robots' hardware as well as their software.

On the one hand, their embodiment enables them to manipulate the physical world (Hood, Lemaignan, and Dillenbourg 2015; Litoiu and Scassellati 2015; Zaga et al. 2015) which is advantageous especially for young children; moving images presented on screen attract children's attention regardless of the content (Beentjes and van der Molen 1997; Böhme-Dürr 1999). It can be assumed that this effect is even stronger when information is presented by a moving robot engaging in children's physical environment. On the other hand, the capability of social robots to communicate verbally can facilitate their usability, especially for children who cannot yet read (Fong, Nourbakhsh, and Dautenhahn 2002). Taken together, hardware and software offer a multimodal and adaptive learning environment which is regarded as beneficial for young children because they need visual as well as auditory cues to elaborate information (Hoffner, Cantor, and Thorson 1988). For instance, on a cognitive level information can be repeated as often as required, and on an affective level low engagement can be counteracted by different strategies of personal referencing such as calling a child by his/her name (Baxter et al. 2017; Blanson Henkemans et al. 2013; Coninx et al. 2015; Janssen et al. 2011).

The positive learning effects referred to in studies on cHRI are explained by children behaving more socially towards an embodied robot than towards a virtual agent (Kennedy, Baxter, and Belpaeme 2015; Leyzberg et al. 2011; Li 2015). Furthermore, children show more engagement and joy (Kidd and Breazeal 2004; Köse et al. 2015; Li 2015; Powers et al. 2007; Wainer et al. 2007) and also obey a physically present robot more than a tele-present robot (Bainbridge et al. 2011; Powers et al. 2007) supposedly because they perceive embodied robots as interaction partners rather than as mere toys. Against this background, it is argued that robots are able to take over social roles (Belpaeme et al. 2018).

3. Social Roles of Educational Robots Interacting with Children

To find out which social roles robots take on in early childhood education, how a robot is introduced to the children, how it interacts with the children verbally, non-verbally, and physically, and which role human caregivers have in this interaction, a literature review[1] was conducted in 2018. The review included nine papers presenting 12 embodied social robots[2] applied in early education with children aged between eleven months and six years.

1 The query was conducted in the ACM Digital Library, ERIC, and IEEE Xplore. It was restricted to English articles published in journals or as conference papers between 2000 and 2018. The following exclusion criteria decided on the adoption or rejection of the search results: (1) Robots are applied as teaching tools or visualization material, not as social actors; (2) robots are applied in a higher grade than kindergarten/pre-school and/or children are older than 6 years; (3) the study does not focus on child-robot interaction but e.g. teachers' opinions on robots.

2 One of the papers described the application of two robots – Nao and Pepper by *SoftBank Robotics* – in a pre-school (Causo et al. 2017). While Nao always acted as a teacher during the

3.1 *Robots as Teachers*

In general, most of the robots applied in educational settings are programmed to take on the role of teachers or tutors (Belpaeme et al. 2018). During an assignment, a robot supervises children, gives hints, and adapts the exercises to children's individual performance and interests (Janssen et al. 2011; Okita, Ng-Thow-Hing, and Sarvadevabhatla 2011; Saerbeck et al. 2010). Furthermore, the teacher robot scaffolds the tasks, i.e. "breaking up of complex tasks into smaller tasks" (Mubin et al. 2013) and decreasing the amount of assistance to edit the tasks. Lessons, in which a robot fulfills the role of a teacher, are mostly conducted one-on-one because this is regarded as beneficial for individual learning outcomes. However, there are also approaches using robots as teaching assistants for human teachers within one-to-many interactions in which the robot attracts attention and generates motivation in children (Alemi, Meghdari, and Ghazisaedy 2014; Leite et al. 2015; You et al. 2006).

Referring to early education, robots in the role of teachers are as well mostly common. Here, they are applied to teach vocabulary (Malmir et al. 2013; Movellan et al. 2009), literacy and motor skills (Fridin 2014; Hsiao et al. 2015), geographical and environmental knowledge (Causo et al. 2017), and geometrical thinking (Keren, Ben-David, and Fridin 2012). A typical interaction retrieved from the literature review is started and ended by a robot; physically by standing up and sitting down, and verbally by introducing itself and thanking the children for their participation as well as saying goodbye. The cHRI is mainly one-directional in this setting: The teacher robot has its own instructional material – a screen implemented in its carcase or a tablet standing in front of the child – to visualize the learning content, and it instructs the child verbally what to do. Furthermore, the robot gives verbal (e.g. "Well done!") or physical feedback (e.g. head nodding). The robot teacher's actions do not necessarily require a verbal reaction from the child. Instead, children rather take on a passive role and only conduct the tasks assigned by the robot. Furthermore, children seldomly get accompanied by familiar human caregivers. Their only role is to introduce the robot to the children and get their consent to participate in the activity with a robot. The overall goal of the cHRI with the robot in the role of a teacher seems to lie in the fostering of children's engagement and getting them prepared for the tests conducted after the interaction with the robot. During these tests, children's learning progress is often assessed by human experimenters without the robot's presence.

---

six lessons, Pepper took on the role of a teacher, a peer, and a novice depending on the lesson. Therefore, this paper was counted four times, once for each role.

3.2 *Robots as Peers*

Generally, robots are much less often programmed and designed as peers or companions. If they are, however, robot peers are supposed to work on a task together with a child, motivate and also praise him/her (Baxter et al. 2017; Kanda et al. 2004; Lubold, Walker, and Pon-Barry 2016; Zaga et al. 2015). Although it is argued that peer robots should be on the same intellectual level as the children in order to not intimidate them (Belpaeme et al. 2018), peer robots are often designed to appear as a more knowledgeable partner (Hashimoto et al. 2013; Saerbeck et al. 2010).

Looking at pre-school and kindergarten, robots in the role of peers support second language learning (Hsiao et al. 2015), socialization, engagement (Degiorgi et al. 2017), critical thinking, creativity, and independence (Causo et al. 2017) within a playful setting. They are introduced as new friends by human caregivers, attract children's attention and pretend emotions towards children's attempts of interacting with them, e.g. by smiling, using light signals (Hsiao et al. 2015), and inflating. For instance, the robotic companion Puffy has a soft body and can change its form; it puffs up "when it feels happy or confident, and shrinks when feeling depressed or scared" (Degiorgi et al. 2017: 37). Human caregivers encourage children in this setting to connect with the robot by touching it or speaking to it; the activity with the robot does not start before a child has initiated the interaction. The cHRI in which the robot has the role of a peer is mainly physical. They do role playing games (e.g. cashier and customer), play tag with each other, and the robot sings and dances to music. However, these free play activities only function as a warm-up for task-oriented activities in which children get tested about the actual learning content, e.g. vocabulary (Hsiao et al. 2015). During these activities, the robot switches its role from peer to teacher and presents the learning content on a screen implemented in its carcase.

3.3 *Robots as Learners*

Reviewing the roles of educational robots, it becomes obvious that they have been rarely designed as learners or novices that are taught by children so far (Belpaeme et al. 2018). However, these roles are awarded great potential for the learning process because they allow the children to repeat what they have learned and to use meta-cognitive strategies to teach their knowledge and skills to someone and something else, respectively (Renkl 2006). In the role of a learner, the robot keeps making errors. Being instructed by children the robot improves its knowledge and skills leaving the children with more confidence and improved learning outcomes themselves, e.g. in terms of their motor skills such as handwriting (Hood, Lemaignan, and Dillenbourg 2015).

Regarding early education, robotic learners are applied to second language learning (Matsuzoe, Kuzuoka, and Tanaka 2014; Tanaka and Ghosh 2011) and the support of critical thinking, creativity, and independence (Causo et al. 2017). In these settings, human caregivers retain an active role by either teaching both robots and children or teaching only the children, so they can teach a robot afterwards. The cHRI, however, is mostly physical. For instance, children show flashcards depicting animals to the robot that has to demonstrate the corresponding animal gesture. In a test conducted after the activity, children have to prove their knowledge. Here, children neither interact with the robot any longer nor are they accompanied by a familiar person. Instead, the test is conducted by an experimenter (Matsuzoe, Kuzuoka, and Tanaka 2014).

3.4 *Summary*

In congruence with the findings of Belpaeme et al. (2018), this literature review found that educational robots have rarely been designed in the roles of peers and learners so far. These findings should be regarded with attention because these two roles seem promising from a pedagogical (e.g. Renkl 2006) as well as from a psychological perspective (Bandura 1977). While Belpaeme et al. (2018) looked at the study design of social robots applied in educational settings in more general (learner's age, learning outcomes, number of learners per robot, role of the robot), this literature review focused on social robots applied in early childhood education and analyzed the robots' roles in more detail. In the following, the findings of this analysis will be matched with different instruction styles and learning theories and discussed against the background of concepts in early childhood education.

4. Didactic Concepts in Educational Robotics

The social roles robots are programmed by can be connected to different didactic concepts. These concepts help to explain how the relationship between the instructor and a learner is shaped and what imagery of a learning child and of learning itself is constructed. In the following, focus is laid on the concepts derived from the literature review: learning by listening, learning by playing, and learning by teaching. While the two latter are well-known in educational contexts, "learning by listening" will be referred to in order to emphasize the one-directional teaching style that can be observed in most studies on cHRI.

4.1 *Learning by Listening*

As described above the instruction style of robots in the role of teachers is more directive and often one-directional: The teacher robot has an active role, while children are put into a passive role and only receive information. For instance, robots for language learning practice vocabulary with the children by scaffolding the task and then interrogate them. This approach to learning and the didactic methods used for it can be explained by a rather material understanding of education (Klafki 1959) focusing on the specific content of learning rather than the process of acquiring knowledge and skills. Children's learning process seem to be extrinsically motivated here because the robot decides what to do and how to do it. With respect to the content taught, it must be questioned whether the acquired knowledge and skills can be transferred to other contexts by the children after this kind of interaction. Furthermore, it must be put into question whether the imbalanced relationship between robot and child is valuable.

Referring to the concept of scaffolding, children depend on the support of more experienced persons in order to learn and reach the zone of proximal development (Vygotsky 1981). This support is not only given content-wise by "breaking up [ ] complex tasks into smaller tasks" as stated in the literature of social robotics (Mubin et al. 2013: 5). In educational psychology and pedagogy scaffolding is referred to as supportive behavior of a familiar caregiver that reciprocally adapts to a child's needs and skills by using gestures and language suitable for children or taking certain actions in order to attract children's attention, reactivate existing knowledge etc. (Leutner 2001). In social robotics, this adaption to children's needs is reached by the same strategies. Furthermore, robots are programmed to offer children various activities that are adapted to children's amount of interest and fit the actual curriculum of the educational institution (Malmir et al. 2013). Scaffolding, however, especially fosters goal-oriented learning in which the aim of a learning scenario is precisely defined (Ireson and Blay 1998); in less structured learning contexts as typical for early childhood education in Germany the concept of scaffolding requires a lot of sensitivity from the caregivers in order to identify children's interests and be able to adapt to their actions (König 2009) rather than exploiting their own advantage in knowledge (Ireson and Blay 1998). Furthermore, the concept of scaffolding is associated with cognitivist approaches to learning that are characterized by a more linear process. These approaches, however, do not seem to match the understanding of learning in early childhood education because the didactic concepts of kindergarten and pre-school usually emphasize children's active role in their own learning processes which are characterized as being explorative. This understanding of learning matches a more constructivist approach to learning (Kasüschke 2010).

4.2 *Learning by Playing*

Referring to the peer role of a robot with which children can learn in a collaborative way rather than being directed by them, corresponds with the concept of learning by playing. This concept pays attention to the fact that "[f]rom children's own perspective, play and learning are not always separate in practices during early year" (Samuelsson and Carlsson 2008: 624). Oerter and Montada (2008: 237, 246) describe play as an end in itself. It is not goal-oriented but rather comes along with the intrinsic motivation to engage with the immediate environment (exploratory play), to experience sensual arousal (functional play), and to cope with challenges such as a child's relationships with his/her parents or siblings (symbolic play) (Fröhlich-Gildhoff, Mischo, and Castello 2015).[3] To do so, children repeat joyful actions and reenact experienced situations, e.g. with dolls or other equipment that is attributed another meaning than it actually has (e.g. a wooden block as a telephone). By playing children train functions that are important for their development without even being aware as play does not require focusing on something externally introduced and lets lose the feeling of time (Oerter and Montada 2008: 237). Within studies on cHRI in early childhood education, the dominant types of play are presented as (1) cooperative play, (2) role-playing games, and (3) functional play: (1) For instance, Pepper (SoftBank Robotics) plays the role of a new student in order to foster children's social skills or (2) Pepper takes on the role of a cashier from which the children have to purchase items from the supermarket (Causo et al. 2017). Furthermore, (3) peer robots are programmed to physically interact with the children, e.g. by playing tag and attract their attention by physical signals such as inflation/deflation and lighting up (Degiorgi et al. 2017). However, these types of play are not the key component of the whole cHRI. Instead, they only function as a warm-up for a situation that could be categorized as frontal teaching. As Hsiao et al. stated "the playway-to-learn with which iRobiQ is endowed arouse children's motivation in learning by engaging them in enjoyable, amusing, and active learning environment, thereby leading people to regard it as 'edutainment'" (2015: 287). Although this description seems to correspond with the characteristics of early childhood education, i.e. playful interactions, the analysis of the cHRI leads to the assumption that educational robotics rather uses gamification elements such as providing

3 There is no general classification of play. Different scientists stress different aspects of play that either focus on structural or functional aspects of the action (Warwitz and Rudolf 2014).

an electronic teammate for cooperative play (Kapp 2012; Werbach and Hunter 2015) in order to influence children's learning behavior. Referring to the understanding of play as being an activity without any intended purpose (Huizinga and Flitner 2017), it can be argued that the instruction style of robot peers is based on manipulation and optimization rather than free play activities.

4.3 *Learning by Teaching*

Robots as novices exploit the concept of "learning by teaching" (Jamet et al. 2018; Martin 1989). While an instructor, i.e. the one who wants to transfer knowledge, usually legitimizes his/her role based on their own expertise in the field of interest, "learning by teaching" follows a more constructivist approach to learning, defining it as an "active, self-regulated, constructive, purposeful, situated, and social process" (Reinmann-Rothmeier and Mandl 1998, translated by the author). Here, children are understood as creators of new knowledge that can be taught to others – in this case to robots as their learning counterparts (Chase et al. 2009). Referring to Klafki (1959), this instruction style could be assigned to have a more formal understanding of education. Here, education refers less to the content taught but rather to the methods that must be acquired in order to gain knowledge and skills and to be able to apply them to new contexts. This is also a common concept in early childhood education. For instance, the Montessori method (Ludwig 2015) and the Reggio Emilia approach (Krieg 2004) view children as explorers of their environment, who should be offered a range of materials to do so. Therefore, robots in the role of learners seem to be promising for early childhood education because they could foster children's self-efficacy which is considered a crucial component in learning settings (Bandura 1977). As the learner – in this case the robot – is inanimate, the question could be raised of how effective the teaching of an object might be for children's learning, however. In fact, studies show that learners do not need an animate counterpart in order to train knowledge and memory (Bargh and Schul 1980; Cohen, Kulik, and Kulik 1982; Koh, Lee, and Lim 2018). In software engineering, for instance, programmers learn to debug a code by explaining a rubber duck step by step what every line of the code is for (Hunt and Thomas 2000). Comparing a rubber duck to an educational robot, however, seems misleading because the robot simulates reciprocity while the rubber duck's understanding is based on pure imagination and might even encourage learners to make mistakes because they know it cannot correct them. With a robot, as will be argued in the following, it might not be obvious for young children that it is only a machine rather than an interaction partner with cognitive and emotional skills. Therefore, educational robots could have an intimidating effect. Furthermore, it might be asked to much if children of pre-school age needed not only to understand the content taught by a human teacher but also be able to structure this newly acquired knowledge to teach it further. On the other hand, it is also difficult from a technological perspective to program a robot that is able to adequately interact with children if the situation is structured by them and not by the robot and the programmer, respectively. Against this background, it is not surprising that robots in the role of learners and the corresponding didactic concepts that are known from human "learning by teaching" have rarely been considered in the design of cHRI so far. Against this background, however, the question arises whether educational theories and instruction styles developed for human teaching can be transferred to social robots interacting with children (Belpaeme et al. 2018). Therefore, the following section deals with children's perceptions of social robots and the sociability of robots as categories to assess the adequacy of applying social robots in early childhood education.

5. The Adequacy of Robot's Roles

As the sociability of embodied and artificially intelligent systems evolving through the interplay between a social robot's features mentioned above (embodiment and mobility, reciprocity, and adaptivity) is the main argument for applying social robots to educational settings (Belpaeme et al. 2018), the question arises of which roles seem adequate for the interaction with young children, who have become a popular target group in social robotics (Miller, Nourbakhsh, and Siegwart 2008). To assess the adequacy of applying social robots in early childhood education, in the following, focus is laid on children's perception of social robots and their actual sociability.

5.1 *Children's Perception of Social Robots*

The anthropomorphization of technology can be explained by the media equation theory (Reeves and Nass 1996) which describes the tendency of humans to treat computers as social actors (CASA). Responsible for this perception are technical functions, which are perceived as social cues, such as interactivity, a natural user interface, i.e. voice input and output, and the adoption of certain social roles (Nass and Moon 2000). With robots this effect might be even stronger because of their physical presence and the capability to take over social roles, as explained above. Furthermore, their interactivity lets social robots be perceived as interlocutors people might develop a pseudo-social relationship to (Krotz 2007; Höflich 2013). In this context, Sherley Turkle speaks of a "robotic moment" (Turkle 2012), i.e. the willingness to engage with robots.

Children move between a psychological view of robots, believing that they are alive because they can speak, and a technological view in which it is argued that robots are not alive because they lack intentionality (Turkle 2005; Ackermann 1991; Levy and Mioduser 2008; Müller and Schulte 2017). Regarding the ascribed intelligence of robots, children think that they are about as smart as cats, and assume that they have free will, emotions, and memory (Beran et al. 2011; Bernstein and Crowley 2008). In terms of gender, it has been found that boys perceive robots more often as being male, while girls perceive robots as male or female equally (Oros et al. 2014).

With respect to the psychological understanding, children are sensitive to social cues and events (Dornes 1998; Mahler et al. 1993) and can feel empathy for others from early childhood on (Bischof-Köhler 1989). Against this background, the question arises of whom or what robots are perceived as by children; even more because empathy is built upon the ability to create distance between oneself and another subject (Ciaramicoli 2001) – or in this case another object – and to assess the actions of one's counterpart. In this context, Sharkey (2016: 291) regards cHRI as a risk. This can be explained by the fact that young children are much more prone to deception than older children or adults (Bruck and Ceci 1999), especially when information are presented to them visually (Beentjes and van der Molen 1997; Böhme-Dürr 1999). Only from the age of pre-school on children develop an understanding for the difference between reality and fiction (Charlton 2004). Therefore, children should be accompanied when using media, e.g. to learn the difference between narrative content and advertisement on television (Klemm and Schulte 2001). Transferred to robots one could argue that it is not necessary to accompany children because of their intuitive user interface that lets children interact with robots on their own. On the other hand, their appearance disguises the mechanisms it is built for, e.g. analyzing children's performance during the interaction and adapting to it. Furthermore, not only is interaction with a robot a new situation for children but they are also less experienced with interpersonal interaction norms in general. Therefore, Rohlfing, Grimminger, and Wrede (2020) plead for children being accompanied by familiar caregivers when interacting with robots. As analyzed within the literature review, however, children are often left alone with the robot.

5.2 *Sociability of Robots*

Breazeal (2003) differentiates between four stages by which social robots can adapt to increasingly complex environments: While (1) *socially evocative robots* such as the Robo Chameleon by *Silverlit* cause anthropomorphization due to their mere physical presence and some social cues such as moving eyes without actually interacting with users, (2) *socially communicative robots* enable scripted interactions, e.g. by answering questions on the exhibits in a museum like Peter that rolls through the world's biggest computer museum in Paderborn, Germany (HeinzNixdorfMuseum Forum 2013). (3) *Socially receptive robots* use interactions with people to learn from them. For instance, the Luvabella doll by Spin Master is programmed to learn more and more words from children when they interact on a regular basis with it by imitating them. Finally, (4) *sociable robots* are driven to pursue their own goals within the interaction with humans. In this last stage, they should be embodied to create a common space of experience with humans (Bartneck and Forlizzi 2004; Dautenhahn, Ogden, and Quick 2002), convey some intentionality (Bartneck and Forlizzi 2004; Breazeal 2003; Fong, Nourbakhsh, and Dautenhahn 2002), be socially situated, i.e. know whom it interacts with (Fong, Nourbakhsh, and Dautenhahn 2002), what its counterpart is doing (Hegel et al. 2009), respond appropriately, and act autonomously (Bartneck and Forlizzi 2004).

Assuming that the stage of sociable robots possessing any theory of mind[4] has not been reached yet, social robots can hardly be characterized as being rational or affective. It is also questionable whether the inter-subjective context of meaning necessary for social action develops between humans and robots. Referring to Weber's (1984) understanding of social action it must, therefore, be questioned whether interactions between humans and robots are actually social or rather "pseudo-social" as stated by Krotz (2007); they are inanimate and despite the technological advances made in robotics, most of the robots developed for interaction with humans are still remotely controlled within a wizard-of-Oz setting (Riek 2012).

In the future, however, human-machine communication will be even more similar to interpersonal interaction (Krämer, von der Pütten, and Eimler 2012). This might further support the perception of robots as social agents and make them even more interesting for areas in which services are provided with and on people – as exemplified for the context of early childhood education.

4 Theory of mind describes a (human's) capability to ascribe certain feelings, needs, intentions, expectations, and attitudes to oneself and others (Resch 1999). Having theory of mind can be expressed by, for instance, joint attention of interacting subjects towards another subject or object (Baron-Cohen 1995).

6. Conclusion

Robots are more than technologies to learn *about*. They take on social roles and interact autonomously *with* children according to different instruction styles. In most of the reviewed studies, social robots are applied as teachers. This corresponds with a meta-study on the application of social robots in education (Belpaeme et al. 2018). However, it can be argued that the current design of robots' roles barely matches the targets of early childhood education, i.e. the support of agency, self-determination, and participation (Kasüschke 2010).

Robot teachers do not allow children to actively explore their (learning) environment because they are programmed to cognitivist approaches in which the instruction style is more directive. This passive role of the learners carries the risk that acquired knowledge cannot easily be transferred to new contexts because they only learn by heart. *Robot companions* follow a more constructivist approach. As children can decide whether they want to play with the robot, the learning process is more intrinsically motivated. However, as the robot decides on the activities by a pre-scheduled program, children quickly lose their formerly active roles within the learning process. Keeping in mind that children even anthropomorphize cuddle toys raises the question of whether a relationship with an artificial friend might harm children's development and their bonds with humans (de Graaf 2016; Sharkey 2016). *Robot learners*, finally, require action from the children, which speaks in favor of a constructivist approach. Having to understand not only for themselves but rather to transfer their knowledge requires responsibility and creativity. Although constructivist approaches put learners in an active role, they come with some challenges, especially for unexperienced learners. The lack of guidelines can overstrain learners, especially when they are very young or have special needs – both are popular target groups in educational robotics (Jamet et al. 2018). Furthermore, robot learners carry the risk of emotional attachment and exploiting children's caring nature (Sharkey 2016).

Finally, there is evidence that robots showing too much social behavior might distract children during the interaction which might lead to poorer learning outcomes because they are focused on the robot rather than the content (Kennedy, Baxter, and Belpaeme 2015; Yadollahi et al. 2018). Looking ahead, it also needs to be discussed whether the further development of artificial intelligence might lead to an automation of evaluating children's learning progress and outcomes, social ties (Malmir et al. 2013) or even forecasting disabilities by data received through the interaction with robots (Scassellati

2007). As "friends" they conceal the actual function of many cHRIs, i.e. to monitor children's learning process and adapting to it.[5]

7. Bibliography

Ackermann, Edith K. (1991) "The 'Agency' Model of Transactions: Toward an Understanding of Children's Theory of Control." In: Montangero, Jacques and Tryphon, Anastasia (eds.) *Psychologie Génétique et Sciences Cognitives*, 63–73. Geneva: Fondation Archives Jean Piaget.

Alemi, Minoo, Ali Meghdari, and Maryam Ghazisaedy (2014) "Employing Humanoid Robots for Teaching English Language in Iranian Junior High-Schools." *International Journal of Humanoid Robotics* 11 (03): 1450022. https://doi.org/10.1142/S0219843614500224.

Bainbridge, Wilma A., Justin W. Hart, Elizabeth S. Kim, and Brian Scassellati (2011.) "The Benefits of Interactions with Physically Present Robots over Video-Displayed Agents." *International Journal of Social Robotics* 3 (1): 41–52. https://doi.org/10.1007/s12369-010-0082-7.

Bandura, Albert (1977) "Self-Efficacy: Toward a Unifying Theory of Behavioral Change." *Psychological Review* 84 (2): 191–215. https://doi.org/10.1037/0033-295X.84.2.191.

Bargh, John A., and Yaacov Schul (1980) "On the Cognitive Benefits of Teaching." *Journal of Educational Psychology* 72 (5): 593–604. https://doi.org/10.1037/0022-0663.72.5.593.

Baron-Cohen, Simon (1995) *Mindblindness: An Essay on Autism and Theory of Mind.* Learning, Development, and Conceptual Change. Cambridge, Mass: MIT Press.

Bartneck, Christoph, and Jodi Forlizzi (2004) "A Design-Centred Framework for Social Human-Robot Interaction." In *RO-MAN 2004. 13th IEEE International Workshop on Robot and Human Interactive Communication (IEEE Catalog No.04TH8759)*, 591–94. https://doi.org/10.1109/ROMAN.2004.1374827.

Baxter, Paul, Emily Ashurst, Robin Read, James Kennedy, and Tony Belpaeme (2017) "Robot Education Peers in a Situated Primary School Study: Personalisation Promotes Child Learning." *PLOS ONE* 12 (5): e0178126. https://doi.org/10.1371/journal.pone.0178126.

Beentjes, Johannes W. J., and Walma van der Molen (1997) "The Impact of Television, Radio and Print on Children's Recall and Creative Imagination. A Review of Media Comparison Researc." *Trends in Communication* 1 (2): 31–50.

Belpaeme, Tony, James Kennedy, Paul Baxter, Paul Vogt, Emiel E. J. Krahmer, Stefan Kopp, Kirsten Bergmann, et al. (2015) "L2TOR – Second Language Tutoring Using

5 Acknowledgement: This article was supported by the Digital Society research program funded by the Ministry of Culture and Science of the German State of North Rhine-Westphalia.

Social Robots." In *Proceedings of the ICSR 2015 WONDER Workshop*. https://pub.uni-bielefeld.de/publication/2900267.

Belpaeme, Tony, James Kennedy, Aditi Ramachandran, Brian Scassellati, and Fumihide Tanaka (2018) "Social Robots for Education: A Review." *Science Robotics* 3 (21): eaat5954. https://doi.org/10.1126/scirobotics.aat5954.

Beran, Tanya N., Alejandro Ramirez-Serrano, Roman Kuzyk, Meghann Fior, and Sarah Nugent (2011) "Understanding How Children Understand Robots: Perceived Animism in Child – Robot Interaction." *International Journal of Human-Computer Studies* 69 (7–8): 539–50. https://doi.org/10.1016/j.ijhcs.2011.04.003.

Bernstein, Debra, and Kevin Crowley (2008) "Searching for Signs of Intelligent Life: An Investigation of Young Children's Beliefs about Robot Intelligence." *Journal of the Learning Sciences* 17 (2): 225–247.

Bischof-Köhler, Doris (1989) *Spiegelbild und Empathie*. Bern u.a.: Ludwig-Maximilians-Universität München. https://doi.org/10.5282/ubm/epub.2919.

Blanson Henkemans, Olivier A., Bert P. B. Bierman, Joris Janssen, Mark A. Neerincx, Rosemarijn Looije, Hanneke van der Bosch, and Jeanine A. M. van der Giessen (2013) "Using a Robot to Personalise Health Education for Children with Diabetes Type 1: A Pilot Study." *Patient Education and Counseling* 92 (2): 174–81. https://doi.org/10.1016/j.pec.2013.04.012.

Böhme-Dürr, Karin (1999) "Bildmagnet Fernsehen." *Television*, 12: 19–25.

Breazeal, Cynthia (2003) "Toward Sociable Robots." *Robotics and Autonomous Systems*, Socially Interactive Robots, 42 (3): 167–75. https://doi.org/10.1016/S0921-8890(02)00373-1.

Breazeal, Cynthia (2018, October 17) "Social Robots Will Become Family Members in the Homes of the Future." *Quartz*. Available at: https://qz.com/1383665/social-robots-will-become-family-members-in-the-homes-of-the-future/.

Broekens, Joost, Marcel Heerink, and Henk Rosendal (2009) "Assistive Social Robots in Elderly Care: A Review." *Gerontechnology* 8 (2). https://doi.org/10.4017/gt.2009.08.02.002.00.

Causo, Albert, Phyo Zin Win, Peng Sheng Guo, and I.-Ming Chen (2017) "Deploying Social Robots as Teaching Aid in Pre-School K2 Classes: A Proof-of-Concept Study." *2017 IEEE International Conference on Robotics and Automation (ICRA)*, 4264–69. https://doi.org/10.1109/ICRA.2017.7989490.

Chang, Chih-Wei, Jih-Hsien Lee, Po-Yao Chao, Chin-Yeh Wang, and Gwo-Dong Chen (2010) "Exploring the Possibility of Using Humanoid Robots as Instructional Tools for Teaching a Second Language in Primary School." *Educational Technology & Society* 13 (2): 13–24.

Charlton, Michael (2004) "Entwicklungspsychologische Grundlagen." In: Mangold, Roland, Vorderer, Peter, and Bente, Gary (eds.) *Lehrbuch Der Medienpsychologie*, 129–150. Göttingen: Hofgrede.

Chase, Catherine C., Doris B. Chin, Marily A. Oppezzo, and Daniel L. Schwartz (2009.) "Teachable Agents and the Protégé Effect: Increasing the Effort Towards Learning." *Journal of Science Education and Technology* 18 (4): 334–52. https://doi.org/10.1007/s10956-009-9180-4.

Church, William Joseph, Tony Ford, Natasha Perova, and Chris Rogers (2010) "Physics With Robotics – Using LEGO MINDSTORMS In High School Education." In *2010 AAAI Spring Symposium Series*. https://www.aaai.org/ocs/index.php/SSS/SSS10/paper/view/1062.

Ciaramicoli, Arthur P (2001) *Der Empathie-Faktor: Mitgefühl, Toleranz, Verständnis*. 2nd edn. München: Deutscher Taschenbuchverlag.

Cohen, Peter A, James A. Kulik, and Chen-Lin C. Kulik (1982) "Educational Outcomes of Tutoring: A Meta-Analysis of Findings." *American Educational Research Journal* 19 (2): 237–248. https://doi.org/10.3102/00028312019002237.

Coninx, Alexandre, Paul Baxter, Elettra Oleari, Sara Bellini, Bert Bierman, Olivier Blanson Henkemans, Lola Cañamero, et al (2015) "Towards Long-Term Social Child-Robot Interaction: Using Multi-Activity Switching to Engage Young Users." *Journal of Human-Robot Interaction* 5 (1): 32–67. https://doi.org/10.5898/JHRI.5.1.Coninx.

Dautenhahn, Kerstin, Bernard Ogden, and Tom Quick (2002) "From Embodied to Socially Embedded Agents – Implications for Interaction-Aware Robots." *Cognitive Systems Research*, Situated and Embodied Cognition, 3 (3): 397–428. https://doi.org/10.1016/S1389-0417(02)00050-5.

Degiorgi, Marzia, Franca Garzotto, Mirko Gelsomini, Giulia Leonardi, Simone Penati, Noëlie Ramuzat, Jacopo Silvestri, Francesco Clasadonte, and Yosuke Kinoe (2017) "Puffy – An Inflatable Robotic Companion for Pre-Schoolers." In *2017 26th IEEE International Symposium on Robot and Human Interactive Communication (RO-MAN)*, 35–41. https://doi.org/10.1109/ROMAN.2017.8172277.

Dornes, Martin (1998) "Müssen Wir Margret Mahlers Theorie Revidieren?" In Burian, Wilhelm (ed.) *Der Beobachtete Und Der Rekonstruierte Säugling*, 41–76. Göttingen: Vandenhoeck & Ruprecht.

European Union (2019) "ANIMATAS. Advancing Intuitive Human-Machine Interaction with Human-like Social Capabilities for Education in Schools." 211560. Brussels: European Commission. Available at: https://cordis.europa.eu/project/rcn/211560_en.html.

Fong, Terrence, Illah Nourbakhsh, and Kerstin Dautenhahn (2002) *A Survey of Socially Interactive Robots*.

Fridin, Marina (2014) "Storytelling by a Kindergarten Social Assistive Robot: A Tool for Constructive Learning in Preschool Education." *Computers & Education* 70 (January): 53–64. https://doi.org/10.1016/j.compedu.2013.07.043.

Fröhlich-Gildhoff, Klaus, Christoph Mischo, and Armin Castello (2015) *Entwicklungspsychologie für Fachkräfte in der Frühpädagogik*. 3rd edn. Köln: Link.

Fukuta, Hiroyoshi, Noriaki Kuwahara, and Kazunari Morimoto (2017) "A Study of Utilizing Communication Robots for Teaching Preschoolers a Good Manner." In Duffy, Vincent G. (ed.) *Digital Human Modeling. Applications in Health, Safety, Ergonomics, and Risk Management: Ergonomics and Design*, 421–434. Lecture Notes in Computer Science. Springer International Publishing.

Gordon, Goren, Samuel Spaulding, Jacqueline Kory Westlund, Jin Joo Lee, Luke Plummer, Marayna Martinez, Madhurima Das, and Cynthia Breazeal (2016) "Affective Personalization of a Social Robot Tutor for Children's Second Language Skills." In *Thirtieth AAAI Conference on Artificial Intelligence*. https://www.aaai.org/ocs/index.php/AAAI/AAAI16/paper/view/11759.

Graaf, Maartje M. A. de (2016) "An Ethical Evaluation of Human – Robot Relationships." *International Journal of Social Robotics* 8 (4): 589–98. https://doi.org/10.1007/s12369-016-0368-5.

Han, Jeong-Hye, Mi-Heon Jo, Vicki Jones, and Jun-H. Jo (2008) "Comparative Study on the Educational Use of Home Robots for Children." *Journal of Information Processing Systems* 4 (4): 159–68. https://doi.org/10.3745/JIPS.2008.4.4.159.

Hashimoto, Takuya, Hiroshi Kobayashi, Alex Polishuk, and Igor Verner (2013) "Elementary Science Lesson Delivered by Robot." In *2013 8th ACM/IEEE International Conference on Human-Robot Interaction (HRI)*, 133–34. https://doi.org/10.1109/HRI.2013.6483537.

Hegel, Frank, Claudia Muhl, Britta Wrede, Martina Hielscher-Fastabend, and Gerhard Sagerer (2009) "Understanding Social Robots." In *Second International Conference on Advances in Computer-Human Interactions (ACHI)*. https://pub.uni-bielefeld.de/publication/2907234.

HeinzNixdorfMuseum (2013) *PETER: Paderborns Erster Toller ErklärRoboter Auf Tour Im HNF*. Available at: https://www.youtube.com/watch?v=IlzJL2yzxpE.

Highfield, Kate, Joanne Mulligan, and John Hedberg (2008) "Early Mathematics Learning through Exploration with Programmable Toys." In *Proceedings Of The Joint Meeting Of Pme 32 And Pme-Na Xxx, Vol 3*, 169–76. Cinvestav-UMSNH. https://researchers.mq.edu.au/en/publications/early-mathematics-learning-through-exploration-with-programmable-.

Hoffner, Cynthia, Joanne Cantor, and Esther Thorson (1988) "Children's Understanding of a Televised Narrative. Developmental Differences in Processing Video and Audio Content." *Communication Research*, no. 15: 227–45.

Höflich, Joachim R (2013) "Relationships to Social Robots: Towards a Triadic Analysis of Media-Oriented Behavior." *Intervalla*, 35–48.

Hood, Deanna, Séverin Lemaignan, and Pierre Dillenbourg (2015) "When Children Teach a Robot to Write: An Autonomous Teachable Humanoid Which Uses Simulated Handwriting." In *Proceedings of the Tenth Annual ACM/IEEE International*

Conference on Human-Robot Interaction, 83–90. HRI '15. New York, NY, USA: ACM. https://doi.org/10.1145/2696454.2696479.

Hsiao, Hsien-Sheng, Cheng-Sian Chang, Chien-Yu Lin, and Hsiu-Ling Hsu (2015) "'iRobiQ': The Influence of Bidirectional Interaction on Kindergarteners' Reading Motivation, Literacy, and Behavior." *Interactive Learning Environments* 23 (3): 269–92. https://doi.org/10.1080/10494820.2012.745435.

Huang, Chien-Ming, and Bilge Mutlu (2013a) "Modeling and Evaluating Narrative Gestures for Humanlike Robots." In *Robotics: Science and Systems IX*. http://citeseerx.ist.psu.edu/viewdoc/citations;jsessionid=C96530BC2D16617242C005CDCF209FE0?doi=10.1.1.657.3848.

Huang, Chien-Ming, and Bilge Mutlu (2013b) "The Repertoire of Robot Behavior: Enabling Robots to Achieve Interaction Goals through Social Behavior." *Journal of Human-Robot Interaction* 2 (2): 80–102.

Huizinga, Johan, and Andreas Flitner (2017) *Homo Ludens: vom Ursprung der Kultur im Spiel*. 25th edn. Hamburg: Rowohlt Taschenbuch Verlag.

Hunt, Andrew, and David Thomas (2000) *The Pragmatic Programmer: From Journeyman to Master*. Reading, Mass: Addison-Wesley.

Ireson, Judith, and Josepha Blay (1998) "Constructing Activity: Participation by Adults and Children." *Learning and Instruction* 9 (1): 19–36. https://doi.org/10.1016/S0959-4752(98)00022-X.

Jamet, Frank, Olivier Masson, Baptiste Jacquet, Jean-Louis Stilgenbauer, and Jean Baratgin (2018) "Learning by Teaching with Humanoid Robot: A New Powerful Experimental Tool to Improve Children's Learning Ability." *Journal of Robotics* 2018 (November): 1–11. https://doi.org/10.1155/2018/4578762.

Janssen, Joris B., Chrissy C. van der Wal, Mark A. Neerincx, and Rosemarijn Looije (2011) "Motivating Children to Learn Arithmetic with an Adaptive Robot Game." In *Social Robotics*, 153–162. Lecture Notes in Computer Science. Berlin, Heidelberg: Springer. https://doi.org/10.1007/978-3-642-25504-5_16.

Kanda, Takayuki, Takayuki Hirano, Daniel Eaton, and Hiroshi Ishiguro (2004) "Interactive Robots As Social Partners and Peer Tutors for Children: A Field Trial." *Hum.-Comput. Interact.* 19 (1): 61–84. https://doi.org/10.1207/s15327051hci1901&2_4.

Kanero, Junko, Vasfiye Geçkin, Cansu Oranç, Ezgi Mamus, Aylin C. Küntay, and Tilbe Göksun (2018) "Social Robots for Early Language Learning: Current Evidence and Future Directions." *Child Development Perspectives*, January. https://doi.org/10.1111/cdep.12277.

Kapp, Karl M (2012) *The Gamification of Learning and Instruction: Game-Based Methods and Strategies for Training and Education*. San Francisco: Pfeiffer.

Kasüschke, Dagmar, ed., (2010) *Didaktik in der Pädagogik der frühen Kindheit*. 1st edn. Köln u.a.: Link.

Kennedy, James, Paul Baxter, and Tony Belpaeme (2015) "The Robot Who Tried Too Hard: Social Behaviour of a Robot Tutor Can Negatively Affect Child Learning." In *Proceedings of the Tenth Annual ACM/IEEE International Conference on Human-Robot Interaction*, 67–74. HRI '15. New York, NY, USA: ACM. https://doi.org/10.1145/2696454.2696457.

Kennedy, James, Séverin Lemaignan, Caroline Montassier, Pauline Lavalade, Bahar Irfan, Fotios Papadopoulos, Emmanuel Senft, and Tony Belpaeme (2017) "Child Speech Recognition in Human-Robot Interaction: Evaluations and Recommendations." In *Proceedings of the 2017 ACM/IEEE International Conference on Human-Robot Interaction – HRI '17*, 82–90. Vienna, Austria: ACM Press. https://doi.org/10.1145/2909824.3020229.

Keren, Guy, Adi Ben-David, and Marina Fridin (2012) "Kindergarten Assistive Robotics (KAR) as a Tool for Spatial Cognition Development in Pre-School Education." In *2012 IEEE/RSJ International Conference on Intelligent Robots and Systems*, 1084–89. https://doi.org/10.1109/IROS.2012.6385645.

Kidd, Cory D., and Cynthia Breazeal (2004) "Effect of a Robot on User Perceptions." In *2004 IEEE/RSJ International Conference on Intelligent Robots and Systems (IROS) (IEEE Cat. No.04CH37566)*, 4:3559–64 vol.4. https://doi.org/10.1109/IROS.2004.1389967.

Klafki, Wolfgang (1959) "Kategoriale Bildung. Zur Bildungstheoretischen Deutung Der Modernen Didaktik." *Zeitschrift Für Pädagogik* 5: 386–412.

Klemm, Michael, and Dirk Schulte (2001) "Eltern Und Kinder." In Holly, Werner, Püschel, Ulrich and Bergmann, Jörg (eds.) *Der Sprechende Zuschauer. Wie Wir Uns Fernsehen Kommunikativ Aneignen*, 262–286. Opladen: Westdeutscher Verlag.

Koh, Aloysius Wei Lun, Sze Chi Lee, and Stephen Wee Hun Lim (2018) "The Learning Benefits of Teaching: A Retrieval Practice Hypothesis." *Applied Cognitive Psychology* 32 (3): 401–10. https://doi.org/10.1002/acp.3410.

König, Anke (2009) *Interaktionsprozesse zwischen ErzieherInnen und Kindern: eine Videostudie aus dem Kindergartenalltag*. 1st. edn. VS Research. Wiesbaden: VS, Verl. für Sozialwiss.

Köse, Hatice, Pınar Uluer, Neziha Akalın, Rabia Yorgancı, Ahmet Özkul, and Gökhan Ince (2015) "The Effect of Embodiment in Sign Language Tutoring with Assistive Humanoid Robots." *International Journal of Social Robotics* 7 (4): 537–48. https://doi.org/10.1007/s12369-015-0311-1.

Kozima, Hideki, Cocoro Nakagawa, and Yuriko Yasuda (2007) "Children – Robot Interaction: A Pilot Study in Autism Therapy." In Claes von Hofsten and Kerstin Rosander (eds.) *From Action to Cognition*, 164:385–400. Amsterdam: Elsevier. https://doi.org/10.1016/S0079-6123(07)64021-7.

Krämer, Nicole, Astrid von der Pütten, Sabrina Eimler (2012) "Human-Agent and Human-Robot Interaction Theory: Similarities to and Differences from Human-Human

Interaction."In Zacarias, Marielba and de Oliveira, José Valente (eds.) Human-Computer Interaction: The Agency Perspective, 215–240. Berlin, Heidelberg: Springer.

Krieg, Elsbeth, ed., (2004) *Lernen von Reggio: Theorie und Praxis der Reggio-Pädagogik im Kindergarten.* 2nd edn. Lage: Jacobs.

Krotz, Friedrich (2007) *Mediatisierung: Fallstudien zum Wandel von Kommunikation.* 1st edn. Wiesbaden: VS, Verl. für Sozialwiss.

Leite, Iolanda, Marissa McCoy, Daniel Ullman, Nicole Salomons, and Brian Scassellati (2015) "Comparing Models of Disengagement in Individual and Group Interactions." In *Proceedings of the Tenth Annual ACM/IEEE International Conference on Human-Robot Interaction*, 99–105. HRI '15. New York, NY, USA: ACM. https://doi.org/10.1145/2696454.2696466.

Lemaignan, Séverin, Fernando Garcia, Alexis Jacq, and Pierre Dillenbourg (2016) "From Real-Time Attention Assessment to 'with-Me-Ness' in Human-Robot Interaction." In *2016 11th ACM/IEEE International Conference on Human-Robot Interaction (HRI)*, 157–64. https://doi.org/10.1109/HRI.2016.7451747.

Leutner, Detlev (2001) "Instruktionspsychologie." In Rost, Detlef H. (ed.) *Handwörterbuch Pädagogische Psychologie*, 2nd edn., 267–275. Weinheim: Beltz.

Levy, Sharona T., and David Mioduser (2008) "Does It 'Want' or 'Was It Programmed to ...'? Kindergarten Children's Explanations of an Autonomous Robot's Adaptive Functioning." *International Journal of Technology and Design Education* 18 (4): 337–59. https://doi.org/10.1007/s10798-007-9032-6.

Leyzberg, Dan, Eleanor Avrunin, Jenny Liu, and Brian Scassellati (2011) "Robots That Express Emotion Elicit Better Human Teaching." In *Proceedings of the 6th International Conference on Human-Robot Interaction*, 347–354. HRI '11. New York, NY, USA: ACM. https://doi.org/10.1145/1957656.1957789.

Li, Jamy (2015) "The Benefit of Being Physically Present: A Survey of Experimental Works Comparing Copresent Robots, Telepresent Robots and Virtual Agents." *International Journal of Human-Computer Studies* 77 (May): 23–37. https://doi.org/10.1016/j.ijhcs.2015.01.001.

Litoiu, Alexandru, and Brian Scassellati (2015) "Robotic Coaching of Complex Physical Skills." In *Proceedings of the Tenth Annual ACM/IEEE International Conference on Human-Robot Interaction Extended Abstracts*, 211–212. HRI'15 Extended Abstracts. New York, NY, USA: ACM. https://doi.org/10.1145/2701973.2702726.

Lubold, Nichola, E.rin Walker, and Heather Pon-Barry (2016) "Effects of Voice-Adaptation and Social Dialogue on Perceptions of a Robotic Learning Companion." In *2016 11th ACM/IEEE International Conference on Human-Robot Interaction (HRI)*, 255–62. https://doi.org/10.1109/HRI.2016.7451760.

Ludwig, Harald, ed., (2015) *Praxishandbuch der Montessori-Methode.* 3rd edn. Freiburg im Breisgau: Herder.

Mahler, Margaret S., Fred Pine, Anni Bergman, and Hilde Weller, eds., (1993) *Die psychische Geburt des Menschen: Symbiose und Individuation.* Frankfurt am Main: Fischer-Taschenbuch-Verl.

Malmir, Mohsen, Debora Forster, Kendall Youngstrom, Lydia Morrison, and Javier R. Movellan (2013) "Home Alone: Social Robots for Digital Ethnography of Toddler Behavior." In *2013 IEEE International Conference on Computer Vision Workshops*, 762–68. https://doi.org/10.1109/ICCVW.2013.104.

Martin, Jean-Pol (1989) "Quand Les e´le`ves Font La Classe." *Français Dans Le Monde*, no. 224: 51–55.

Matsuzoe, Shizuko, Hideaki Kuzuoka, and Fumihide Tanaka (2014) "Learning English Words with the Aid of an Autonomous Care-Receiving Robot in a Children's Group Activity." In *The 23rd IEEE International Symposium on Robot and Human Interactive Communication*, 802–7. https://doi.org/10.1109/ROMAN.2014.6926351.

Miller, David P., Illah R. Nourbakhsh, and Roland Siegwart (2008) "Robots for Education." In Siciliano, Bruno and Khatib, Oussama (eds.) *Springer Handbook of Robotics*, 1283–1301. Berlin, Heidelberg: Springer. https://doi.org/10.1007/978-3-540-30301-5_56.

Movellan, Javier, Micah Eckhardt, Marjo Virnes, and Angelica Rodriguez (2009) "Sociable Robot Improves Toddler Vocabulary Skills." In 307. ACM Press. https://doi.org/10.1145/1514095.1514189.

Mubin, Omar, Catherine J. Stevens, Suleman Shahid, Abdullah Al Mahmud, and Jian-Jie Dong (2013) "A Review of the Applicability of Robots in Education." *Technology for Education and Learning* 1 (1). https://doi.org/10.2316/Journal.209.2013.1.209-0015.

Müller, Kathrin, and Carsten Schulte (2017) "Ein Modell Zur Analyse von Vorstellungen über Roboter und Ihrer Funktionsweise." In Diethelm, Ira (ed.) *Informatische Bildung zum Verstehen und Gestalten der digitalen Welt*, 109–118. Bonn: Gesellschaft für Informatik e.V.

Nass, Clifford, and Youngme Moon (2000) "Machines and Mindlessness: Social Responses to Computers." *Journal of Social Issues* 56 (1): 81–103. https://doi.org/10.1111/0022-4537.00153.

Oerter, Rolf, and Leo Montada, eds., (2008) *Entwicklungspsychologie.* 6th edn. Weinheim: Beltz PVU.

Okita, Sandra Y., Victor Ng-Thow-Hing, and Ravi K. Sarvadevabhatla (2011) "Multimodal Approach to Affective Human-Robot Interaction Design with Children." *ACM Transactions on Interactive Intelligent Systems* 1 (1): 1–29. https://doi.org/10.1145/2030365.2030370.

Oros, Marina, Milutin Nikolić, Branislav Borovac, and Ivan Jerković (2014) "Children's Preference of Appearance and Parents' Attitudes towards Assistive Robots." In *2014 IEEE-RAS International Conference on Humanoid Robots*, 360–65. https://doi.org/10.1109/HUMANOIDS.2014.7041385.

Powers, Aaron, Sara Kiesler, Susan Fussell, Susan Fussell, and Cristen Torrey (2007) "Comparing a Computer Agent with a Humanoid Robot." In *Proceedings of the ACM/IEEE International Conference on Human-Robot Interaction*, 145–152. HRI '07. New York, NY, USA: ACM. https://doi.org/10.1145/1228716.1228736.

Reeves, Byron, and Clifford I. Nass (1996) *The Media Equation: How People Treat Computers, Television, and New Media like Real People and Places*. New York: Cambridge University Press.

Reinmann-Rothmeier, Gabi, and Heinz Mandl (1998) "Wissensvermittlung. Ansätze Zur Förderung des Wissenserwerbs." In Klix, Friedhart and Spada, Hans (eds.) *Wissen*, 457–500. Göttingen: Hofgrede.

Renkl, Alexander (2006) "Lernen Durch Lehren." In Rost, Detlef H. (ed.) *Handwörterbuch Pädagogische Psychologie*, 3rd edn., 416–420. Weinheim: BELTZ.

Resch, Franz, ed., (1999) *Entwicklungspsychopathologie des Kindes- und Jugendalters: ein Lehrbuch*. 2nd edn. Weinheim: Beltz.

Riek, Laurel D. (2012) "Wizard of Oz Studies in HRI: A Systematic Review and New Reporting Guidelines." *Journal of Human-Robot Interaction* 1 (1): 119–136.

Rohlfing, Katharina, Angela Grimminger, and Britta Wrede (2020) "The Role of Caregivers in Keeping a Child-Robot Interaction Going." In Rohlfing, Katharina and Müller-Bauers, Claudia (eds.) *International Perspectives on Digital Media and Early Literacy: The Impact of Digital Devices on Learning, Language Acquisition and Social Interaction*. London: Routledge.

Saerbeck, Martin, Tom Schut, Christoph Bartneck, and Maddy D. Janse (2010) "Expressive Robots in Education: Varying the Degree of Social Supportive Behavior of a Robotic Tutor." In 1613. ACM Press. https://doi.org/10.1145/1753326.1753567.

Samuelsson, Ingrid Pramling, and Maj Asplund Carlsson (2008) "The Playing Learning Child: Towards a Pedagogy of Early Childhood." *Scandinavian Journal of Educational Research* 52 (6): 623–641. https://doi.org/10.1080/00313830802497265.

Scassellati, Brian (2007) "How Social Robots Will Help Us to Diagnose, Treat, and Understand Autism." In *Robotics Research*, edited by Sebastian Thrun, Rodney Brooks, and Hugh Durrant-Whyte, 552–563. Springer Tracts in Advanced Robotics. Springer Berlin Heidelberg.

Schodde, Thorsten, Kirsten Bergmann, and Stefan Kopp (2017) "Adaptive Robot Language Tutoring Based on Bayesian Knowledge Tracing and Predictive Decision-Making." In *Proceedings of the 2017 ACM/IEEE International Conference on Human-Robot Interaction*, 128–136. HRI '17. New York, NY, USA: ACM. https://doi.org/10.1145/2909824.3020222.

Sharkey, Amanda J. C. (2016) "Should We Welcome Robot Teachers?" *Ethics and Information Technology* 18 (4): 283–297. https://doi.org/10.1007/s10676-016-9387-z.

Tanaka, Fumihide, and Madhumita Ghosh (2011) "The Implementation of Care-Receiving Robot at an English Learning School for Children." In *2011 6th ACM/IEEE*

International Conference on Human-Robot Interaction (HRI), 265–266. https://doi.org/10.1145/1957656.1957763.

Tanaka, Fumihide, and Shizuko Matsuzoe (2012) "Children Teach a Care-Receiving Robot to Promote Their Learning: Field Experiments in a Classroom for Vocabulary Learning." *Journal of Human-Robot Interaction*, August, 78–95. https://doi.org/10.5898/JHRI.1.1.Tanaka.

Turkle, Sherry (2005) *The Second Self: Computers and the Human Spirit*. Cambridge: MIT Press.

Turkle, Sherry (2012) *Verloren unter 100 Freunden: wie wir in der digitalen Welt seelisch verkümmern*. 1st edn. München: Riemann.

Verified Market Research (2019) "Global Educational Robot Market By Type, By Education Level, By Geographic Scope And Forecast To 2026." 4048. Verified Market Research. Available at: https://www.verifiedmarketresearch.com/product/global-educational-robot-market-size-and-forecast-to-2025/.

Vygotsky, Lev Semenovič (1981) *Mind in Society: The Development of Higher Psychological Processes*. Cambridge: Harvard Univ. Press.

Wainer, Joshua, David J. Feil-Seifer, Dylon A. Shell, and Maja J. Mataric (2006) "The Role of Physical Embodiment in Human-Robot Interaction." In *ROMAN 2006 – The 15th IEEE International Symposium on Robot and Human Interactive Communication*, 117–22. https://doi.org/10.1109/ROMAN.2006.314404.

Wainer, Joshua, David J. Feil-Seifer, Dylon A. Shell, and Maja J. Mataric (2007) "Embodiment and Human-Robot Interaction: A Task-Based Perspective." In *RO-MAN 2007 – The 16th IEEE International Symposium on Robot and Human Interactive Communication*, 872–77. https://doi.org/10.1109/ROMAN.2007.4415207.

Warwitz, Siegbert A., and Anita Rudolf (2014) *Vom Sinn des Spielens: Reflexionen und Spielideen*. 3rd edn. Baltmannsweiler: Schneider-Verl. Hohengehren.

Weber, Max. 1984. *Soziologische Grundbegriffe*. 6th edn. Tübingen: Mohr.

Werbach, Kevin, and Dan Hunte (2015) *The Gamification Toolkit – Dynamics, Mechanics, and Components for the Win*. Philadelphia: Wharton Digital Press.

Westlund, Jacqueline Kory, and Cynthia Breazeal (2015) "The Interplay of Robot Language Level with Children's Language Learning During Storytelling." In *Proceedings of the Tenth Annual ACM/IEEE International Conference on Human-Robot Interaction Extended Abstracts*, 65–66. HRI'15 Extended Abstracts. New York, NY, USA: ACM. https://doi.org/10.1145/2701973.2701989.

Yadollahi, Elmira, Wafa Johal, Ana Paiva, and Pierre Dillenbourg (2018) "When Deictic Gestures in a Robot Can Harm Child-Robot Collaboration." In *Proceedings of the 17th ACM Conference on Interaction Design and Children*, 195–206. IDC '18. New York, NY, USA: ACM. https://doi.org/10.1145/3202185.3202743.

You, Zhen-Jia, Chi-Yuh Shen, Chih-Wei Chang, Baw-Jhiune Liu, and Gwo-Dong Chen (2006) "A Robot as a Teaching Assistant in an English Class." In *Sixth IEEE Interna-*

tional Conference on Advanced Learning Technologies (*ICALT'06*), 87–91. https://doi.org/10.1109/ICALT.2006.1652373.

Zaga, Cristina, Manja Lohse, Khiet P. Truong, and Vanessa Evers (2015) "The Effect of a Robot's Social Character on Children's Task Engagement: Peer Versus Tutor." In Tapus, Adriana, André, Elisabeth, Martin, Jean-Clause, Ferland, François, and Ammi, Mehdi (eds.) *Social Robotics*, 704–713. Lecture Notes in Computer Science. Springer International Publishing.

PART V

Society

Artificial Intelligence and Polygenic Scoring

The Risk of Personal Eugenics

Johannes Grössl

Abstract

This paper discusses social and ethical challenges that arise through the application of artificial intelligence on the genetic analysis of humans. AI can be used to calculate individuals' polygenic scores which statistically predict character traits and behavioral dispositions. This technology might foster the selection of unborn children according to the likelihood of engaging in socially compliant behavior or developing a certain level of cognitive abilities. Through exercising moral and cognitive enhancements on individuals and groups, AI-enhanced personal eugenics runs the risk of undermining or destroying essential characteristics of humanity such as moral responsibility and autonomy. It is shown that an argument for the rejection of this technology requires certain philosophical or theological assumptions concerning the essence or teleology of human nature.

1. Ethical Challenges of AI Applications

The term 'artificial intelligence' (AI) refers to complex self-learning algorithms, i.e. computer programs that can solve problems surprisingly well. AI can recognize structures and correlations within large amounts of data, which even the best human mathematicians cannot determine. What is crucial about the application of AI is that we cannot understand retrospectively how a complex algorithm has arrived at its result. Intelligent algorithms are already in use today, for example in marketing: Social networks can forecast which advertisements are likely to be clicked on, based on usage behavior; streaming services know which video will arouse further interest. An influence on voting behavior is also conceivable, as the *Cambridge Analytica affair* and the increasing use of *social bots* on Facebook have shown. In this article, I will discuss an ethical-social challenge that arises in a similar way through the application of artificial intelligence, but in an area that challenges our self-understanding of being human much more than a purely *external* manipulation of behavior would: the application of artificial intelligence in the genetic analysis of humans.

 | DOI:10.30965/9783957437488_022

2. Polygenic Scoring

Companies such as *23andme* or *MyHeritage* offer to sequence and analyze one's genome for less than 300 dollars, and prices will continue to fall. *23andme* claims that more than ten million customers[1] have already taken advantage of this service: They learn in which regions of the world their ancestors lived, whether they are more likely to contract certain diseases, whether they are more likely to become overweight or depressed. If a customer deals adequately with the statistical character of this information, she does not note this result as inevitable, but recognizes the special challenge resulting from this finding. Anyone who is aware of an increased predisposition to the development of depression, for instance, takes appropriate precautions and may even be treated preventatively. Such a prognosis, based on genome analysis, works because almost all physical and psychological characteristics of people are *partly* hereditary. However, very few traits are encoded in a single gene segment, leading to a simple dominant, recessive, or autosomal dominant inheritance, such as, for example, Huntington's disease. The expression of quantitative characteristics is, on the contrary, a complex process of interaction and feedback between DNA, proteins, and cell plasma. Research on the relationships between complex gene structures and phenotypic traits is therefore part of so-called *quantitative genetics*: The probability of a trait developing depends in some cases on thousands of gene segments, each of which leads to a very small statistical shift.[2] Summing up the influence of the respective relevant gene segments (SNPs), an algorithm can calculate a so-called *polygenic score*, which provides information about the probability of the development of a certain trait. The more influential segments that can be found and assessed, the more precise is the prediction.[3] This is somewhat comparable with insurance companies and banks, which calculate the individual risk or the probability of credit default with the help of age, address, number of children, and many other variables. The best way to picture such probabilistic prediction is as a shift in the normal distribution. For example, in studies regarding academic assessment in Britain, a high polygenic score in academic success (top 10%) can predict that it is

1 Source: https://mediacenter.23andme.com/company/about-us/ (Oct 30, 2019). According to *MIT Technology Review*, more than 26 million customers had added their DNA to four leading commercial ancestry and health databases by 2019. Cf. Regalado 2019.

2 Cf. Plomin 2018, 120–133. Ibid., 70: "[G]enetic influence is caused by thousands of DNA differences of extremely small effect size; this is called polygenicity."

3 Cf. Ibid., 134–147.

about twice as likely that a certain person scores in the top 15% in her GCSE score, and even four times as likely to score in the top 2% (see Figure 21.1).[4]

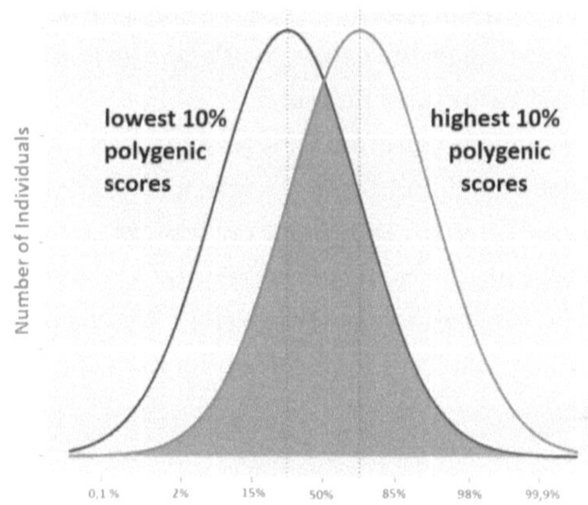

FIGURE 21.1

Classical hereditary research already points to a certain genetic part that is worth researching. Provided that they live together as a family, identical twins have approximately the same environment and genes, while fraternal twins also have the same environment but their genes vary by 50%. Here, one compares the correlation of a property in the monozygotic pair with the correlation of the same property in the dizygotic pair; From the difference of these correlations, one can calculate the hereditary portion.[5] Of course, such a method is prone to error: The shared environment is not fully identical even among twins, and the individual environment can change significantly with increasing age. The behavioral geneticist Robert Plomin argues, however, that corresponding studies in recent years increasingly show that the impact of the shared environment was overestimated (and the genetic influence in the choice of the individual environment was underestimated) for a long time. This means that the values determined by twin studies indicate lower rather than upper limits of the heritability of the respective trait.[6] Even more notably, his studies show that in a liberal society many individual environmental factors contain a strong genetic component, as we choose our environment according to our interests

4 Source: Plomin 2018, 158. The better the predictive power of a polygenic score, the more the curves will drift apart.

5 Cf. Ibid., 17–31.

6 Cf. Plomin 2018, 95: "What look like systematic environmental effects in fact reflect genetic differences. For example, the socioeconomic status of parents is correlated with their children's educational and occupational outcomes. This correlation has been interpreted as if it is caused environmentally. [... However:] Socioeconomic status of parents is a measure of their educational and occupational outcomes, which are both substantially heritable."

and abilities:[7] This means that the percentage of the genetic influence of many properties tends to be even larger than suggested from prior findings.[8] In contemporary research, polygenic scoring is used to determine this proportion even more precisely, in order to be able to find out the genetically determined proportion of the variance of a property solely from the analysis of the DNA of a born or unborn person.[9]

It can be assumed that this method will continue to be perfected, also because more and more people will make use of the services already offered. About 80% of *23andme*'s customers consent to the use of their data for research purposes. The crucial question is what effects an increasing spread and improvement of this technology will have on our society, as well as on individual persons and on our understanding of humanity.

3. Predicting and Manipulating Behavior

Based on the current state of research on polygenic scoring, which I have outlined, and an optimistic assumption on how this research will develop and interconnect with information technology over the next years, I will stipulate and defend two theses. My first thesis is that artificial intelligence can (and will improve in its ability to) recognize complex statistical correlations between genes, abilities, and behavior due to the large amounts of data available, highly adaptive learning, and enormous computing capacities. Not only can an ever-increasing variance of characteristics be directly explained with the help of AI.[10] The existing data may also be cross-checked with other information collected

7 Plomin 2018, 185: "Most measures of the environment used in psychology show substantial genetic influence. […] [W]hat looks like systematic environmental effects are often due to us choosing environments correlated with our genetic propensities."

8 It is interesting to note that the measured values are often misjudged by laypersons. For example, many people believe, probably due to media coverage of this topic, that breast cancer has a high genetic proportion, but only 10% of the variance can be genetically explained. The disposition to mental illnesses, on the other hand, is often underestimated. The empirical data pointing to the heredity of intelligence and other success factors are often ignored, presumably because they conflict with a society's ideal of equal opportunity and with the hope that the world is somehow fair or, if we were to establish a fair political or economic system, could at least become so.

9 It is important to note that the percentage of heritability is not a culturally independent variable. It depends on the homogeneity of the population and the nature of social structures. This means that we can only calculate the genetically conditioned portion *in a certain society*. Evidence for this are findings that show that the genetic portion *increases* with increasing equality of opportunity and freedom from discrimination in a society.

10 Cf. Paré et. al. 2017.

on the genetically measured individuals: income, career choice, buying behavior, personality traits, hobbies, and much more. When social networks gain access to the DNA of their users, or – even more explosive – when states gain access to data from both social networks and genetic studies, artificial intelligence can be used to predict people's behavior in certain situations more and more effectively. It also makes it easier for people to manipulate each other. Already today, an algorithm can estimate the Big-5 personality traits analysing 250 likes on Facebook with better accuracy than the spouse of this person can.[11] The method of *polygenic scoring* can therefore be perfected by AI and *big data*.

My second thesis concerns a probable application of this method, its use for personal eugenics. Through exercising cognitive and moral enhancements on individuals and groups, AI-enhanced personal eugenics runs the risk of undermining or destroying essential characteristics of humanity. What the essential characteristics of humanity are, however, is not a scientific question, but must be determined within the framework of a philosophical or theological anthropology.

I would like to turn to my prognosis that new eugenics can be expected. Most dystopian novels and films on this subject, such as *Gattaca*, address the question of how a totalitarian state would use these technologies. The advocate of eugenics can always answer that such technologies should, of course, be used in a free society; autocrats, they respond, will use any new technology for their own benefit or to maintain the political system. However, ethical challenges may even arise in a free society. In the literature on digital anthropology[12] to date, this is only dealt with marginally: What influence does the possibility of associating our DNA with behavioral and competence dispositions have on demographic development? Or, more specifically, does the possibility of analyzing the DNA of unborn children bring about the danger of a new, non-directed, personal eugenics? The sociologists Dalton Conley and Jason Fletcher discuss this effect in their book *The Genome Factor*:

> When the predictive power of these single polygenic scores starts reaching double digits, a serious policy debate about artificial selection (aka personal eugenics) will be needed [...].[13]

---

11 Cf. Kosinski 2013. For the comparison with the predictive power of spouses , cf. the documentation "Do Not Track" (2015) by Brett Gaylor, in which Michal Kosinski, the author of the study, was interviewed.

12 Cf. Horst/Miller 2012.

13 Conley/Fletcher 2017, 57.

They refer to the fact that soon a variance of over 10% of IQ differences can be explained. According to Robert Plomin, we have already reached 10% of the predictive power of academic success.[14] Although this is still far from a complete explanation of the genetic proportion of 50–60%, there is already no better predictor.[15] The only reason those interested in knowing the likelihood of one's child completing a doctorate are not able to attain this information is because most companies have chosen not to reveal it. *23andme* explains their policy the following way:

> Educational attainment is something we have looked at previously but are not currently pursuing for our product for several reasons, one being the pitfalls of potential misinterpretation of such a report.[16]

It is evident that the companies are not simply concerned that people cannot appropriately deal with the statistical nature of the information. One could find ways to educate and council affected persons. It is more likely that these companies are sensitive to what this information can be used for. Parents naturally want their children to have the best chances in life. The so-called *helicopter mentality* does not only occur at primary school age, but much earlier, even before birth: In many regions of the world, more than 90% of parents decide not to continue with a pregnancy if a serious genetic disorder is diagnosed.[17] As has been shown, over a few years, not only monogenic but also polygenic values can be determined with the help of AI and *big data*; more sophisticated prenatal tests are likely to be developed based on this technology. This means that, regarding any polygenic trait, we can already know during pregnancy how much higher the probability of a child falling below a minimum value is, compared to the average population. Using this information, companies are able to develop non-invasive prenatal tests (NIPTs) which select, based not only on monogenic traits, mutations or genetic disorders (such as Trisomy 21

---

14 In the UK measured by GCSE scores, in the USA frequently measured by the college admission test SAT. Cf. Plomin 2018, 168: „For educational attainment, we have seen that the polygenic score […] can predict 9 per cent of the variance in tests of school achievement at the age of sixteen."

15 According to the current state of research, the heritability of IQ is at least 50%, of academic achievement at least 60%. Cf. Plomin 2018, 6.

16 Official statement of 23andMe. Cited in Regalado 2018.

17 This number results from the rate of abortion for children with Trisomy 21. In some countries it almost reaches 100%; for the situation in Britain see Gee 2016. Non-invasive prenatal tests can easily diagnose this anomaly today. For children with this disorder, the future average IQ is about 50, i.e. the probability of above-average aptitude (IQ>100) is only about 0.1%.

or Klinefelter syndrome[18]), but also on physical traits, personality traits, even on intelligence.

4. Cognitive Enhancements

Why should parents be interested in increasing their children's intelligence? Scientifically, intelligence is a clearly defined concept, precisely measurable and a good predictor of success in professional life.[19] Although there are other factors that correlate with academic and professional success, these also have a significant genetic component. This includes not only the ability to delay gratification,[20] but low consensus orientation (*agreeableness*), high *conscientiousness* and high determination (*grit*).[21] But even if parents attach importance to other factors, they can measure these individually chosen factors and make their decision dependent on them. Artificial intelligence can help to put together ideal combinations from different polygenic values, from which a high probability of a 'good' or 'happy' life of the future living person can be derived in the expected environment of the family and society.

Is such personal eugenics realistic? It has already been discussed politically in the context of preimplantation diagnostics (PID). PID has, in most countries, only been permitted so far in exceptional cases, namely in cases of known serious hereditary diseases. But a softening of the criteria is likely – only some countries will try to uphold original standards. Due to the effort and risk of *in vitro fertilization*, however, this method is not suitable for the masses. Only a method of non-invasive prenatal testing (NIPT) would achieve this. It is to be expected that there will be an increasing number of abortions of children with 'worse chances': a tendency that can already be seen today in the case of disabilities. Here, too, a slow softening of the criteria is to be expected. Targeted DNA manipulation is already possible, which has been publicly recognized

18 NIPTs for various trisomy disorders and sex chromosome anomalies are already available. In some countries (among those Germany), these tests are sometimes even paid for by public health insurance.

19 Cf. Strenze 2007, 409. Plomin (2018:155) emphasizes that according to the current state of research the polygenic score for academic achievement is an even better predictor for academic achievement than IQ is.

20 Cf. Mischel et. al. 1989.

21 Both conscientiousness and agreeableness are, like all 5 basic character traits, 40 to 60 percent heritable. Cf. Plomin 2018, 6; 74; Jang et. al. 1996: "Broad genetic influence on the five dimensions of Neuroticism, Extraversion, Openness, Agreeableness, and Conscientiousness was estimated at 41%, 53%, 61%, 41%, and 44%, respectively."

when the news about the first designer child from China appeared.[22] Although the CRISPR/Cas9 method is not well suited for manipulating polygenic inherited traits, new methods for complete DNA resequencing may soon be established, also reducing the likelihood of off-target effects. Furthermore, with the help of AI, the likelihood of unwanted side-effects of polygenic manipulation due to *pleiotropy* (the fact that each DNA difference affects more than one trait) can subsequently be reduced.

There are good reasons to assume that these technologies will not only be developed, but that they will soon be available on the market. Some countries will have little moral inhibition in using them, which can be concluded based on other policies and prior efforts: Singapore enacted a eugenics law in 1986, China in 1994.[23] On the one hand, countries with practiced polygenic eugenics will probably be economically more successful than those refraining from this technology. This will lead to adjustment pressure on other countries. On the other hand, the great desire of parents to offer their children the best possible start in life will initially lead to 'eugenics tourism', and ultimately to political pressure to adapt.

A total ban, although surely intended by several global agencies, may fail because of the problem of arbitrary demarcation: Can we find an argument that makes it reasonable to allow a child with Down's syndrome to be aborted, but not a child with low intelligence?[24] Due to the lack of feasibility, and the pressure to adapt already mentioned, countries will have two alternatives to deal with the challenge: either exposing eugenics to the free market or regulating it themselves. The same applies to other, i.e. non-genetic, forms of cognitive enhancements, whether medicinal[25] or technological. If the free market solution is chosen, cognitive enhancements will only be available to those who can afford them. This will most likely lead to an increasing division of society into two classes. Even if, in a welfare state, we specifically promote the disadvantaged in

---

22 In November 2018, Chinese scientist He Jiankui claimed that two babies whose DNA was modified using the CRISPR/Cas9 method had been born.

23 In China, the "Maternal and Infant Health Care Law" was introduced. It forbids carriers of hereditary diseases to reproduce; cf. Dickson 1994. China's one-child policy had, through the disproportionate abortion of female embryos, also strong eugenic effects on society.

24 In Germany, the "eugenic indication" as a possible reason to abortion was abolished in 1995, following protests of disability advocate groups. Although individual choices still have eugenic consequences, they are legally legitimized by referring to the psychological health of the mother, who sees herself unfit to raise a disabled child.

25 The most widespread medical cognitive enhancement is Ritalin; also, illegal drugs such as Cocaine belong to this category.

our own country, we risk splitting *world society* into two classes or, eventually, even two species.[26]

Even with a less drastic course, an important psychological argument is often put forward, something we can already observe today: When a society segregates people *before* birth because of their inadequacy, it tends increasingly to discriminate against or even segregate "inadequate" people *after* birth. However, slippery slope arguments can often be overturned; advocates will call for appropriate precautionary or compensatory measures.

5. Anthropological Constraints: Theology and Philosophy

Since arguments based on utilitarianism and involving fears of slippery slopes are hard to evaluate, one may ask whether deontological arguments against genetic enhancement can be put forward: arguments based on certain anthropological assumptions. This question can be raised from a theological or philosophical point of view. Let us begin with a theological perspective: According to the Christian conception of humanity, every human being possesses the ability to enter into a loving relationship with God.[27] This ability is a facet of our being created in the image of God and one of the foundations of the infinite value of the human individual. A perfect loving relationship presupposes, however, that a person can also decide against this relationship. Thus, ideally, every person should have the ability to freely accept or reject the Christian faith.[28] Using this argument, the Dominican monk Bartholemé de las Casas had already established fundamental aspects of human rights in the 16th century. But it also requires a certain degree of spirituality as well as the ability to reflect and relate in order to decide freely for or against the Christian faith or a relationship with God. Thus, from a Christian perspective, it seems to be a desirable goal that as many people as possible should be *able* to enter into a free, loving relationship with God. A great danger in polygenic eugenics lies in a reduction of spiritual dispositions. This is already a proven genetic

26 It is even conceivable that, by subsequent genetic engineering, humanity eventually splits apart into two distinct species. In the fictional Star Trek universe, this scenario leads to so-called "Eugenic Wars" in the 21st century.

27 Catholic theologian Karl Rahner called the ability to believe in God *potentia oboedientialis*; for him, this ability is innate to all human beings and an essential part of human nature. Cf. Rahner 1969.

28 A note for theologians: If one favors a Thomistic doctrine of grace and wants to avoid semi-pelagianism, it is enough to defend the possibility of a free decision between *rejecting* and *not rejecting* faith.

disposition, which Dean Hamer boldly calls "God genes". It enables a restriction of the faith autonomy of future individuals.[29] Although an expansion of faith autonomy using this technology is possible, too, one could argue that the risk of destroying this important value is so high that we should, rather, not use this technology to potentially enhance it.

A similar argument can be advanced using a philosophical rather than a theological anthropology. According to the Enlightenment concept of humanity, specifically in the Kantian tradition, we regard acting according to universalizable principles as a prerequisite of autonomy. If autonomy grounds human dignity, as suggested by Kantian ethics, it would even be a worthwhile goal to grant as many people as possible the ability to act according to universalizable principles.[30] This ability seems to be very unequally distributed: According to Lawrence Kohlberg's frequently repeated studies, only 5–10% of people in our society actualize this potential.[31] It is questionable whether education and upbringing alone, as proposed by the optimistic Enlightenment anthropology, are sufficient to increase this percentage significantly, especially if moral reasoning capability strongly correlates with cognitive powers. "Humanist" transhumanists might thus strive for making people better moral agents, thereby actualizing the potentials of human nature in more and more individuals. However, the same mechanisms involving artificial intelligence and polygenic scoring can be used to diminish autonomy. A critic of enhancements, due to the risk involved, might argue on the following lines:

(1) Human nature necessarily includes the disposition to become a moral agent.
(2) The disposition to become a moral agent requires a certain level of cognitive powers (*moral reasoning capability*), the power to choose between alternatives (*freedom of will*), and the feasibility of these options (*freedom of action*)
(3) It is likely that polygenic eugenics, even if originally aimed at increasing cognitive powers or the scope of free will, is used to reduce or destroy a person's moral reasoning capabilities or her freedom of will.

29 Cf. Hamer 2005.
30 Increased ability for moral reasoning can be achieved by cognitive enhancements, since there seems to be a strong connection between IQ and moral development; cf. Trautner 1997, 443.
31 Bründler et. al. 2004, 70: "According to Kohlberg, less than ten percent of adults each attain level 6, which comes very close to Kant's 'categorical imperative' […]. On the other hand, more than ten percent seem to remain at level 1 [= authority orientation]. Level 5 is most frequently reached (28 percent) [= orientation on the articles of association]".

(4) A technology that is likely to be used to destroy human nature must be strictly controlled or forbidden.
(5) Polygenic eugenics must be strictly controlled or forbidden.

In this article, a proper evaluation of the magnitude of this risk cannot be evaluated. But what can be evaluated is that necessary qualities of being human – whether they are philosophically or theologically justified – are not a priori incompatible with an improvement through cognitive or volitionary enhancements. One needs to balance the risks with some important advantages: Higher intelligence, greater willpower, and improved empathy increase one's scope of freedom and might even improve moral judgement. Adding the theological perspective, improving theological reasoning ability, and enhancing spirituality may even, especially in a secular culture, make people more likely to engage with religious traditions and possibly enter freely into a relationship with God.[32]

The anthropological discussion reminds us that, besides negative consequences in society, there exists the danger that a point of improvement will be exceeded where freedom is no longer increased but effectively reduced. On the one hand, an improvement of certain abilities can increase the moral autonomy of a future individual. *Cognitive enhancements* increase intelligence; *volitionary enhancements* increase strength of will (or reduce the power of natural drives, leading to a similar effect).[33] On the other hand, a change in behavioral dispositions – *moral enhancements* or *behavioral enhancements* – can limit the moral autonomy of a future individual. This would not be compatible with the anthropology of Enlightenment philosophy.[34] Moral enhancements could even be applied to undermine the autonomy of faith. This would not be compatible with a Christian-theological anthropology. The danger in AI-enhanced polygenic eugenics, from these two perspectives, lies mainly in behavior control: Even a well-intended moral enhancement program[35] would destroy human nature if it undermined human freedom of will and responsibility. Even in the disputed case of volitionary enhancements, one could argue

---

32 From a Christian point of view, one could even use technology to reduce pride and arrogance to support people on their way to faith.

33 Similar to cognitive abilities, research suggests that willpower has also a strong genetic component. Important work on willpower and self-control was done by Roy Baumeister and Tom Denson; on the genetics of willpower see Goldman 2012.

34 For a discussion of different variants of moral enhancements, see Ach 2018, 191–197. He argues that *collective moral enhancements* are only conceivable with a restriction of individual freedom, *parental moral enhancements* only by taking away a "right to an open future" from the child. Cf. ibid., 194.

35 As an elaborate example for such an agenda see Hughes 2015.

from the perspective of Aristotelian virtue ethics that a self-acquired, hard-earned virtue is more valuable than an innate or artificially produced virtue – even when one has freely decided to acquire it.[36]

6. Conclusion

I consider the concern to be justified that AI and polygenic scoring will be used to not only morally empower and educate people, but to manipulate and control them. In his book *Unfit for the Future*, transhumanist Julian Savulescu calls for such enhancements in order to meet global ethical challenges such as climate change.[37] Those who are solely concerned with minimizing suffering and maximizing pleasure for present and future generations will definitely use this technology to minimize suffering – regardless of whether the autonomy of the respective persons can be guaranteed. Such utilitarian thinking is widespread: Freedom is often restricted to create prosperity or security; in democracies, such measures are sometimes even supported by the majority of the population.[38] It can be expected that, in addition to the already highly problematic *social scoring*, a polygenic scoring value for socially compliant behavior will also be calculated; based on this, individual eugenics might be pursued by many people without exerting any coercion.[39] The danger, therefore, is that instead of improving ethical judgement, there will be a pure improvement in what Kant would call action *according to* duty (in contrast to action *out of* duty), which will lead to the destruction of human autonomy.[40]

---

36 Cf. Rath, Beth, On the Impossibility of Engineering Moral Virtue, Talk at the Conference "Transhumanism, Posthumanism, and Super-Naturalism", University of Oxford, July 6, 2018: https://www.ianramseycentre.info/rath-on-the-impossibility-of-engineering-moral-virtue (Oct 30, 2019).

37 Persson/Savulescu 2012.

38 However, we also have an interest in truth and self-determination. Robert Nozick argues that we would not exchange this for the sake of happiness (cf. Nozick 1974). As the figure of *Cypher* in Matrix shows, this does not necessarily apply to all humans.

39 Hare 1981, 172: "It is a good thing that we should have these attitudes firmly implanted; a wise utilitarian educator would wish his pupils to be so equipped to face the stresses of the moral life."

40 Interestingly, the theological reflection on the theodicy problem gives us indications that human autonomy is also more important for God than avoidance of suffering. Here there is a correspondence between theological and Kantian philosophical anthropology: Better an imperfect world with more suffering, but with self-determination, than an apparently perfect world without suffering, but without 'real' people.

Too little research has so far been done into the effects of artificial intelligence on personal eugenics, especially against the background fact that this development is difficult to prevent in an open society. Such research must be interdisciplinary, involving the natural sciences, social sciences, philosophy, and theology. The task of philosophical and theological anthropology is to discuss the limits of change through enhancements and genetics. Central to this is the question of which essential characteristics of being human must not, under any circumstances, be endangered. The ethical challenge is to use the available technologies and at the same time ensure social cohesion, moral autonomy and religious autonomy.

In my preliminary ethical assessment, cognitive and volitionary enhancements may be permitted within narrowly defined framework conditions. Moral enhancements, however, must be distinguished from the other types for they are used to *control* or *change* behavior, whether genetically or technologically. Despite the best of intentions, applying moral enhancements could eventually eliminate human freedom and thereby destroy the essence of humanity itself.

7. Bibliography

Ach, J. (2018): "Transhumanismus und Enhancement der Moral," in: Göcke, B. & Meier-Hamidi, F. (eds.), *Designobjekt Mensch,* Freiburg: Herder, 181–198.

Bründler, P., et. al. (2004): *Einführung in die Psychologie und Pädagogik*, Zürich: Compendio Bildungsmedien.

Conley, D. & Fletcher, J. (2017): *The Genome Factor What the Social Genomics Revolution Reveals about Ourselves, Our History, and the Future*, Princeton University Press.

Dickson, D. (1994): Concern grows over China's plans to reduce number of 'inferior births', *Nature* 367 (6458):3.

Dumfart, B. & Neubauer, A. (2016): "Conscientiousness Is the Most Powerful Noncognitive Predictor of School Achievement in Adolescents," in: *Journal of Individual Differences* 37.1, 8–15.

Grössl, J. (2018): "Verbesserung oder Zerstörung der menschlichen Natur? Eine theologische Evaluation des Transhumanismus," in: Göcke, B. / Meier-Hamidi, F. (ed.), *Designobjekt Mensch*, Freiburg: Herder, 339–361.

Hare, R. M. (1981): *Moral Thinking. Its Levels, Method, and Point,* Oxford: Clarendon Press.

Horst, H. / Miller, D. (2012) (eds.): *Digital Anthropology*, Bloomsbury T&T Clark.

Hamer, D. (2005): *The God Gene: How Faith Is Hardwired Into Our Genes,* New York: Anchor Books.

Kohlberg, L. et. al. (1983): *Moral stages. A current formulation and a response to critics*, Basel – New York: Karger.
Kosinski, M. & Stillwell, D. & Graepel, T. (2013): "Private traits and attributes are predictable from digital records of human behavior," *Proceedings of the National Academy of Sciences of the United States of America* 110 (15).
Jang, K. L. & Livesley, W. J. & Vernon, P. A. (1996): "Heritability of the big five personality dimensions and their facets: a twin study", in: Journal of Psychology 64 (3), 577–591.
Mischel, W., et. al. (1989): Delay of gratification in Children, *Science* 244, 933–938.
Mischel, W., et. al. (2011): "'Willpower' over the life span: decomposing self-regulation," *Social cognitive and affective neuroscience* 6 (2), 252–256.
Nozick, R. (1974): *Anarchy, State, and Utopia*, New York.
Gee, A. (2016): A world without Down's syndrome?, BBC News Magazine, https://www.bbc.com/news/magazine-37500189 (Nov 13, 2019).
Goldman, David (2012): *Our Genes, Our Choices: How Genotype and Gene Interactions Affect Behavior*, Amsterdam: Elsevier Science.
Hughes, J. (2015): Enhancing Virtues. Fairness, Ethical Technology, http://ieet.org/index.php/IEET2/more/hughes20150301 (Mar 1, 2019).
Paré, Guillaume & Mao, Shihong & Deng, Wei Q. (2017): A machine-learning heuristic to improve gene score prediction of polygenic traits. *Scientific Reports* 7, 12665.
Persson, I. & Savulescu, J. (2012): *Unfit for the Future: The Need for Moral Enhancement*. Oxford University Press.
Plomin, R. (2018): *Blueprint: How DNA Makes Us Who We Are*, Milton Keynes: Random House.
Rahner, K. (1969), Art. 'Potentia oboedientialis', in: *Sacramentum mundi* 3, 1245–1249.
Regalado, A. (2018): "DNA tests for IQ are coming, but it might not be smart to take one," *Technology Review* (April 2), www.technologyreview.com/s/610339.
Regalado, A. (2019): "More than 26 million people have taken an at-home ancestry test", Technology Review (February 11), www.technologyreview.com/s/612880.
Strenze, T. (2007): "Intelligence and socioeconomic success: A meta-analytic review of longitudinal research," *Intelligence* 35 (2007), 401–426.

Could Machines Replace Human Scientists?

Digitalization and Scientific Discoveries

Jan G. Michel

Abstract

The focus of this article is a question that has been neglected in debates about digitalization: Could machines replace human scientists? To provide an intelligible answer to it, we need to answer a further question: What is it that makes (or constitutes) a scientist? I offer an answer to this question by proposing a new demarcation criterion for science which I call "the discoverability criterion". I proceed as follows: (1) I explain why the target question of this article is important, and (2) show that it leads to a variant of the demarcation problem of science. (3) By arguing that it is probably an essential feature of science that we can make scientific discoveries, I suggest a novel way of dealing with this problem by proposing a new demarcation criterion. Before introducing it, (4) I analyze an exemplary case of a real scientific discovery, and (5) argue that scientific discovery processes have a general underlying structure. (6) I introduce my discoverability criterion for science and present my master argument that helps us understand which criteria have to be fulfilled in order to decide whether machines can replace human scientists or not. (7) I conclude by answering the article's target question and bringing forward a take-home message.

1. Introduction

Artificial intelligence (AI) seems to be on everyone's mind these days. This is, from my perspective at least, hardly suprising, mainly for two reasons.[1]

First of all, it is no surprise to find that AI *as such* seems to be on everyone's mind when one takes into account the extraordinary fascination that already

1 Preliminary remark: One of the ideas behind the anthology in which this article appears is that it should be accessible to readers from a wide range of disciplines comprising, among others, computer science, linguistics, mathematics, philosophy, psychology, sociology, and theology. In order to conform to this idea in my contribution, I will try to refrain from using technical vocabulary and keep the presupposed background knowledge at a minimum. As I see it, this can be best achieved by maintaining the character of my original lecture from which this article has evolved.

 | DOI:10.30965/9783957437488_023

the very idea of an artificial being has exuded on humankind throughout history.[2] In other words: AI has, seemingly, never been out of fashion.

Secondly, it is no surprise to find that AI is on everyone's mind *these days* when one takes into account that we live in an age that is also known as "computer age", "information age", or "digital age", an age directly following the so-called digital revolution that started in the late 1950s. Even though it is not an easy task to pin down what exactly is meant by "information" or by "digital" here, it seems still intelligible to say that we currently live in an age in which information has become a resource or a commodity that is quickly and widely disseminated and easily available especially through the use of computer technology and AI in general.

Ever since the introduction of the term "artificial intelligence",[3] philosophical debates about AI – and especially about GOFAI ("Good Old-Fashioned Artificial Intelligence", Haugeland 1985) – have centered around questions like the following: Is it possible to build a machine that thinks rationally and that acts intelligently? A machine that can really understand a language? And what about the mind and consciousness? Could there be a sentient machine, a machine with feelings or emotions? If so, would such a machine also have a free will?

Like many others, I have grappled with these and other related questions in the philosophy of mind for quite some time now.[4] In this article, however, I will approach the topic of AI from a different perspective. Apart from well-known and long-standing questions of principle like the ones listed above, fresh topics and new questions concerning AI have evolved, partly due to the zeitgeist of our digital age. Many of these topics and questions reveal an obvious practical relevance, such as the topic of – as it is sometimes called – digitalization (or computerization) which clearly belongs to the most important public topics

2 This dates back, at least, to the times of ancient Egypt. For a very short overview with examples of artificial beings in history (but also in mythology) see Michel and Backmann 2009b: 191.

3 According to common belief, the term "artificial intelligence" was first used during a workshop at Dartmouth College in 1956. This, however, is not entirely correct since the term "artificial intelligence" was already part of the proposal for the Dartmouth workshop. This proposal was written by John McCarthy, Marvin L. Minsky, Nathaniel Rochester, and Claude E. Shannon, and is dated August 31, 1955; it can be accessed online via http://raysolomonoff.com/dartmouth/boxa/dart564props.pdf. What seems to be entirely correct, however, is that AI as an academic discipline and field of research had its official beginning with the Dartmouth workshop in 1956. See Bringsjord and Govindarajulu 2018 for an overview.

4 See Michel 2009, Michel 2011, Michel 2013, Michel 2016, Michel 2018a, Michel 2018b, Michel 2018c, Michel and Backmann 2009a, Michel and Backmann 2009b, Michel and Kober 2011, Michel and Münster 2013, Michel, Boström, and Pohl 2016.

at the moment. Questions like the following lie at the center of debates about digitalization: What will be the impact of automation, robotics, and AI on our job market in the (near) future? What should and will happen to those whose jobs are replaced by machines? And which jobs can be replaced in the first place?

Now, this is the background against which this article is set, and I have two aims here, one general and one more specific: The general aim I have is to illuminate and also to give a tentative answer to a question that has been completely neglected in current debates about digitalization, namely: Could machines replace human scientists? In order to provide an intelligible answer to this question, however, it is necessary to answer a second question, namely: What exactly is it that makes (or constitutes) a scientist? Now, the more specific aim I have in this article is to make a new contribution to a good answer to this second question, and I will do so by proposing a new demarcation criterion for science which I call "the discoverability criterion for science".[5]

I proceed in seven steps: In a first step, I motivate the general topic of this article by briefly showing that, and to what extent, it is a neglected but nevertheless important and also pressing question to ask whether human scientists could be replaced by machines. In a second step, I show that any attempt at answering this question leads to a variant of one of the most prominent yet unresolved problems in the philosophy of science, namely the so-called demarcation problem of science. By arguing that it is an important and maybe even essential feature of science that we can make scientific discoveries, I suggest, in a third step, a novel way of dealing with the demarcation problem by proposing a new demarcation criterion. Before introducing this criterion, however, I analyze an exemplary case of a real scientific discovery in a fourth step, and argue, in a fifth step, that scientific discovery processes share a common structure. Against this background, I introduce the discoverability criterion for science in a sixth step, and also present my master argument that helps us understand which criteria have to be fulfilled in order to decide whether machines can replace human scientists or not. In a seventh step, I conclude

5 I would like to thank Mitchell S. Green who suggested that I could also speak of a *symptom* of science. I like this suggestion for two reasons: (1) I have the impression that speaking of a symptom somehow hints at the possibility of other symptoms. With regard to science, this is a view I embrace – I do not want to claim that there is only one symptom of science. (2) Speaking of symptoms of science connects the topics I am dealing with here with the metaphorical field of "diagnosing a disease". In the context of the so-called demarcation problem of science, this is also something I embrace. There are, however, several reasons why I will continue to use "criterion" in this article, maybe first and foremost because it seems to be more neutral and better suited for my purposes here.

by tentatively answering the article's target question and bringing forward a take-home message with further questions to be dealt with.

2. Motivating the Topic: Could Scientists be Replaced by Machines?

In our so-called age of digitalization, many human workers live in fear of losing their jobs to machines. And as current research on the consequences of digitalization for the working world documents, this is not an irrational fear, but rather a well-justified one. In Germany, for instance, the so-called Institute for Employment Research – a special office of the Federal Employment Agency (Bundesagentur für Arbeit) – has published detailed research reports on the substitution potentials of jobs in Germany. Let us begin by taking a brief look at some of the results of this research.[6]

Even though the experts who have carried out these studies do not expect machines to replace entire professions, they still think that many human workers should better prepare for unpleasant developments in the near future, especially workers in segments in which the majority of the overall professional activity is already carried out by machines. They point out that the average substitution potential is highest in the segment of manufacturing professions with more than 70% while in all other segments the weighted average substitution potential is below 50%. In a nutshell, one of the main results is that 4.4 million human workers in Germany could theoretically be replaced by machines today, that means around 15% of all current employees subject to social insurance contributions in Germany.

It may, however, not come as a big surprise that digitalization affects different professional segments differently. To illustrate: In the segment of chemical manufacturing professions, about 90% of the professional activities can be carried out by machines while only about 5% of the professional activities in the segment of social professions (such as teachers) can be replaced by machines. The situation is similar when it comes to the comparison of the substitutability of, say, bankers, on the one hand, and nurses, on the other hand.

Against this background and in view of the public debates about digitalization, it is a pressing question to ask whether also human scientists could be

6 For the following see, inter alia, Arnold et al. 2017, Arntz et al. 2016, Arntz et al. 2017, Dengler and Matthes 2015a, Dengler and Matthes 2015b, Dengler and Matthes 2016, Dengler and Matthes 2017, Dengler and Matthes 2018a, Dengler and Matthes 2018b, Dengler and Matthes 2019, Frey and Osborn 2013, Kruppe et al. 2017, Lehmer and Matthes 2017a, Lehmer and Matthes 2017b. Note that, although most of these studies focus on Germany, the overall situation is similar in other industrialized or developed countries.

replaced by machines – a question that the aforementioned studies have not touched upon. The question is pressing for several reasons, and I would like to mention only two of them, an obvious one and a less obvious one: An obvious reason why the question is so pressing is that – according to figures provided by the German Federal Ministry of Education and Research (cf. BMBF 2017: 23) – about 600,000 people working in science in Germany are affected by it. A less obvious, but not less important reason why the question is pressing is that we can expect the question to lead us to significant insights concerning the overall (human) practice of science as such, in general, and the role that machines can play in it, in particular.

3. Introducing a Variant of the Old Demarcation Problem

In order to find out whether human scientists could be replaced by machines, we need to specify what it is that makes a scientist – what is it that distinguishes a scientist from a non-scientist? This question can be regarded as a variant of one of the most prominent as well as most bothering problems in philosophy of science, namely the so-called demarcation problem of science. Simply put, it amounts to the question of what distinguishes science[7] from non-science.[8]

But, as often with philosophical problems, things are not so simple. First of all, the problem is not new, and its history and its development are rather complex. Even though the label "demarcation problem" was only coined in the 1930s (Popper 1934: 7–11), the problem designated by this label has a long-standing history and can already be found in the works of Aristotle (Laudan 1983: 112f.). More importantly, the demarcation problem does not only comprise the question what distinguishes science from non-science, but also what distinguishes science from pseudo-science and proto-science as well as from such diverse areas as religion, art, or literature. And it is also legitimate to ask what exactly distinguishes empirical science from non-empirical enterprises such as mathematics and logic or simply good science from bad science.

Since these problems are all interrelated and only come in a variety of guises, I prefer not to interpret the expression "demarcation problem" as referring to *the* demarcation problem (as if there was only one), but rather to *the family* of demarcation problems. Among others, the demarcation problem family

7 My considerations in this paper are based on a broad reading of the term "science" (similar to the German term "Wissenschaft").

8 See also Michel 2019 for the following overview.

has the following questions as family members: Why do we have no problem with counting physics and biology as clear cases of (natural) sciences, while we resist to do so with astrology, homeopathy, parapsychology, or ufology? What are the criteria or features that demarcate science from non-science? Many philosophers and scientists have tried to solve the demarcation problem, but it is fair to say that we are still far from reaching a consensus.

So, the question whether human scientists could be replaced by machines leads us to the question: What is it that makes a scientist? Seriously trying to answer this question will end up in an attempt at developing good criteria to demarcate scientists from non-scientists. This, however, is nothing but a modification or a new variant of the old demarcation problem of science.

4. A New Proposal: The Role of Discoveries in Science

Prima facie, it may seem plausible to say: If you want to know what makes a scientist, just see what they do! However, considering the variety and diversity of professional activities that scientists (have to) carry out in their daily routine – among them research and teaching as well as different organizational, technical, communicative, and, not to forget, administrative tasks, to name just a few –, it is hardly surprising that it is not an easy task to pin down what exactly it is that makes a scientist. This also shows that we will probably not find a general demarcation criterion for science as a whole just by looking at the contingent and varying everyday activities of today's scientists.

We should rather ask whether there are any general features that are constitutive of scientific practice across the board. Of course, this question is not new, and some well-known and highly sophisticated answers to it have been brought forward and discussed over the years. There have, for instance, been debates – more or less in chronological order – about verification, falsification, testability, scientific methodology, inductive logic, the social role of science, reduction, unity, pluralism, and systematicity.[9] Unfortunately, we have to admit that none of these proposals has yet proven successful in providing a satisfactory solution to the demarcation problem.

It may seem bold in the light of this diagnosis, but I would still like to give it another try and propose a new demarcation criterion, a criterion that I prefer to call "the discoverability criterion for science". The origins of this criterion

9 For these debates, see Carnap 1945, Kuhn 1962, 1970, Popper 1963, Ziman 1968, Lakatos 1978, Feyerabend 1975, Thagard 1978, Thagard 1988, Laudan 1982, Ruse 1982, Kitcher 1983, Kitcher 1993, Dupré 1993, Hoyningen-Huene 2013, Seidel 2014.

can be traced back to an insight captured by the following claim: *It is at least an important if not essential feature of science that we can make discoveries in science.* Against the background of this claim, the question arises what exactly it is that distinguishes discoveries in science from discoveries in other areas. As I argue and elaborate on elsewhere,[10] making a scientific discovery is much more than just seeing or finding something. To put it in a nutshell, I argue that scientific discoveries differ from other discoveries in that they have a general underlying structure with characteristic structural features, some of which being indispensable for the overall scientific discovery process. In order to illustrate what that means, let us have a quick look at an exemplary case before introducing the discoverability criterion for science.

5. An Exemplary Case of a Real Scientific Discovery

Even though I cannot go much into detail here, I would still like to illustrate my basic idea with the help of a real scientific discovery case. For this, I chose the case of a recent discovery in biology: a species of deep-sea anglerfish, *Lasiognathus dinema*. Of course, I could have picked any other example, but, for several reasons, I found the discovery of *Lasiognathus dinema* to be a textbook or ideal paradigmatic case of a scientific discovery.[11] Moreover, my structural analysis of this biological discovery can be generalized and applied to scientific discoveries in other disciplines or fields of research. My following presentation is mainly based on personal communication with *Lasiognathus dinema*'s discoverer, renowned biologist Theodore W. Pietsch from the University of Washington.[12]

Let us begin with an overview of the background events that led to the discovery of *Lasiognathus dinema:* In 2010, a blowout on board of the Deepwater Horizon, a drilling rig in the Gulf of Mexico, caused an explosion and a fire that could not be extinguished. The Deepwater Horizon sank, leaving the well gushing at the seabed and causing the largest offshore oil spill in history so far. This disaster led to a subsequent Natural Resource Damage Assessment conducted by the National Oceanic and Atmospheric Administration. During this, a few hundred specimens of deep-sea fish were collected in a research trawl that samples marine life at depths between 800 and 1,300m – "[i]n effect,

10 Cf. Michel 2019, Michel forthcoming-a, Michel forthcoming-b, Michel forthcoming-c.

11 Cf. Michel 2019 where I analyze this example in greater detail. See also the articles in Michel forthcoming-d.

12 In addition, see Pietsch and Sutton 2015 as well as my more detailed presentation of this discovery in Michel 2019.

they haul up a lot of dead stuff and see what they have."[13] The whole material collected was then brought to the Virginia Institute of Marine Science at Gloucester Point.

Knowing about his expertise on anglerfish, Theodore Pietsch's colleagues asked him to take a look at their material. Responding to their invitation, Pietsch traveled from Seattle to Gloucester Point where he sat for a week in front of a microscope and assigned names to the specimens. After identifying about 400 specimens and dozens of common things, Pietsch saw something that stood out from all the rest. In the end, he and his colleague Tracey Sutton found three new exemplars of fish. Pietsch made several detailed sketches of them, took measurements and pictures, wrote out a detailed description, and prepared a manuscript for publication which he submitted to a respected ichthyology and herpetology journal in October 2014. After responding to comments from outside referees, the paper was accepted for publication in January 2015 and appeared in print in June 2015. In their paper, Pietsch and Sutton described their findings, declared it a new biological species, and proposed a name for it: "*Lasiognathus dinema*". All of the three exemplars they found have been deposited at the University of Washington Burke Museum of Natural History and Culture.

6. The Structure of Scientific Discoveries

Against the background of this example, we see that a certain body of background knowledge – exemplified by Pietsch's biological expertise – is a necessary precondition for making scientific discoveries. Moreover, we see that the overall scientific discovery process comprises several steps. First of all, something (here: a sufficient number of exemplars) has to be found, and the finder somehow needs to come to the suspicion that this something is new – "new" meaning here: so far not covered by the relevant theories and taxonomies (here: that these are exemplars of a new species). We can say that finding something is an indispensable feature of scientific discoveries.[14]

In addition, the example case illustrates that there are also features that are important but not indispensable to the overall scientific discovery process, such as drawing sketches, taking measurements and pictures, as well as the

13 See www.news-journalonline.com/news/20150815/scientist-discovers-ugly-new-species-of-fish-in-gulf-of-mexico.

14 And for that matter, I think that finding something is an indispensable feature of any discovery.

preparation of detailed descriptions. Depending on the exact circumstances, discussions with colleagues or checking the literature again can contribute to the success of a discovery. Although important for the overall discovery process in question, each of the steps mentioned in this case may be dispensable or unnecessary in other scientific discovery cases.

Considering our current conception of science (biology, in our case), it is necessary that a manuscript be submitted to a relevant and recognized journal (or other publishing outlet).[15] In order to be accepted and published, the manuscript submitted must contain a sufficiently detailed description meeting certain demands and sometimes also pictures or diagrams. Moreover, it is often the case that the manuscript must contain a proposal for a name of the new findings. If there are any, this has to be in accord with the nomenclature rules of the given field of research. By submitting the manuscript, the authors declare to introduce a new discovery into their field of research. Sometimes, the paper enters a review process in which reviewers recommend the paper to be rejected, revised, or accepted. In the revision case, it is the authors' task to act accordingly. Once the paper is accepted for publication, this is tantamount to the satisfaction of the authors' declarations and proposals in the paper, the most important being: The findings described in the paper now count as the discovery of a new species in biology. It is only by means of an author's (public) declaration that the new findings can be accepted or recognized as new species discoveries by other scientists.[16] Therefore, we can say that the submitting of a relevant manuscript taken together with its subsequent acceptance for publication is an indispensable feature of the scientific discovery process. In short, a scientific discovery differs from other discoveries in that the discovery in question has to be accepted or recognized as a scientific discovery by a relevant scientific community.

To give a very rough summary: Against a certain body of background knowledge, scientific discovery processes comprise several structural features, some of which being indispensable, others, albeit important, being only contingent. We found that the indispensable features comprise (a) the initial finding of something new and (b) a scientific community's acceptance of the finding as a new discovery. In other words: We can only have a case of a scientific discovery

---

15 I would like to thank Michael Ohl and Theo Michael Schmitt for helpful discussions and good advice.

16 As my presentation already indicates, it is declarations or declarative speech acts that are the focus of my analysis. Unfortunately, for reasons of space, I cannot go into details here, but see Michel 2019 and Michel forthcoming-b for detailed analyses in which I rely on and even extend the speech act theoretical account of John Searle (1975, 1995, 2007, 2010a, 2010b, 2010c, 2010d).

if (a) and (b) are satisfied – something new has to be found and then accepted as a new scientific discovery, otherwise we are not dealing with a case of a scientific discovery.

7. Scientific Discoverability and the Master Argument

Let us return to the variant of the demarcation problem of science we are faced with here. At the beginning, I proposed to introduce a new criterion that may help to tell science and non-science apart, namely the discoverability criterion for science. Now, in order to see what this criterion says, consider the following line of thought that I have proposed here: I claimed that it seems to be an essential feature of science that we can make discoveries in science. I went on to argue that, once we acknowledge this and take a closer look at clear cases of scientific discoveries, we find that they have a general underlying structure with certain structural features some of which being indispensable for the overall scientific discovery process. The indispensable structural features include both the initial finding of something as well as a relevant scientific community's acceptance of this finding as a new discovery in a given field of research. As I see it, something can only be a case of a scientific discovery if it satisfies a sufficient number of these features. In the light of this, I propose the following criterion for science:[17]

> *The discoverability criterion for science:* If a discipline is a science, scientific discoveries can be made in it. That means: If scientific discoveries cannot be made in a discipline, the discipline in question is not a science.

Now, as pointed out above, we are dealing with a variant of the demarcation problem – the question is not how to distinguish science from non-science but, rather, how to distinguish scientists from non-scientists. It is, however, unproblematic to modify the discoverability criterion for this purpose:

> *The discoverability criterion for scientists:* If an entity[18] is a scientist, it can make scientific discoveries. That means: If an entity cannot make scientific discoveries, the entity in question is not a scientist.

17 The following may be a first and somewhat loose formulation.

18 Speaking of an entity, I try not to be chauvinistic in the following sense: I do not want to preclude machines, animals, or others from being scientists by definition already. Moreover, "entity" could also refer to a group or a team (e.g., of human researchers) – a point I find important for several reasons (but I cannot go into detail here).

With all this in mind, we arrive at the master argument of this article. The argument comprises seven steps. I will present it in an informal way.

1. The target question is whether machines could replace human scientists. Against the background of my reasoning so far, we can say that machines can replace human scientists if and only if the machines in question are capable of satisfying the essential features of human scientific practice.
2. I have claimed that it is an essential feature of human scientific practice that scientific discoveries can be made.
3. If the ability to make scientific discoveries is an essential part of human scientific practice, then everything that can make scientific discoveries is a possible candidate for counting as a scientist (or a team of scientists).
4. Scientific discoveries are no atomic phenomena but, rather, have a general structure with certain features, some of which are indispensable.
5. A machine can make scientific discoveries if and only if it satisfies the indispensable structural features of scientific discoveries as well as a sufficient number of the other features (this depends on the context).
6. If a machine can make scientific discoveries, it is a possible candidate for counting as a scientist.
7. If a machine is a possible candidate for counting as a scientist, the machine in question could replace human scientists (at least in principle).

8. An Answer and a Take-Home Message

Now, could machines replace human scientists? Yes, they could – but only if they are capable of making scientific discoveries. To pin down what exactly this means is, as we have seen, quite a challenge. I have, however, accepted the challenge and made a proposal which includes a structural analysis of scientific discoveries. Among other things, I came to the conclusion that the overall scientific discovery process comprises both finding something and a relevant scientific community's acceptance or recognition of the finding as a new discovery in a given field of research. We also saw that a finding can only be accepted or recognized as a new scientific discovery with the help of several different declarations (i.e., declarative speech acts).

In the light of my analysis, I have the impression that, with regard to scientific discoveries, it is both too cumbersome to appropriately integrate machines into the relevant scientific practices and too demanding for today's machines to satisfy the core structural features of scientific discovery processes. Rather, machines serve as important tools in science already but cannot (yet) replace human scientists. Although much more could be said here, we arrive at some good news: human scientists do not have to fear soon to be replaced by machines.

And there is also a take-home message. Against the background of my approach, we are confronted with questions that have not been central in AI debates, an example being: Can machines perform speech acts, especially declarative speech acts? This, again, is not an easy task, since, in order to answer this question, we need to know much more about (the preconditions of) communication in general and communicative practice in science in particular.[19]

9. Bibliography

Arnold, Daniel, Lutz Bellmann, Bernhard Christoph, Katharina Dengler, Oliver Falck, Florian Lehmer, Tobias Maier, Britta Matthes, Andreas Mazat, Susanne Steffes, Bastian Stockinger, Enzo Weber, Marc Ingo Wolter, Stefanie Wolter, Gerd Zika (2017). "Digitalisierung." In: *Arbeitsmarkt Kompakt: Analysen, Daten, Fakten (IAB-Bibliothek)*, Joachim Möller and Ulrich Walwei (eds.), 110–128. Nürnberg.

Arntz, Melanie, Terry Gregory, and Ulrich Zierahn (2017). "Revisiting the Risk of Automation." *Economics Letters* 159, 157–160.

Arntz, Melanie, Terry Gregory, Florian Lehmer, Britta Matthes, and Ulrich Zierahn (2016). "Arbeitswelt 4.0 – Stand der Digitalisierung in Deutschland: Dienstleister haben die Nase vorn." *IAB-Kurzbericht 22/2016*, available at: http://doku.iab.de/kurzber/2016/kb2216.pdf.

BMBF (Bundesministerium für Bildung und Forschung/Federal Ministry of Education and Research) (2017). "Education and Research in Figures 2017." Available at: http://www.datenportal.bmbf.de/portal/en/education_and_research_in_figures_2017.pdf.

Bringsjord, Selmer, and Naveen Sundar Govindarajulu (2018). "Artificial Intelligence." In: *The Stanford Encyclopedia of Philosophy* (Fall 2018 Edition), Edward N. Zalta (ed.). Available at: https://plato.stanford.edu/archives/fall2018/entries/artificial-intelligence.

Carnap, Rudolf (1945). "On Inductive Logic." *Philosophy of Science*, 12(2), 72–97.

19 Acknowledgments: The material this paper is based on has been presented on different occasions. I am grateful for the helpful comments from my audiences and for many good discussions. Most of all, I would like to thank Kim Boström, Benedikt Göcke, Mitch Green, David Gunkel, Bill Lycan, Michael Ohl, Ted Pietsch, Selma Šabanovic, Theo Michael Schmitt and Dragos-Christian Vasilescu. Research for this article was funded by the German Research Foundation (Deutsche Forschungsgemeinschaft, project number 295845819).

Dengler, Katharina, and Britta Matthes (2015a). "Folgen der Digitalisierung für die Arbeitswelt: Substituierbarkeitspotenziale von Berufen in Deutschland." *IAB-Forschungsbericht 11/2015*, available at: http://doku.iab.de/forschungsbericht/2015/fb1115.pdf.

Dengler, Katharina, and Britta Matthes (2015b). "Folgen der Digitalisierung für die Arbeitswelt: In kaum einem Beruf ist der Mensch vollständig ersetzbar." *IAB-Kurzbericht 24/2015*, available at: http://doku.iab.de/kurzber/2015/kb2415.pdf.

Dengler, Katharina, and Britta Matthes (2016). "Auswirkungen der Digitalisierung auf die Arbeitswelt: Substituierbarkeitspotenziale nach Geschlecht." *IAB-Bericht 24/2016*, available at: http://doku.iab.de/aktuell/2016/aktueller_bericht_1624.pdf.

Dengler, Katharina, and Britta Matthes (2017). "Folgen der Digitalisierung für die Arbeitswelt: Welche Berufe sich potenziell durch Computer ersetzen lassen." In: *Arbeitsmarkt Kompakt: Analysen, Daten, Fakten (IAB-Bibliothek)*, Joachim Möller and Ulrich Walwei (eds.), 116–117. Nürnberg.

Dengler, Katharina, and Britta Matthes (2018a). "The Impacts of Digital Transformation on the Labour Market: Substitution Potentials of Occupations in Germany." *Technological Forecasting and Social Change* 137, 304–316.

Dengler, Katharina, and Britta Matthes (2018b). "Substituierbarkeitspotenziale von Berufen: Wenige Berufsbilder halten mit der Digitalisierung Schritt." *IAB-Kurzbericht 04/2018*, available at: http://doku.iab.de/kurzber/2018/kb0418.pdf.

Dengler, Katharina, and Britta Matthes (2019). "Digitalisierung in Deutschland: Substituierbarkeitspotenziale von Berufen und die möglichen Folgen für die Beschäftigung." In: *Bildung 2.1 für Arbeit 4.0?*, Rolf Dobischat et al. (eds.), 49–62. Wiesbaden.

Dupré, John (1993). *The Disorder of Things: Metaphysical Foundations of the Disunity of Science*. Cambridge, Mass.

Feyerabend, Paul (1975). *Against Method: Outline of an Anarchist Theory of Knowledge*. London.

Frey, Carl Benedikt, and Michael A. Osborne (2013). "The Future of Employment: How Susceptible are Jobs to Computerisation?" *Technological Forecasting and Social Change* 114, 254–280.

Haugeland, John (1985). *Artificial Intelligence: The Very Idea*. Cambridge, Mass.

Hoyningen-Huene, Paul (2013). *Systematicity: The Nature of Science*. Oxford.

Kitcher, Philip (1983). *Abusing Science: The Case Against Creationism*. Cambridge, Mass.

Kitcher, Philip (1993). *The Advancement of Science: Science without Legend, Objectivity without Illusions*. New York.

Kruppe, Thomas, Ute Leber, and Britta Matthes (2017). "Sicherung der Beschäftigungsfähigkeit in Zeiten des digitalen Umbruchs." *IAB-Stellungnahme 7/2017*, available at: http://doku.iab.de/stellungnahme/2017/sn0717.pdf.

Kuhn, Thomas (1962). *The Structure of Scientific Revolutions*. Chicago.

Kuhn, Thomas (1970). "The Logic of Discovery or the Psychology of Research?" In: *Criticism and the Growth of Knowledge*, Imre Lakatos and Alan E. Musgrave (eds.), 1–24. Cambridge.

Lakatos, Imre (1978). "Science and Pseudoscience." In his *Philosophical Papers*, vol. 1, John Worrall and Gregory Currie (eds.), 1–7. Cambridge, Mass.

Laudan, Larry (1982). "Science at the Bar: Causes for Concern." *Science, Technology & Human Values* 7 (41), 16–19.

Laudan, Larry (1983). "The Demise of the Demarcation Problem." In: *Physics, Philosophy and Psychoanalysis: Essays in Honor of Adolf Grünbaum. Boston Studies in the Philosophy of Science*, vol. 76, Robert S. Cohen & Larry Laudan (eds.), 111–127. Dordrecht.

Lehmer, Florian, and Britta Matthes (2017a). "Auswirkungen der Digitalisierung auf die Beschäftigungsentwicklung in Deutschland." *IAB-Bericht 5/2017*, available at: http://doku.iab.de/aktuell/2017/aktueller_bericht_1705.pdf.

Lehmer, Florian, and Britta Matthes (2017b) "Stand der Digitalisierung" In: *Arbeitsmarkt Kompakt: Analysen, Daten, Fakten (IAB-Bibliothek)*, Joachim Möller and Ulrich Walwei (eds.), 112–113. Nürnberg.

Michel, Jan G. (2009). "Überlegungen zu einem panpsychistischen Physikalismus". In: *Physikalismus, Willensfreiheit, Künstliche Intelligenz*, Jan G. Michel and Marius Backmann (eds.), 43–50. Paderborn.

Michel, Jan G. (2011). *Der qualitative Charakter bewusster Erlebnisse: Physikalismus und phänomenale Eigenschaften in der analytischen Philosophie des Geistes*. Paderborn.

Michel, Jan G. (2013). "Mit Gedankenexperimenten argumentieren: Eine Fallstudie in der Philosophie des Geistes". In: *Die Suche nach dem Geist*, Jan G. Michel and Gernot Münster (eds.), 81–120. Münster.

Michel, Jan G. (2016). "The Extended Mind: Was ist eigentlich mit 'Mind' gemeint? Zum Verhältnis von Geist und Kognition." In: *Ist der Geist im Kopf?*, Jan G. Michel, Kim J. Boström, and Michael Pohl (eds.), 57–82. Münster.

Michel, Jan G. (2018a). "Materialism." Overview article in the online encyclopedia of the digital humanities project *Special Divine Action* at Oxford University, available at: https://sda.bodleian.ox.ac.uk/sda/#!/themes/article/277.

Michel, Jan G. (2018b). "Determinism." Overview article in the online encyclopedia of the digital humanities project *Special Divine Action* at Oxford University, available at: https://sda.bodleian.ox.ac.uk/sda/#!/themes/article/278.

Michel, Jan G. (2018c). "Reductionism." Overview article in the online encyclopedia of the digital humanities project *Special Divine Action* at Oxford University, available at: https://sda.bodleian.ox.ac.uk/sda/#!/themes/article/379.

Michel, Jan G. (2019). "How Are Species Discovered? Declarative Speech Acts in Biology." *Grazer Philosophische Studien* 96 (3), 419–441.

Michel, Jan G. (forthcoming-a). "How Do Science and Fiction Relate? The Case of Scientific Discoveries."

Michel, Jan G. (forthcoming-b). "Extending Declaratives: The Case of Scientific Discoveries."

Michel, Jan G. (forthcoming-c). *The Structure of Scientific Discoveries.*

Michel, Jan G. (ed.) (forthcoming-d). *Making Scientific Discoveries: Interdisciplinary Reflections.* Paderborn.

Michel, Jan G., and Marius Backmann (eds.) (2009a). *Physikalismus, Willensfreiheit, Künstliche Intelligenz.* Paderborn.

Michel, Jan G., and Marius Backmann (2009b). "Menschen, Maschinen und Möglichkeiten." In: *Physikalismus, Willensfreiheit, Künstliche Intelligenz,* Jan G. Michel and Marius Backmann (eds.), 191–198. Paderborn.

Michel, Jan G., Kim J. Boström, and Michael Pohl (eds.) (2016). *Ist der Geist im Kopf?* Münster.

Michel, Jan G., and Michael Kober (2011b). *John Searle.* Paderborn.

Michel, Jan G., and Gernot Münster (eds.) (2013). *Die Suche nach dem Geist.* Münster.

Pietsch, Theodore W., and Tracey T. Sutton (2015). "A New Species of the Ceratioid Anglerfish Genus *Lasiognathus* Regan (Lophiiformes: Oneirodidae) from the Northern Gulf of Mexico." *Copeia* 103 (2), 429–432.

Popper, Karl R. (1934). *Logik der Forschung: Zur Erkenntnistheorie der modernen Naturwissenschaften.* Wien: Springer. Available at: https://monoskop.org/images/e/ec/Popper_Karl_Logik_der_Forschung.pdf. Translated into English as *The Logic of Scientific Discovery*, London, 1959.

Popper, Karl R. (1963). *Conjectures and Refutations.* New York.

Ruse, Michael (1982). "Pro Judice." *Science, Technology & Human Values* 7(41), 19–23.

Searle, John R. (1975). "A Taxonomy of Illocutionary Acts." In his *Expression and Meaning: Studies in the Theory of Speech Acts*, 1–29. Cambridge, 1979.

Searle, John R. (1995). *The Construction of Social Reality.* New York.

Searle, John R. (2007). "What Is Language: Some Preliminary Remarks." In: *John Searle's Philosophy of Language: Force, Meaning and Mind,* Savas L. Tsohatzidis (ed.), 15–45. Cambridge.

Searle, John R. (2010a). *Making the Social World: The Structure of Human Civilization.* Oxford.

Searle, John R. (2010b). "The Basic Reality and the Human Reality." In: *John R. Searle: Thinking about the Real World,* Jan G. Michel et al. (eds.), 19–44. Frankfurt am Main.

Searle, John R. (2010c). "Reply to 'Searle's Biological Naturalism: A Typology'." In: *John R. Searle: Thinking about the Real World,* Jan G. Michel et al. (eds.), 210–214. Frankfurt am Main.

Searle, John R. (2010d). "Reply to '51 Years on: Searle on Proper Names Revisited'." In: *John R. Searle: Thinking about the Real World,* Jan G. Michel et al. (eds.), 220–222. Frankfurt am Main.

Seidel, Markus (2014). "Rezension von Paul Hoyningen-Huene: *Systematicity. The Nature of Science.*" *Zeitschrift für philosophische Literatur* 2 (4), 33–38.

Thagard, Paul (1978). "Why Astrology Is a Pseudoscience." *PSA: Proceedings of the Biennial Meeting of the Philosophy of Science Association, Volume 1: Contributed Papers*, 223–234.

Thagard, Paul (1988). *Computational Philosophy of Science*. Cambridge, Mass.

Ziman, John (1968). *Public Knowledge: An Essay Concerning the Social Dimension of Science*. New York.

Organizational and Societal Implications of the Adoption of Machine Invention Systems

Dragos-Cristian Vasilescu, Michael Filzmoser

Abstract

In recent years, creative systems capable of producing new and innovative solutions in fields such as quantum physics, fine arts, robotics or defense and security have emerged. These systems, called machine invention systems, have the potential to revolutionize the current standard invention process. Because of their potential, there are widespread implications to consider, on a societal and organizational level alike: changes in the workforce structure, intellectual property rights and ethical concerns regarding such autonomous systems. On the organizational side, integration of such machine systems in the largely human-driven innovation process requires careful consideration. Delegation of decisions to human agents, technology assessments, and concepts like the complementarity approach that is used to strengthen the interplay between human and machine in innovation processes seem to be a few of the solutions needed to address the changes incurred by such machine invention systems. Future work in this field will address the integration of machine-based invention from a different perspective as well, focusing on establishing necessary characteristics needed in order to be more easily accepted by human users and customers.

1. Introduction

Due to technological lifecycles happening more rapidly (Linstone 2011), new technologies emerge at a faster pace than ever before. At the same time, by reducing the time-to-market and shortening the test phases until full implementation of new technologies (Stark 2015), societal changes due to technological changes are happening at an accelerated pace as well. Therefore, understanding what impact these emerging technologies have on the society earlier in the assessment process is increasing in importance (Lee 2008; Jovane, et al. 2008; Brey 2012; Weaver, et al. 2017).

 | DOI:10.30965/9783957437488_024

With the advent of machine invention systems (Filzmoser and Koeszegi 2018; Vasilescu and Filzmoser 2018), a type of machine learning system that is able to create new and unique results, actionable models and pattern-free solutions, also known as computationally creative systems (Colton, et al. 2009) – the need to understand how these emerging partly autonomous technologies will impact our society is now greater than ever. As these systems become more prevalent and their outputs used on a larger scale, organizational and societal implications have to be considered. Changes in business processes through the integration of human-machine systems are happening largely in production systems (Hermann, et al. 2016) and have initiated polarizing discussions about the role of machines and future of work (Milgrom and Roberts 1990; Demerouti, et al. 2014; Frey and Osborne 2017). The discussion sparked by creative systems is even more pronounced as until recently, creativity was considered to be a human-only feature (Frey and Osborne 2017). At the same time, a lot of current academic research is dedicated to finding solutions to problems raised by autonomous systems; with ethical considerations at the forefront (Brey 2005; Lin, et al., eds. 2012; Lanfranchi 2017).

Based on these underlying considerations, this chapter focuses on providing solutions to some of the organizational and societal difficulties caused by the introduction of an emerging technology, in this case, machine invention systems. In order to provide better answers, we divide our research focus into three research questions:

(i) What are the societal implications of this new technology?

(ii) What are the organizational implications of the adoption of machine invention systems?

(iii) How can the adoption of this new technology be conducted in order to minimize the resistance to change and to insure a smooth transition from current systems and processes?

To answer these questions, we use a combination of literature research and methods from organization theory and design. While the first two research questions are addressed in the subsequent sections, the last one will find its solutions distributed across the chapter. A discussion of the solutions found in this case will take place in the conclusion.

The remainder of this chapter is organized as follows: the next section provides a brief description of machine invention systems together with practical applications. Section 3 discusses opportunities and challenges presented by these systems from a societal perspective. The next two sections on ethics and accountability and technology acceptance bridge the topics of societal and organizational implications. The section on process approach considers the implementation of machine invention technology from an organizational perspective. The concluding section finishes the chapter by discussing the provided solutions, addressing the limitations of current research and by providing an outlook on future work.

2. Machine Invention Systems

Machine invention systems as described in (Vasilescu and Filzmoser 2018) have been constantly expanding their reach. Figure 23.1 shows a block diagram of the functionality of machine invention systems.

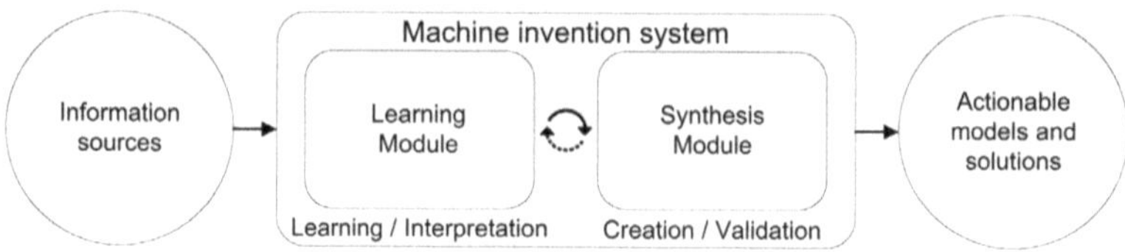

FIGURE 23.1 Main components of a machine invention system.

These systems use a combination of algorithms to derive their creative outputs. Besides unsupervised and deep learning we can find examples of reinforcement learning, transduction and learning to learn processes as well in the various machine invention systems currently implemented. In a first step, the learning module learns and interprets the input data set or, if applicable, generates data, by means of simulation, from which it can learn. In order achieve this result, depending on the application, the system will use a series of machine learning algorithms simultaneously. Finding patterns or hidden links within the data is the main task of the machine learning algorithms that are found in this module.

In the synthesis module, the emerging patterns are conceptualized, sets of rules are derived, models are created and validated, and potential results are predicted. While all machine learning algorithms have feed-forward capabilities, not all allow for feedback loops. The differentiation occurs when analyzing how the algorithm learns and how it passes information from one perceptron (or neuron) to another. If there is no backward link to other neurons that serve as input for the one being analyzed, then the algorithm only possesses feed-forward capabilities (Ming 2015). One such example of a feed-forward algorithm would be the convolutional neural networks used in creating new imagery (Gatys, et al. 2015). There are, however, special cases of convolutional neural networks that allow for feedback loops as well (Stollenga, et al. 2014).

As a rule, the presence of feedback loops between the synthesis module and the machine learning portion increases the quality of the delivered output.[1] Because not all the output in the current state-of-the-art is in a ready-to-be-used format by a customer, the common denominator lies in the invention aspect of the output, rather than coining all output as innovations. Therefore, the concept of machine invention describes the output, and thus the system, more accurately than machine innovation.

Machine invention systems are already used in practice. For example, there are applications in which machine systems create unique forms of art. In this field, Kulitta (Quick 2014; Quick 2015) is developing new musical scores based on a structural/abstract generation, musical interpretation, and learning algorithms. In visual arts, convolutional neural networks can process images by the same principles of abstraction and object recognition. Using content representation, these neural networks then create a new image with the same recognized objects, using a variety of styles learned by abstracting different painting techniques (Gatys, et al. 2015). In October 2018, a unique portrait created by a generative adversarial network (GAN) portraying a fictitious person was sold for a large amount of money by a renowned auction house (Obvious Art 2018). This case has significantly increased the visibility of machine invention as it has been one of the first instances of a work of art known to be created only through machine learning algorithms that was sold.

Machine invention systems can also help us understand and model different parts of our environment. Especially in the field of quantum physics, where phenomena seem at first counterintuitive for researchers, a computer algorithm called Melvin (Krenn, et al. 2016) is able to find and optimize the arrangement of quantum building blocks that result in viable solutions. In order to do so, the algorithm is learning by identifying useful groups of elements that ultimately lead to new suggested experiments. These experiments help researchers understand and expand their knowledge of quantum effects and help in fringe areas like quantum cryptography, or in quantum entanglement.

Even if machine invention seems to have a lot of potential, the implementation of this technology faces considerable resistance. The next section discusses major challenges and opportunities of machine invention systems.

---

1 The delimitation between the two modules of the machine invention systems does not have to be apparent in practice however. Such a separation has only been shown for the purpose of better explaining the model.

3. Systemic Opportunities and Challenges

The development of machine invention systems could be seen as incremental in nature and a logical next step from simple machine learning systems; however, it fundamentally challenges the understanding of the invention process itself. Based on the work of Norman and Verganti, we postulate that this development represents a radical change for our society (Norman and Verganti 2014). In order to further justify this claim, we use the work of Dahlin and Behrens to identify whether the criteria for a radical innovation are met (Dahlin and Behrens 2005). The technology described above meets all the requirements, by (1) being a novel and unique approach and (2) by having a large impact on the development of future technology. Being a radical innovation, there is a larger resistance to change to be expected than in the case of incremental innovations (Clifton and Wharton 1971).

Not only is future technology affected by this development, but the very core of what we consider human-centric activities. Innovation was a task always attributed to animals (Reader, et al. 2016), but with the advent of machine invention systems, this basic paradigm is being challenged. An intersection of machines in this sector was to be expected, however, based on their extreme evolution in the past 50 years (Barnet 2004).

The lurking issue here seems to be the loss of jobs to new automation methods that reduce the areas of human labor (Frey and Osborne 2017). Currently, the jobs being slowly replaced are typically repetitive, manual and sometimes dangerous. Phasing out such jobs is not a negative development in itself, but causes a shift toward a more skilled workforce. Evidence, however, suggests that even though routine jobs are already declining, there is a definite positive trend toward low-level service-oriented jobs (Cully 1999), based on the model prediction by Reich (1992) who postulated that both "symbolic-analysts" and service-oriented jobs will gain importance in the 21st-century workforce. One possible reason for such a phenomenon lies in the inherent value placed on human interaction as opposed to technology-only interactions (Ganguli and Roy 2011).

A significant challenge faced by machine invention systems is the fact that the expected output is, in the best case, an innovative model, application or actionable experiment and therefore should be protected by patents, copyrights or other protection methods of intellectual property rights. Current policy making is still operating on the premise of human-only inventions and innovations. A change in global policy making is necessary, as technologies become more autonomous. Such a difficulty has become apparent in the case of accidents involving driverless cars (Ganesh 2017) where the accountability cannot be clearly assigned as in the traditional driver-based system. This topic will be discussed further in the following section.

The human invention process is chaotic and often fruitless. A machine attempting to innovate on the other hand has a more methodological approach, by virtue of its nature. Therefore, advancements made in the direction of machine invention would have the potential to create significantly different inventions in comparison with human inventions. With the advance of machine learning and their innate capability of assessing and connecting large amounts of data, their potential to realize previously unrelated connections is far beyond human capability. From creating new connections in data, there are only a few more steps to deriving new inventions. The potential for different inventions together with a larger number of inventions altogether can be a tremendous asset for all fields, industries and economy in general.

4. Ethics and Accountability

A perhaps more important discussion regarding the generation of innovative solutions through machine systems lies in the ethical issues arising. With the assignment of tasks and decision rights to machine agents[2], the questions of economic, legal and moral accountability for the execution of the assigned tasks and decisions and their consequences emerge. It is assumed that organizational control ensures the accountability of employees within companies. Organizational control is achieved by designing and applying appropriate hierarchical structures and procedures, and by setting targets. Through monitoring performance as well as the conformity to rules and procedures, the management exercises control over the organization's members. Employees who do not conform can be held accountable. However, human agents can only be held accountable for consequences of their actions if they can understand, predict and influence work processes (Boos, et al. 2013). Although machine agents can also interact with their environment to pursue and achieve predefined targets, accountability for outcomes cannot simply be transferred to these machine agents that, at least at the current developmental state of the underlying technologies, do not possess a self-generated belief system, do not feel motivation, or build relationships and trust. Machine agents also do not have the status of a natural or legal person required for legal liability. Thus,

2 Such as robots, digital systems and artificial intelligent agents. For the purpose of this chapter, the terms of machine agent and artificial agent will be used interchangeably.

when distributing tasks and decision rights to artificial agents, human agents still need be held accountable.

For example, a digital system equipped with such machine invention capabilities could generate solutions to the most efficient way of taking life in specific scenarios; solutions that would completely defy human logic and instincts, and therefore cannot be combated by the typical means at our disposal, except through other machine invention systems. Such ethical concerns should be thoroughly discussed and solutions should be systemically applied to all autonomous-prone technologies. There are already models to assess the ethical implications of emerging technologies, such as the ethical technology assessment, the techno-ethical scenario or the ETICA approach (Busby, et al. 2008; Brey 2012), but there is no globally-accepted axiom or set of principles regarding the ethics of autonomous technologies yet.

Recently, an increasing trend toward ethical considerations can be found in the literature (Lin, et al., eds. 2012; Lanfranchi 2017; Nørskov 2017) and in practice (DeepMind 2018). Moreover, country-wide recommendations have started to appear, with Germany being one of the first countries to release a report on ethical guidelines for automated and connected driving (BMVI 2017).

As in the case of autonomous driving, such ethical principles can be implemented to filter out dangerous inventions generated by the machine invention systems. The ethical enhancement of systems capable of machine invention can be envisioned as a two-step process: first, the global ethical principles and the application-specific principles (if necessary) are defined, and second, building the system in such a way that these principles will always be applied. One such design principle is called 'value sensitive design' (Mingers and Walsham 2010; Friedman, et al. 2013).

One way of reducing the resistance to implementing such a technology, even without implementing ethical enhancements to machine invention systems, is to delegate the final decision and control tasks back to human agents: in essence, moving the accountability of accepting or rejecting the inventions produced from the machine decision-making process into a human decision-making process.

In the domain of robotics and artificial intelligence, the delegation of decision rights could be realized by the application of algorithmic selection. Algorithmic selection suggests alternatives in decision situations based on the analysis of big data by means of machine learning. Currently, the technology is implemented in many internet services like recommender systems or news aggregation, but it is also in use to support credit rating or promotion and employment decisions (Laux and Liermann 2005; Petrina 2008).

When evaluating the possibility of delegating decision rights to human agents, several advantages and disadvantages must be considered. Without delegation the decision lies solely with the machine invention system. On the other hand, in the case of delegation, the decision suffers from a lack of overview and questionable motivation of the human counterpart. If making a decision simultaneously bears certain consequences like additional work or overtime, then a conflict of interests must be expected. Users of artificial intelligence technologies, for example for big data analyses or algorithmic selection, often do not understand the internal functionality of the programs, so that the decision quality remains questionable. Delegation of decisions to machine agents is reasonable in case of routine decisions that are frequently recurring, so there exists relevant historic data and the decision can be analyzed and modeled. In the case of machine invention, such a joint decision system can be applied in part as well. However, due to the nature of the output of machine invention systems, there would be significantly fewer decisions that could be taken by artificial agents, as the output is explicitly novel.

A viable concept to deal with this accountability and delegation issue is proxy agency. Traditionally, we distinguish between direct agency (of an individual) and collective agency, when individuals participate in networks with synergistic goals (Bandura 2001). Proxy agency is a third form of a mediated agency, where a human agent utilizes the resources or know-how of another (artificial) agent to act on their behalf. Thus, the human agent remains accountable as if they were acting by themselves. Proxy agency augments the possibilities of the human agent to reach their goals beyond their own capabilities.

However, the proxy agency concept is only suited to resolve the accountability issue if the human agent can understand the work processes of these systems. As a consequence, (i) the decision process of the system has to be explainable or justifiable, and (ii) the human agent must have the ability to accept or reject the final output of such a system. In this case, the first condition is harder to meet, while the second one can be applied without much difficulty.

5. Technology Acceptance

The advent of new concepts like digital work, flexible work and new ways of working (Scandura and Lankau 1997; Demerouti, et al. 2014) is already shaping the work of tomorrow. In addition to these, changes to a type of work that was thought innately human, namely the research and development of new content, models and experiments, just increases the resistance toward implementation of such a technology even further. Such a societal resistance to change was already observed in global shifts in technology, for example during the industrial revolution (Hudson 2014). However, the long-term changes in that case were tremendously positive, with the reduction of mortality at work, the diversification of types of jobs, emergence of new industries, higher paying wages or salaries and a better quality of life, to name a few (Hartwell 1961).

Due to the above reasons, availability of knowledge, the innovation diffusion and technology substitution and adoption rates (Miranda and Lima 2013), machine invention systems are predicted to exhibit only a slow implementation rate initially. The currently still relatively low number of applications using machine invention attests to this. This adoption process also coincides with the natural changes in the workforce, with the newer generations being educated from the beginning to be able to use new technologies. Nonetheless, systemic changes in the society have to be considered carefully. One way to help steer the entire process is through policy making (Wennberg, et al. 2014).

Another approach that can be used to promote the acceptance of machine invention technologies in a localized fashion is the use of technology acceptance evaluations. Such evaluations and assessments with system developers, organizational decision makers and designers would need to be performed in order to elicit perceived benefits and drawbacks of the changes. Afterwards, the perceived issues should be addressed in the change management. An established model to conduct technology acceptance evaluations in the domain of information systems is the technology acceptance model (TAM) developed by Davis (1989). The model and its further developments use standardized questionnaires to collect information about the perceived usefulness and perceived ease of use, which are the major factors influencing the attitude towards using a new technology. In the case of autonomous systems, the TAM model would have to be adjusted to reflect the usefulness, ease of operating (rather than using) and perceived benefits of implementing such a technology.

6. Changes in the Business Processes

With respect to collaboration of different roles in business processes there exists a continuum between integration and fragmentation of processes. As illustrated in Figure 23.2, a specific instance of a process can be handled by just one worker, a so-called case worker, which constitutes case management. On the other hand, there could be several roles involved in a fragmented process, up to the number of tasks the process consists of.

FIGURE 23.2
Case management vs. fragmented process.

The same principle holds for the assignment of roles to human and machine agents. For example, the invention process can be divided into several tasks, such as researching the topic, finding potential gaps, finding solutions in these gaps, selecting potential solutions, testing and prototyping. These tasks can be performed either by a single researcher or can be divided between several individuals.

The activities could, however, also be assumed by artificial agents, in this case machine invention systems, that combine all these tasks into one system. Beside the two extremes of full manual work and full automation, there is of course the possibility of combining the work of human and artificial agents. The question arises however – which level of automation is best for business processes. Organization theory provides a clear answer (Davenport 1993; Frey and Osborne 2017): the more integration, the better. Case management saves waiting times for unnecessary set-up, wind-down and control activities, and for humans it also leads to an increased identification with the final product. Therefore working on a case becomes an intrinsic motivator for the worker.

One of the drawbacks of integrated processes is the social isolation of case workers. In an optimized integrated process, the interactions are reduced as far as possible. This negative aspect is irrelevant for machine agents. Another drawback however, is critical also when it comes to determining the degree of automation: case workers, and therefore also machines that assume the role of case workers, that handle whole processes, need more skills and competencies than workers specialized on fewer tasks in a fragmented process. This could be problematic if the systems are not sufficiently flexible or versatile to fulfill this requirement.

The combination of different artificial agents similar to the combination of several human agents, on the other hand, might cause the same kind of inefficiencies mentioned: waiting and set-up times, error-prone interfaces and so on. Finally, a combination of human and artificial agents is also not advisable as this undermines the amortization of the high investments in new technologies. Unless shift working is applied, which might not be applicable in all areas, machine agents cannot economize on one of their major advantages: namely that they can work 24 hours a day, seven days a week without breaks, vacations or illness. Instead, these systems would be restricted to the working hours of their human co-workers.

With respect to the process integration, a fully automated or fully human process, provided there is only one execution unit, would be optimal based on the above arguments. However, there might be processes where these analyses are not valid, and rather a cooperation of machine and human agents will be superior in terms of quality. This will be relevant therefore for processes where the output quality is of higher importance than time or costs associated with a process, for example in the field of innovation. This represents another approach to solving the problem of slow integration, and the perceived threat of machine invention systems. This concept is called the complementarity approach and in this concept each actor, human or otherwise, focuses on the tasks for which they are most qualified or best equipped to handle. This complementarity approach was first described by Milgrom and Roberts (1990; 1995) among others, as one series of tasks benefitting a completely different set of tasks, therefore them being typically chosen to be done together for the added synergistic effects. One such example would be core business processes like manufacturing and sales working together with support processes like acquisition of goods, logistics and finances. The individual tasks in each of these processes is very different, yet they support each other in order to optimize the production and sales output of a company.

While even in innovation processes there is a shift toward shorter time-to-market (Davenport 1993), having a human component as part of the system serves a dual purpose. On one hand it increases acceptance of the solutions generated by machine invention systems by human users and customers and, on the other hand, it would allow a better use of the native capabilities available to both human and machines. Rather than the machine invention systems being perceived as a threat, they could be seen as a complementary system to improve human invention and creativity, while still having a human component; essentially allowing a more in-depth view on a given topic than was previously possible. In that respect, a situation where a near full integration is possible, and only some specific tasks that cannot (or should not: see the section on ethics and accountability) be performed efficiently by the artificial agents seems to be the current optimal result for processes where output quality is the highest priority.

From the examples of AlphaGo (Silver, et al. 2016) and AlphaGo Zero (Silver, et al. 2017) there is something to learn about where to place the human – machine interface. Both systems are designed to play the game of Go. The AlphaGo system had a mixed training, starting with learning from past human games of Go and then continuing by playing against itself in order to improve. At the same time, expert Go players would interpret the moves the system was making and instruct the system to stop using moves that yield little or no benefits. On the other hand, AlphaGo Zero was only given the basic rules of the game and it only learned from playing against itself, without any human interaction. When both systems went head to head, AlphaGo Zero obtained a flawless victory over 100 games played (Silver, et al. 2017). Drawing from this example, the human – machine interface should be best placed either at the beginning of the process, by setting up tasks, rules and limitations for the machine invention system, or at the end, by accepting or rejecting the final output of the system – rather than in other parts of the process. The latter placement was already discussed in the previous sections.

7. Conclusion

Machine invention systems represent a novel technology and phenomenon, with implications for society and organizations not yet fully understood or known. These systems process input data by means of machine learning algorithms and use the derived patterns to develop new, pattern-free creative outcomes. Questions arise about what impact the diffusion and adoption of this technology in the innovation processes of companies will have and how they can be integrated to transform human-only to human-machine systems with the least amount of resistance possible.

This chapter identified that the radical change imposed by machine invention systems requires a reconsideration of the current understanding of how inventions occur and from which sources. Also, there is currently a lack of regulation when it comes to the intellectual property rights of inventions generated by these systems. This also relates to ethics and accountability of machine invention systems. Can machines be held accountable for consequences of actions they perform, but do not understand, as in the case of unusable or defective developments? For this scenario, similar to delegation in organization theory, a supervisor of the system has to be accountable for the consequences of the delegation. In innovation processes this requires a collaboration of human and machine and a final decision by the human agent.

Throughout the sections of this chapter, a series of solutions have been presented to approach the consequences and implications of machine invention systems, some of them requiring no work whatsoever, like the natural change in the workforce, others requiring a great deal of effort from various stakeholders of society, such as developing systems that can make justifiable decisions, based on ethical principles set through policy making. Further, the chapter discussed methods to minimize the resistance to change for such systems, such as the TAM model, delegation of decisions and suggested using complementarity principles when designing business innovation processes. Based on examples, suggestions were also made about where in the innovation process the human-machine interfaces should exist. Finally, in the field of innovation, joint human-machine invention systems could create synergies that would elevate the performance and quality of such processes beyond current capabilities.

Future work will focus on creating a typology of machine invention systems based on their fields of application, together with the used algorithms, in order to serve as a roadmap for future developments and to potentially uncover more application possibilities in currently unexplored areas. Further, experiments are planned to assess what characteristics a machine-based invention should have in order to be accepted by human agents. This will in turn enable machine invention developers to design systems that will be easier to implement and be more readily accepted in our day-to-day environments.

The systems and applications this research is based on represent only a snapshot of the state-of-the art. Considering the relative novelty of the field, only currently known opportunities and challenges have been discussed in this chapter, based on current issues arising from organizations, public society and policy makers. Ongoing research will have to consider the future development of the field. We, however, hope to have laid the foundations for this research with this chapter on organizational and social implications of the adoption of machine invention systems.

8. Bibliography

Bandura A (2001) "Social Cognitive Theory: An Agentic Perspective." *Annual Review of Psychology* 52:1–26.

Barnet B (2004) "Technical Machines and Evolution." *CTheory* 3–16/2004.

BMVI (2017) *German Federal Ministry of Transport and Digital Infrastructure – Report by the Ethics Commission on Automated and Connected Driving*. Avaliable at: https://www.bmvi.de/SharedDocs/EN/publications/report-ethics-commission-automated-and-connected-driving.html. Accessed 14 Jan 2019.

Boos D, Guenter H, Grote G, Kinder K (2013) "Controllable accountabilities: the internet of things and its challenges for organisations." *Behaviour & Information Technology* 32, 449–467.

Brey PAE (2012) "Anticipatory Ethics for Emerging Technologies." *Nanoethics* 6, 1–13.

Busby H, Hervey T, Mohr A (2008) "Ethical EU law? The influence of the European group on ethics in science and new technologies." *European Law Review* 33, 803–842.

Clifton R, Wharton J (1971) *Risk, uncertainty, and the subsistence farmer: Technological innovation and resistance to change in the context of survival. Studies in Economic Anthropology*. George Dalton (Ed) Washington, DC American Anthropological Association, 151–180.

Colton S, de Mántaras RL, Stock O (2009) "Computational creativity: Coming of age." *AI Magazine* 30,11.

Cully M (1999) "A more or less skilled workforce? Changes in the occupational composition of employment, 1993 to 1999." *Australian Bulletin of Labour*, 25, 98–104.

Dahlin KB, Behrens DM (2005) "When is an invention really radical?" *Research Policy*. 34, 717–737.

Davenport TH (1993) *Process innovation: reengineering work through information technology*. Harvard Business Press.

Davis FD (1989) "Perceived Usefulness, Perceived Ease Of Use, And User Accep." *MIS Quarterly* 13, 319–340.

DeepMind (2018) "DeepMind Ethics & Society." In: *DeepMind*. https://deepmind.com/applied/deepmind-ethics-society/. Accessed 15 Feb 2018.

Demerouti E, Derks D, ten Brummelhuis LL, Bakker AB (2014) "New Ways of Working: Impact on Working Conditions, Work – Family Balance, and Well-Being." In: Korunka C, Hoonakker P (eds) *The Impact of ICT on Quality of Working Life*. Dordrecht: Springer Netherlands, 123–141.

Filzmoser M, Koeszegi ST (2018) Integrating Robotics and Artificial Intelligence in Business Processes – An Organization Theoretical Analysis. *Envisioning Robots in Society – Power, Politics, and Public Space: Proceedings of Robophilosophy 2018/TRANSOR 2018* 311, 177–184.

Frey CB, Osborne MA (2017) "The future of employment: How susceptible are jobs to computerisation?" *Technological Forecasting and Social Change* 114, 254–280.

Friedman B, Kahn PH, Borning A, Huldtgren A (2013) "Value sensitive design and information systems." In: *Early engagement and new technologies: Opening up the laboratory*. Springer, 55–95.

Ganesh MI (2017) "Entanglement: Machine learning and human ethics in driver-less car crashes." *APRJA* 6(1).

Ganguli S, Roy SK (2011) "Generic technology-based service quality dimensions in banking: Impact on customer satisfaction and loyalty." *Intl Jnl of Bank Marketing*. 29, 168–189.

Gatys LA, Ecker AS, Bethge M (2015) *A Neural Algorithm of Artistic Style*. arXiv:150806576 [csCV]. doi: 10.1167/16.12.326.

Hartwell RM (1961) "The Rising Standard of Living in England, 1800–1850" *The Economic History Review* 13, 397–416.

Hermann M, Pentek T, Otto B (2016) "Design principles for industrie 4.0 scenarios." In: *System Sciences (HICSS), 2016 49th Hawaii International Conference on. IEEE*, 3928–3937.

Hudson P (2014) *The Industrial Revolution*, 1st ed. Bloomsbury Publishing.

Jovane F, Yoshikawa H, Alting L, Boër CR, Westkamper E, Williams D, Tseng M, Seliger G, Paci AM (2008) "The incoming global technological and industrial revolution towards competitive sustainable manufacturing." *CIRP Annals* 57, 641–659.

Krenn M, Malik M, Fickler R, Lapkiewicz R, Zeilinger A (2016) "Automated Search for new Quantum Experiments." *Physical Review Letters* 116:090405.

Lanfranchi V (2017) "Machine Learning and Social Media in Crisis Management: Agility vs Ethics." *Proceedings of the 14th International Conference on Information Systems for Crisis Response and Management*. IMT Mines Albi-Carmaux (École Mines-Télécom).

Laux H, Liermann F (2005) *Fundamentals of organization: The management of decisions as fundamental problem of business administration*. Berlin.

Lee, EA (2008) "Cyber physical systems: Design challenges". *2008 11th IEEE International Symposium on Object and Component-Oriented Real-Time Distributed Computing (ISORC)*, 363–369.

Lin P, Abney K, Bekey GA, eds (2012) *Robot ethics: the ethical and social implications of robotics*. Cambridge, Mass: MIT Press.

Linstone, HA (2011) "Three eras of technology foresight." Technovation 31, 69–76.

Milgrom P, Roberts J (1990) "The Economics of Modern Manufacturing: Technology, Strategy, and Organization." *The American Economic Review* 80, 511–528.

Milgrom P, Roberts J (1995) "Complementarities and fit strategy, structure, and organizational change in manufacturing." *Journal of accounting and economics* 19, 179–208.

Ming Liang, Xiaolin Hu (2015) "Recurrent convolutional neural network for object recognition." In: *2015 IEEE Conference on Computer Vision and Pattern Recognition (CVPR). IEEE*, Boston, MA, 3367–3375.

Mingers J, Walsham G (2010) "Toward ethical information systems: the contribution of discourse ethics." *Mis Quarterly* 34, 833–854.

Miranda LCM, Lima CAS (2013) "Technology substitution and innovation adoption. The cases of imaging and mobile communication markets." *Technological Forecasting and Social Change* 80(6), 1179–1193.

Norman DA, Verganti R (2014) "Incremental and Radical Innovation: Design Research vs. Technology and Meaning Change." *Design Issues* 30, 78–96.

Nørskov M (2017) *Social Robots: Boundaries, Potential, Challenges*. Taylor & Francis.

Obvious Art (2018) | Edmond de Belamy, from La Famille de Belamy | Prints & Multiples, digital print | Christie's. https://www.christies.com/Lotfinder/lot_details.aspx?hdnSaleID=27814&LN=363&intsaleid=27814&sid=41bfe836-b0c1-4afa-9298-09cc909345ee. Accessed 14 Jan 2019.

Petrina AM (2008) "Robotics: Present and future." *Scientific and Technical Information Processing* 35, 73–79.

Quick D (2014) *Kulitta: A Framework for Automated Music Composition*. Yale University, 247.

Quick D (2015) "Composing with Kulitta." *International Computer Music Conference Proceedings (ICMC)*, 306–309.

Reader SM, Morand-Ferron J, Flynn E (2016) "Animal and human innovation: novel problems and novel solutions." *Philosophical Transactions of the Royal Society B: Biological Sciences* 371, 20150182.

Reich RB (1992) *The work of nations: preparing ourselves for 21st century capitalism*, 1st Vintage Books ed. Vintage Books, New York.

Scandura TA, Lankau MJ (1997) "Relationships of gender, family responsibility and flexible work hours to organizational commitment and job satisfaction." *Journal of Organizational Behavior* 18, 377–391.

Silver D, Huang A, Maddison CJ, et al (2016) "Mastering the game of Go with deep neural networks and tree search." *Nature* 529, 484–489.

Silver D, Schrittwieser J, Simonyan K, et al (2017) "Mastering the game of Go without human knowledge." *Nature* 550, 354–359.

Stark, J (2015) "Product lifecycle management." In: *Product Lifecycle Management*. Volume 1. Springer, 1–29.

Stollenga MF, Masci J, Gomez F, Schmidhuber J (2014) "Deep networks with internal selective attention through feedback connections." *Advances in neural information processing systems* 27, 3545–3553.

Vasilescu D-C, Filzmoser M (2018) "Machine Invention Systems: A New Approach to Innovation." In: *VSS* 2018, 22–23.

Weaver P, Jansen L, Grootveld G van, Spiegel E van, Vergragt P (2017) *Sustainable Technology Development*. Routledge. doi: 10.4324/9781351283243.

Wennberg K, Stadin E, Bergström A (2014) *How policy could handle workplace digitization*. Ratio Working Papers 237.The Ratio Institute.

About the Authors

Agnieszka Wykowska, Dr. habil., Principal Investigator and team leader of the unit “Social cognition in human-robot interaction” at the Istituto Italiano di Tecnologia.
Publications:

(1) Ciardo, Francesca, Frederike Beyer, Davide de Tommaso, and Agnieszka Wykowska. 2020. “Attribution of Intentional Agency Towards Robots Reduces One’s Own Sense of Agency.” *Cognition* 194: 104109.
(2) Willemse, Cesco, and Agnieszka Wykowska. 2019. “In Natural Interaction with Embodied Robots, We Prefer It When They Follow Our Gaze: A Gaze-Contingent Mobile Eyetracking Study.” *Philosophical transactions of the Royal Society of London. Series B, Biological sciences* 374 (1771): 20180036.
(3) Perez-Osorio, Jairo, and Agnieszka Wykowska. 2019. “Adopting the Intentional Stance Toward Natural and Artificial Agents.” *Philosophical Psychology* 5 (7): 1–27.

Andreas Theodorou, Dr., Postdoctoral fellow at the Department of Computer Science, Umeå University.
Publications:

(1) Theodorou, Andreas and Dignum, Virginia. 2020. “Towards ethical and socio-legal governance in AI”, Nature Machine Intelligence 2(1), 10–12.
(2) Theodorou, Andreas, Robert H. Wortham, and Joanna J. Bryson. 2017. “Designing and Implementing Transparency for Real Time Inspection of Autonomous Robots.” *Connection Science* 29 (3): 230–41.
(3) Bryson, Joanna J., and Andreas Theodorou. 2019. “How Society Can Maintain Human-Centric Artificial Intelligence.” In *Human-Centered Digitalization and Services*. Vol. 19, edited by Eveliina Saari and Marja Toivonen, 305–23. Translational systems sciences volume 19. Singapore: Springer.

Andreas Bischof, Dr. phil., Research Group Leader at Chemnitz University of Technology.
Publications:

(1) Pentzold, Christian, and Andreas Bischof. 2019. “Making Affordances Real: Socio-Material Prefiguration, Performed Agency, and Coordinated Activities in Human – Robot Communication.” *Social Media+ Society* 5 (3): 2056305119865472.

(2) Bischof, Andreas. 2018. "The Challenge of Being Self-Aware When Building Robots for Everyday Worlds." In *Developing Support Technologies: Integrating Multiple Perspectives to Create Assistance That People Really Want*. Vol. 23, edited by Athanasios Karafillidis and Robert Weidner, 127–35. Biosystems & Biorobotics 23. Cham: Springer International Publishing.
(3) Bischof, Andreas. 2017. *Soziale Maschinen Bauen: Epistemische Praktiken Der Sozialrobotik:* transcript Verlag.

Arne Maibaum, Dipl.-Soz.-Wiss., PhD fellow at TU Berlin.
Publications:
(1) Hergesell, Jannis, Maibaum, Arne, and Meister, Martin. 2020. *Genese und Folgen der Pflegerobotik. Die Konstitution eines interdisziplinären Forschungsfeldes*. Weinheim/Basel: Juventa.
(2) Hergesell, Jannis, and Maibaum, Arne. 2018. "Interests and Side Effects in Geriatric Care". In *Developing Support Technologies – Integrating Multiple Perspectives to Create Support that People Really Want*, edited by Robert Weidner and Athanasius Karafilidis. Wiesbaden: VS-Verlag.
(3) Maibaum Arne. 2018. "Innovating while Inventing". In *Innovationsphänomene*, edited by Jannis Hergesell, Arne Maibaum, C. Minnetian, and A. Sept. Springer VS, Wiesbaden

Astrid Marieke Rosenthal-von der Pütten, Prof. Dr., Chair Individual and Technology, Department of Society, Technology, and Human Factors, RWTH Aachen.
Publications:
(1) Rosenthal-von der Pütten, Astrid M., Nicole C. Krämer, Stefan Maderwald, Matthias Brand, and Fabian Grabenhorst. 2019. "Neural Mechanisms for Accepting and Rejecting Artificial Social Partners in the Uncanny Valley." *The Journal of Neuroscience* 39 (33): 6555–70. https://doi.org/10.1523/JNEUROSCI.2956-18.2019.
(2) Von der Pütten, Astrid M, Nicole C. Krämer, Jonathan Gratch, and Sin-Hwa Kang. 2010. "It Doesn't Matter What You Are! Explaining Social Effects of Agents and Avatars." *Computers in Human Behavior* 26 (6): 1641–50.
(3) Rosenthal-von der Pütten, Astrid M, Nicole C. Krämer, Laura Hoffmann, Sabrina Sobieraj, and Sabrina C. Eimler. 2013. "An Experimental Study on Emotional Reactions Towards a Robot." *International Journal of Social Robotics* 5 (1): 17–34.

Benedikt Paul Göcke, Prof. Dr. Dr., Chair of Philosophy of Religion and Philosophy of Science, Catholic Faculty of Theology, Ruhr-University Bochum.
Publications:

(1) Göcke, Benedikt Paul. 2018. *The Panentheism of Karl Christian Friedrich Krause (1781–1832): From Transcendental Philosophy to Metaphysics*: Peter Lang.
(2) Göcke, Benedikt Paul. 2014. *A Theory of the Absolute*: Palgrave Macmillan.
(3) Göcke, Benedikt Paul. 2017. "Christian Cyborgs: A Plea for a Moderate Transhumanism." *Faith and Philosophy* 34 (3): 347–64.

Carmen Krämer, Dr. phil., Research Assistant at RWTH Aachen.
Publications:

(1) Krämer, Carmen. 2018. "9 Menschenwürde – Flucht – Medien. Reflexionen zu Ziffer 1 des deutschen Pressekodexes im Kontext der Medienberichterstattung über Flucht." In *Migration, Integration, Inklusion: Medienethische Herausforderungen Und Potenziale Für Die Digitale Mediengesellschaft*, edited by Nina Köberer, Marlis Prinzig, and Michael Schröder. 1st ed., 151–62. Baden-Baden: Nomos Verlagsgesellschaft mbH & Co. KG.
(2) Kellerwessel, Wulf, and Carmen Krämer. 2018. "Flucht, Immigration Und Aufenthalt – Normative Perspektiven (Kellerwessel, Carmen Krämer)." *PhLA* 71 (2): 162–98.
(3) Krämer, Carmen. 2016. "Peter Singer und die Anfänge der Tierethik in der Moderne." In *Mensch – Tier – Ethik im interdisziplinären Diskurs*, edited by Annette Förster, Wulf Kellerwessel, and Carmen Kramer. Ethik interdisziplinär Band 22. Berlin, Münster: LIT.

Daniel Neumann, Mag., Research Associate at the Department of Philosophy, Alpen-Adria-Universität Klagenfurt.
Publications:

(1) Daniel Neumann, "Ist der geistige Automat frei? Göttliches und anarchisches Denken in Spinozas Ethik", in: Thomas Kisser, Katrin Wille (Hg.), *Spinozismus als Modell. Deleuze und Spinoza* (Paderborn: Wilhelm Fink 2019).
(2) Daniel Neumann, "Kleine Geschichte denkender Automaten", in: Tyyne Claudia Pollmann (Hg.), *visions4people. Artistic Research Meets Psychiatry* (Berlin: Jovis 2019).

(3) Daniel Neumann, "Die unsterbliche, mechanische Seele des Körpers. Descartes' Maschine in der Kontroverse", in: *Von Menschen und Maschinen. Proceedings der 3. Tagung des Nachwuchsnetzwerks „INSIST"*, 05.– 07. Oktober 2018 (in Vorbereitung).

David J. Gunkel, PhD, Distinguished Teaching Professor, Northern Illinois University (USA).
Publications:

(1) Gunkel, David J. 2012. *The Machine Question: Critical Perspectives on AI, Robots, and Ethics:* MIT Press.
(2) Gunkel, David J. 2016. *Of Remixology: Ethics and Aesthetics After Remix:* MIT Press.
(3) Gunkel, David J. 2018. *Robot Rights:* MIT Press.

Davide Ghiglino, PhD fellow at the Istituto Italiano di Tecnologia and University of Genova.
Publications:

(1) Ghiglino, Davide, De Tommaso, Davide, Wykowska, Agnieszka. 2018. Attributing Human-likeness to an Avatar: the Role of Time and Space in the Perception of Biological Motion. In: *Social Robotics. 10th International Conference. ICSR 2018*. Vol. 11357, edited by Shuzhi S. Ge, 400–409. Lecture Notes in Computer Science 11357. Cham: Springer.
(2) Marchesi, Serena, Ghiglino, Davide, Ciardo Francesca, Perez-Osorio, Jairo, Baykara, Ebru, and Wykowska, Agnieszka. 2019. "Do We Adopt the Intentional Stance Toward Humanoid Robots?" *Frontiers in Psychology* 10:450.
(3) Ghiglino, Davide, Willemse, Cesco, De Tommaso, Davide, Bossi, Francesco, & Wykowska, Agnieszka. 2020. "At first sight: robots' subtle eye movement parameters affect human attentional engagement, spontaneous attunement and perceived human-likeness". *Paladyn, Journal of Behavioral Robotics* 11.1:31–39.

Diana Löffler, Dr., Post-Doc at Ubiquitous Design/Experience & Interaction Group, University of Siegen.
Publications:

(1) Löffler, Diana, Jörn Hurtienne, and Ilona Nord. 2019. "Blessing Robot BlessU2: A Discursive Design Study to Understand the Implications of Social Robots in Religious Contexts." *International Journal of Social Robotics* 22 (5): 353.

(2) Löffler, Diana, Nina Schmidt, and Robert Tscharn. 2018. "Multimodal Expression of Artificial Emotion in Social Robots Using Color, Motion and Sound." In *HRI'18, March 5–8, 2018, Chicago, IL, USA*, edited by Takayuki Kanda, Selma Ŝabanović, Guy Hoffman, and Adriana Tapus, 334–43. New York, New York, USA: ACM Press.
(3) Dörrenbächer, J., Löffler, Diana, and Hassenzahl, Marc. 2020. "Becoming a Robot – Overcoming Anthropomorphism with Techno-Mimesis". *Accepted for CHI 2020.*

Dragos-Cristian Vasilescu, Dipl.-Ing., Teaching and Research Associate at the Institute of Management Science, TU Wien.
Publications:
(1) Vasilescu, D.-C. and Filzmoser, M. 2017. "Machine Invention Systems: A Fundamental Paradigm Shift in the Field of Innovation and its Societal Implications". In *Proceedings of Innovation and Sustainability International Scientific Conference, third edition*, 6–13. Bucharest: Niculescu.
(2) Vasilescu, D.-C. and Filzmoser, M. 2018. "Machine Invention Systems: A New Approach to Innovation". Vienna Young Scientists Symposium 2018, Book-of-Abstracts.com, Austria edited by P. Hans, G. Artner, J. Grames, H. Krebs, H. Mansouri Khosravi, and T. Rouhi. 22–23.

Gábor L. Ambrus, Dr., Post-doctoral researcher at Charles University, Faculty of Protestant Theology, Prague.
Publications:
(1) Ambrus, Gábor. 2019. "Creators and Creatures: The Creation Account in Genesis and the Idea of the Artificial Humanoid." *Zygon* 54 (3): 557–74.
(2) Ambrus, Gábor. 2018. "Information, Language and the Human Self in the Play of Biblical Revelation." *The Heythrop Journal.*
(3) Ambrus, Gábor. 2016. "In the Beginning Was the Word: Theological Reflections on Language and Technical Media in the Context of the Gospel of John." *Acta Universitatis Carolinae Theologica* 6 (2): 135–51.

Ilona Nord, Prof. Dr., Chair of Protestant Theology and Religious Education, University of Würzburg.
Publications:
(1) Nord, Ilona, and Kristin Merle, eds. 2020. *Mediatisierung Religiöser Kultur: Praktisch-Theologische Standortbestimmungen Im Interdisziplinären Kontext.* 1. Auflage. Veröffentlichungen der Wissenschaftlichen Gesellschaft für Theologie (VWGTh) 58. Leipzig: Evangelische Verlagsanstalt.

(2) Nord, Ilona and Zipernovsky, Hanna, eds. 2017. *Religious Education in a Mediatized World*. Stuttgart: Kohlhammer
(3) Nord, Ilona. 2008. *Realitäten Des Glaubens: Zur Virtuellen Dimension Christlicher Religiosität:* Berlin/NewYork: Walter De Gruyter.

Jan G. Michel, Dr., Post-Doc in the Emmy Noether Research Group "Theologie als Wissenschaft?!", Chair of Philosophy of Religion and Philosophy of Science, Ruhr-University Bochum.
Publications:
(1) Backmann, Marius, and Jan G. Michel, eds. 2009. *Physikalismus, Willensfreiheit, Künstliche Intelligenz.* Paderborn: Mentis.
(2) Michel, Jan G., ed. (forthcoming). Making Scientific Discoveries: Interdisciplinary Reflections. Paderborn: Mentis/Brill.
(3) Michel, Jan G. 2019. "How Are Species Discovered? Declarative Speech Acts in Biology." *Grazer Philosophische Studien* 96 (3): 419–41.

Johannes Grössl, Dr., Assistant Professor for Fundamental Theology and Comparative Religion at University of Würzburg, Germany.
Publications:
(1) Grössl, Johannes. 2015. *Die Freiheit des Menschen als Risiko Gottes: Der offene Theismus als Konzeption der Vereinbarkeit von göttlicher Allwissenheit und menschlicher Freiheit.* Studien zur systematischen Theologie, Ethik und Philosophie 3. Münster: Aschendorff.
(2) Stump, Eleonore, Georg Gasser, and Johannes Grössl, eds. 2015. *Göttliches Vorherwissen Und Menschliche Freiheit: Beiträge Aus Der Aktuellen Analytischen Religionsphilosophie:* Kohlhammer Verlag.
(3) Grössl, Johannes. 2018. "Verbesserung oder Zerstörung der menschlichen Natur? Eine theologische Evaluation des Transhumanismus." In *Designobjekt Mensch: Die Agenda des Transhumanismus auf dem Prüfstand*, edited by Benedikt P. Göcke and Frank Meier-Hamidi, 339–361. Freiburg, Basel, Wien: Herder.

Jörn Hurtienne, Prof. Dr., Chair of Psychological Ergonomics, University of Würzburg.
Publications:
(1) Hurtienne, Jörn. 2017. "How Cognitive Linguistics Inspires HCI: Image Schemas and Image-Schematic Metaphors." *International Journal of Human – Computer Interaction* 33 (1): 1–20.

(2) Klüber, S., Löffler, Diana, Hassenzahl, Marc, Nord, Ilona, and Hurtienne, Jörn. 2020. "Designing Ritual Artifacts for Technology-Mediated Relationship Transitions". *Accepted for TEI '20.*
(3) Löffler, Diana, Jörn Hurtienne, and Ilona Nord. 2019. "Blessing Robot BlessU2: A Discursive Design Study to Understand the Implications of Social Robots in Religious Contexts." *Internatinal Journal of Social Robotics* 22 (5): 353.

Julia Alessandra Harzheim, Forum Scientiarum Tübingen & Institut für Ethik und Geschichte der Medizin Tübingen.
Publications:
(1) Teichmann, Fabian and Harzheim, Julia Alessandra. 2019. „Sterbehilfe: Recht und Gesellschaft". *ZBJV* 155: 377–401. Bern.

Kilian Karger, M.A., PhD fellow at the Munich School of Philosophy, Scholarship Holder from Hanns Seidel Stiftung.

Leonie Seng, M.A., Scientific Assistant at Ludwigsburg University of Education.
Publications:
(1) Maisenhölder, Patrick, and Leonie Seng. 2019. "The Serious Side of Science Fiction – on the Usage of Medial Depictions of Artificial Intelligence for Ethical Reflection Using Detroit: Become Human as an Example." In *12th International Conference for Education, Research and Innovation (ICERI) 2019*, 3321–27.
(2) Seng, Leonie. 2019. "Mein Haus, mein Auto, mein Roboter? Eine (medien-)ethische Beurteilung der Angst vor Robotern und künstlicher Intelligenz" In *Maschinenethik: Normative Grenzen autonomer Systeme*, edited by Matthias Rath, Friedrich Krotz, and Matthias Karmasin, 57–72. Wiesbaden: Springer Fachmedien.
(3) Seng, Leonie. 2018. "Maschinenethik und Künstliche Intelligenz." In *Handbuch Maschinenethik*. Vol. 22, edited by Oliver Bendel. Living reference work, 1–21. Springer Reference Geisteswissenschaften.

Lukas Brand, Mag.-Theol., Research Assistant at the Chair for Philosophy of Religion and Philosophy of Science at Ruhr-Universität Bochum, Germany.
Publications:
(1) Brand, Lukas. 2018. *Künstliche Tugend: Roboter Als Moralische Akteure:* Verlag Friedrich Pustet.

(2) Brand, Lukas. 2020. Die Maschine als Mensch. Ökumenische Rundschau 64 (1), 50–59.

Michael Filzmoser, PhD, Associate Professor at the Institut of Management Science, TU Wien.
Publications:
(1) Filzmoser, Michael. 2010. *Simulation of Automated Negotiation.* Berlin: Springer.

Rebecca Davnall, PhD., Lecturer in Philosophy, Olaf Stapledon Centre for Speculative Futures, University of Liverpool.
Publications:
(1) Davnall, Rebecca. 2019. "Solving the Single-Vehicle Self-Driving Car Trolley Problem Using Risk Theory and Vehicle Dynamics." *Science and engineering ethics.*

Scarlet Siebert, M.Sc., PhD candidate at TH Köln University of Applied Sciences, Institute for Media Research and Media Education.
Publications:
(1) Siebert, Scarlet, Nils Tolksdorf, Katharina Rohlfing, and Isabel Zorn. 2019. "Raising Robotic Natives? Persuasive Potentials of Social Robots in Early Education." *The Journal of Communication and Media Studies* 4 (4): 21–35.

Stefan Reining, Dr., Postdoctoral Research Associate, Chair of Information Systems and Strategic IT Management, University of Duisburg-Essen
Publications:
(1) Reining, Stefan. 2016. "On the Supposed Dilemma of Conciliationism." *Episteme* 13 (3): 305–28. https://doi.org/10.1017/epi.2015.47.
(2) Reining, Stefan. 2016. "Peerhood in Deep Religious Disagreements." *Religious Studies* 52 (3): 403–19.
(3) Reining, Stefan. 2018. "Reconsidering the Case for Colour Relativism." *Metaphysica* 19 (1): 57–86.

Swantje Luthe, Dipl. theol., Research assistant at the Chair of Protestant Theology and Religious Education, University of Würzburg.
Publications:
(1) Luthe, Swantje, Nord, Ilona, Löffler, Diana, and Hurtienne, Jörn. 2019. „Segensroboter „BlessU-2". Forschungsimpulse für die Praktische

Theologie angesichts der Entwicklung sozialer Roboter". *Pastoraltheologie* 108 (3): 107–123.
(2) Luthe, Swantje. 2017. „Priestertum aller Glaubenden? Digitale Formen von Andacht und Partizipationsmöglichkeiten". *Praktische Theologie* 52 (1): 31–38.
(3) Luthe, Swantje. 2016. „Trauerarbeit online – Facebook als Generator für Erinnerungen". In *Tod und Trauer im Netz: Mediale Kommunikationen in der Bestattungskultur*, edited by Ilona Nord and Thomas Klie, Stuttgart.

Tobias Müller, Dr., Lecturer in Philosophy at the Munich School of Philosophy.
Publications:
(1) Müller, Tobias. in press. *Das Rätsel des Bewusstseins. Eine Theorie der Subjektivität im Zeitalter der Naturwissenschaften*: Alber Verlag.
(2) Müller, Tobias. in press. Anthropologische Grundbestimmungen und ihre Bedeutung für die aktuellen Entwicklungen von KI, Robotik und Assistenzsystemen. In *Roboter als (Er-)Lösung? Orientierung der Pflege von morgen am christlichen Menschenbild in Forschung – Technik – Praxis*, edited by Stephan Mokry and Maximilian T. L. Rückert, Paderborn: Bonifatius Verlag.
(3) Müller, Tobias. in press. Zur Anthropologie des Transhumanismus. In *Schöne neue Welt, oder was kommt nach dem Menschen? Der Transhumanismus als Utopie und als technologisches Projekt*, edited by H. Watzka and S. Herzberg.

Vanessa Schäffner, M.A., M.Sc., Ph.D. Candidate at Munich School of Philosophy.
Publications:
(1) Schäffner, Vanessa. 2018. "Caught up in Ethical Dilemmas: An Adapted Consequentialist Perspective on Self-Driving Vehicles." In *Envisioning Robots in Society –Power, Politics, and Public Space, Proceedings of Robophilosophy 2018/TRANSOR 2018*, edited by M. Coeckelbergh, J. Loh, M. Funk, J. Seibt, and M. Nørskov. Frontiers in Artificial Intelligence and Applications. Amsterdam: IOS Press.
(2) Schäffner, Vanessa. 2020. "Wenn Ethik zum Programm wird: Eine risikoethische Analyse moralischer Dilemmata des autonomen Fahrens." *Zeitschrift für Ethik und Moralphilosophie* (2020). Available at: https://doi.org/10.1007/s42048-020-00061-9.

Walther Ch. Zimmerli, Prof. Dr. DPhil h.c. (University of Stellenbosch), Honorary Professor Humboldt-University Berlin, Senior Visiting Fellow Institute for Human Sciences (IWM) Vienna, DSI-Fellow Digital Society Initiative University of Zürich.

Publications:

(1) Zimmerli, Walther Ch., and Stefan Wolf, eds. 1994. *Künstliche Intelligenz: Philosophische Probleme.* Stuttgart: Reclam.

(2) Zimmerli, Walther Ch. 2014. "From Fiction to Science. A German-Japanese Era Project." In *Robotics in Germany and Japan: Philosophical and Technical Perspectives*, edited by Michael Funk and Bernhard Irrgang. 1st, New ed. Dresdener Studien zur Philosophie der Technologie 5. Frankfurt a.M: Peter Lang GmbH Internationaler Verlag der Wissenschaften. Philosophy of Technology Studies / Dresdner Studien zur Philosophie der Technologie 5. Frankfurt a.M: Peter Lang GmbH Internationaler Verlag

(3) Zimmerli, Walther Ch. 2020. "Macht Information Sinn? Reflexionen zur Iteration von Unterschied und Nichtwissen." In *Information und Wissen. Beiträge zum transdisziplinären Diskurs*, edited by K. Pelikan and Th. Roelcke, 69–83. Berlin: Peter Lang.

Index